Self-Organizing Neural Networks

Studies in Fuzziness and Soft Computing

Editor-in-chief
Prof. Janusz Kacprzyk
Systems Research Institute
Polish Academy of Sciences
ul. Newelska 6
01-447 Warsaw, Poland
E-mail: kacprzyk@ibspan.waw.pl
http://www.springer.de/cgi-bin/search_book.pl?series=2941

Udo Seiffert · Lakhmi C. Jain
Editors

Self-Organizing Neural Networks

Recent Advances and Applications

With 119 Figures
and 27 Tables

Springer-Verlag Berlin Heidelberg GmbH

Dr. Udo Seiffert
Faculty of Electrical Engineering
Technical Computer Science Group
University of Magdeburg
Universitätsplatz 2
39106 Magdeburg
Germany
seiffert@iesk.et.uni-magdeburg.de

Professor Lakhmi C. Jain
Knowledge-Based Intelligent
Engineering Systems Centre
University of South Australia
Adelaide, Mawson Lakes
South Australia 5095
Lakhmi.Jain@unisa.edu.au

ISSN 1434-9922
ISBN 978-3-662-00343-5   ISBN 978-3-7908-1810-9 (eBook)
DOI 10.1007/978-3-7908-1810-9

Cataloging-in-Publication Data applied for
Die Deutsche Bibliothek – CIP-Einheitsaufnahme
Self-organizing neural networks: recent advances and applications; with 27 tables / Udo Seiffert; Lakhmi C. Jain, ed. – Heidelberg; New York: Physica-Verl., 2002
  (Studies in fuzziness and soft computing; Vol. 78)
  ISBN 978-3-662-00343-5

© Physica-Verlag Heidelberg 2002
Originally published by Physica-Verlag Heidelberg New York in 2002

Hardcover Design: Erich Kirchner, Heidelberg

SPIN 10843052   88/2202-5 4 3 2 1 0 – Printed on acid-free paper

Preface

The Self-Organizing Map (SOM) is one of the most frequently used architectures for unsupervised artificial neural networks. Introduced by Teuvo Kohonen in the 1980s, SOMs have been developed as a very powerful method for visualization and unsupervised classification tasks by an active and innovative community of international researchers.

A number of extensions and modifications have been developed during the last two decades. The reason is surely not that the original algorithm was imperfect or inadequate. It is rather the universal applicability and easy handling of the SOM. Compared to many other network paradigms, only a few parameters need to be arranged and thus also for a beginner the network leads to useful and reliable results. Nevertheless there is scope for improvements and sophisticated new developments as this book impressively demonstrates. The number of published applications utilizing the SOM appears to be unending.

As the title of this book indicates, the reader will benefit from some of the latest theoretical developments and will become acquainted with a number of challenging real-world applications. Our aim in producing this book has been to provide an up-to-date treatment of the field of self-organizing neural networks, which will be accessible to researchers, practitioners and graduated students from diverse disciplines in academics and industry.

We are very grateful to the father of the SOMs, Professor Teuvo Kohonen for supporting this book and contributing the first chapter.

We would like to express our sincere appreciation to all authors for contributing to this volume. It has been a great pleasure for us to work on this project. Furthermore we would like to take this opportunity to thank Professor Janusz Kaprzyk for including this book in his series and the Springer Verlag Company for their excellent editorial guidance. We also thank Dr. Neil Allen for his valuable input and Werner Liebscher for his assistance in processing the electronic files of this book.

Udo Seiffert, Magdeburg, Germany

Lakhmi C. Jain, Adelaide, Australia

April 2001

Contents

Contributors

Bernd Brückner
Leibniz Institute for Neurobiology Magdeburg, Germany
brueckner@ifn-magdeburg.de

Timo D. Hämäläinen
Digital and Computer Systems Laboratory
Tampere University of Technology, Finland
timo.d.hamalainen@tut.fi

Lakhmi C. Jain
University of South Australia, Adelaide, Australia
lakhmi.jain@unisa.edu.au

Hans A. Kestler
Department of Neural Information Processing
University of Ulm, Germany
hkestler@neuro.informatik.uni-ulm.de

Teuvo Kohonen
Neural Networks Research Centre
Helsinki University of Technology, Finland

Timo Kostiainen
Laboratory of Computational Engineering
Helsinki University of Technology, Finland
timo.kostiainen@hut.fi

Jouko Lampinen
Laboratory of Computational Engineering
Helsinki University of Technology, Finland
jouko.lampinen@hut.fi

Erzsébet Merényi
Department of Electrical and Computer Engineering
Rice University, Texas, U.S.A.
erzsebet@ece.rice.edu

Thomas Natschläger
Institute for Theoretical Computer Science
Graz University of Technology, Austria
tnatschl@igi.tu-graz.ac.at

Günther Palm
Department of Neural Information Processing
University of Ulm, Germany
palm@neuro.informatik.uni-ulm.de

Daniel Polani
Institute for Neuro- and Bioinformatics
University of Lübeck, Germany
polani@inb.mu-luebeck.de

Marina Resta
Facoltà di Economia
Università degli Studi di Genova, Italy
marina.resta@tiscalinet.it

Berthold Ruf
Institute for Theoretical Computer Science
Graz University of Technology, Austria
bruf@igi.tu-graz.ac.at

Michael Schmitt
Lehrstuhl Mathematik und Informatik
Ruhr-Universität Bochum, Germany
mschmitt@lmi.ruhr-uni-bochum.de

Friedhelm Schwenker
Department of Neural Information Processing
University of Ulm, Germany
fschwenker@acm.org

Udo Seiffert
Technical Computer Science
University of Magdeburg, Germany
seiffert@iesk.et.uni-magdeburg.de

Thomas Villmann
Clinic of Psychotherapy
University Leipzig, Germany
villmann@informatik.uni-leipzig.de

Thomas Wesarg
Leibniz Institute for Neurobiology Magdeburg, Germany
wesarg@ifn-magdeburg.de

1 Overture

Teuvo Kohonen

Abstract. An introduction to and overview of the Self-Organizing Map (SOM) methods is presented in this chapter.

1.1 Introduction

The Self-Organizing Map (SOM) [1], [2], [3] is a computational mapping principle that forms an ordered nonlinear projection of high-dimensional input data items onto a low-dimensional, usually 2D regular grid. The grid points are also called *nodes*. The display produced onto the nodes can be regarded as a similarity graph. Consider that some distance measure $d(X_i, X_j)$ between any kind of items X_i and X_j can be defined: pairs of items that have a small mutual distance will then be mapped onto the same or nearby nodes. On the other hand, unlike in some other projection methods such as multidimensional scaling (MDS) [4], [5], larger distances between items will in general not be preserved in the SOM projection, not even approximately.

The SOM also differs from the other projection methods in that the set of all occurring input items is approximated by a much smaller number of *models M_i* that are associated with the nodes and have the same format as the X_j; an arbitrary input item is then approximated by the model M_i for which the distance $d(X_j, M_i)$ is smallest. In this way, finding the best-matching model in the display needs significantly fewer comparisons than if every input item is compared with all the others.

Fig. 1.1 exemplifies a set of models that represent acoustic spectra of speech. The occurring spectra have been quantized into 96 models, which are arranged as a hexagonal grid. Models adjacent in the grid are mutually more similar than the more distant ones.

By virtue of the models used for the representation of projections, the SOM, unlike MDS, has a useful characteristic property of allocating more grid points for input items that occur more frequently, improving their resolution. This "magnification factor" is similar as in the biological brain maps where more cortical area is allocated for those representations of sensory events or sensory features that occur more frequently in observations or are otherwise more important. In this way the SOM, like the biological nervous systems, tends to optimize the resources used for representation.

Now consider that the X_i are generally clustered; then the probability for the occurrence of input values between the clusters is small, and there is no need to allocate space on the SOM for the data space between the clusters. In the SOM, the clusters will automatically be mapped closely nearby. In order to distinguish the clusters in the display and to demarcate their borders, one may use, e.g., special coloring methods such as the U-matrix [6], [7], where the clusters are indicated

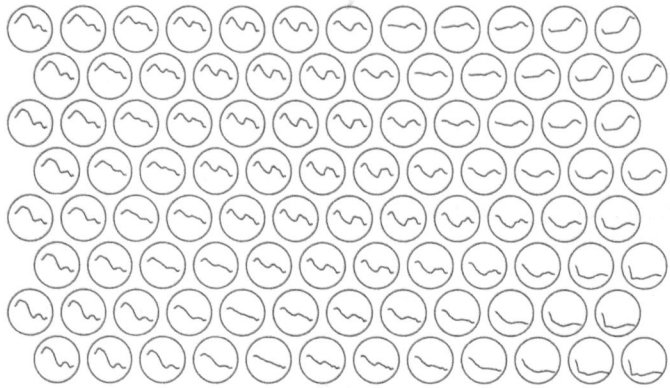

Fig. 1.1. In this exemplary application, each processing element in the hexagonal grid holds a model of a short-time spectrum of natural speech (Finnish). Note that neighboring models are mutually similar.

by light shades and the borders with darker shades, respectively. Alternatively one can combine the SOM algorithm with some more traditional clustering method that draws the lines between the clusters on the SOM display.

Fig. 1.2 illustrates the U-matrix. The SOM here forms the map of 126 countries according to 39 socioeconomic indicators that describe the household economy, health services and educational facilities of the countries, and thus forms a similarity graph of the countries based on their poverty characteristics. One can discern roughly half a dozen clusters, the borders of which are automatically demarcated by shades of gray: when the vectorial distance of the neighboring models in the original 39-dimensional indicator space is small, the interstitial areas between the labeled models are indicated by light shades of gray, whereas for areas between neighboring models that differ more from each other, darker shades are used.

The SOM was originally introduced for vectorial input data, and the mapping algorithm was defined as an incremental, stepwise correction process, which is still used in most SOM publications. If $\mathbf{x}(t)$ is an n-dimensional vectorial input item identified by the variable t (a running index of the samples, and also the index of the iteration step), and if $\{\mathbf{m}_i(t)\}$ is a spatially ordered set of vectorial models arranged as a grid, where i is the index of the node (grid point), then the original SOM algorithm reads:

Step 1. Define the $\mathbf{m}_i(t)$ closest to $\mathbf{x}(t)$:

$$c = \arg\min_i\{\|\mathbf{x}(t) - \mathbf{m}_i(t)\|\} . \qquad (1.1)$$

It is customary to call the best-matching node the *winner*. In this algorithm the winner is found by direct comparison of $\mathbf{x}(t)$ with all the $\mathbf{m}_i(t)$; in special adaptive networks (cf. [26]) the function for the determination of c is called the winner-take-all (WTA) function.

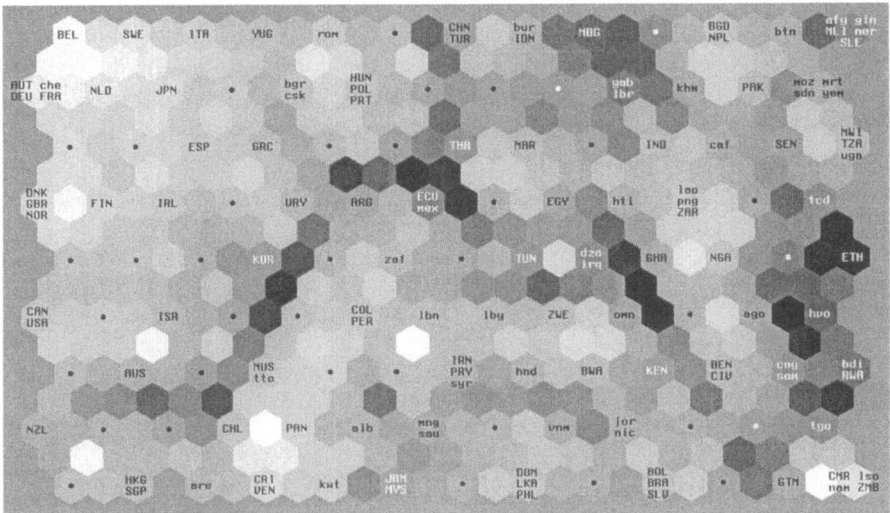

Fig. 1.2. "Poverty map" of 126 countries with the clustering shown by shades of gray.

Step 2. Correct the $\mathbf{m}_c(t)$ and the $\mathbf{m}_i(t)$ in the neighborhood of $\mathbf{m}_c(t)$ on the grid towards $\mathbf{x}(t)$:

$$\mathbf{m}_i(t + 1) = \mathbf{m}_i(t) + h_{ci}(t)[\mathbf{x}(t) - \mathbf{m}_i(t)] \ . \qquad (1.2)$$

Here $h_{ci}(t)$ is called the *neighborhood function*. It is similar to a smoothing kernel that has its maximum at the grid point $i = c$ and its value decreases monotonically with increasing spatial distance between grid points i and c. Also $h_{ci}(t)$ usually decreases monotonically with the sample index t, and becomes narrower. It can be proven mathematically, although this will not be very easy, that in many cases, when $t \to \infty$, the asymptotic values of the $\mathbf{m}_i$ constitute the wanted ordered projection (cf., e.g., [8]).

As the SOM process has turned out very difficult to describe mathematically, it cannot be discussed in this short chapter in more detail. Some of its aspects will hopefully become clearer below and from the rest of this book; also a few special monographies written on the SOM are useful as references, e.g., [3], [12], [31], [32], and [33].

A newcomer can always in the first place rely on software packages, some of which (e.g. SOM_PAK and Matlab SOM Toolbox) can be downloaded freely from the web address *http://www.cis.hut.fi/research/*.

1.2   A More General Setting for the Computation of the SOMs

The algorithm defined by eqs. (1.1) and (1.2) will now be generalized for items that may not be vectorial. As the incremental correction rule (1.2) is then not applicable, we must resort to a batch computation process that directly tends to find the models in such a way that they quantize the set of occurring input samples in an optimal way.

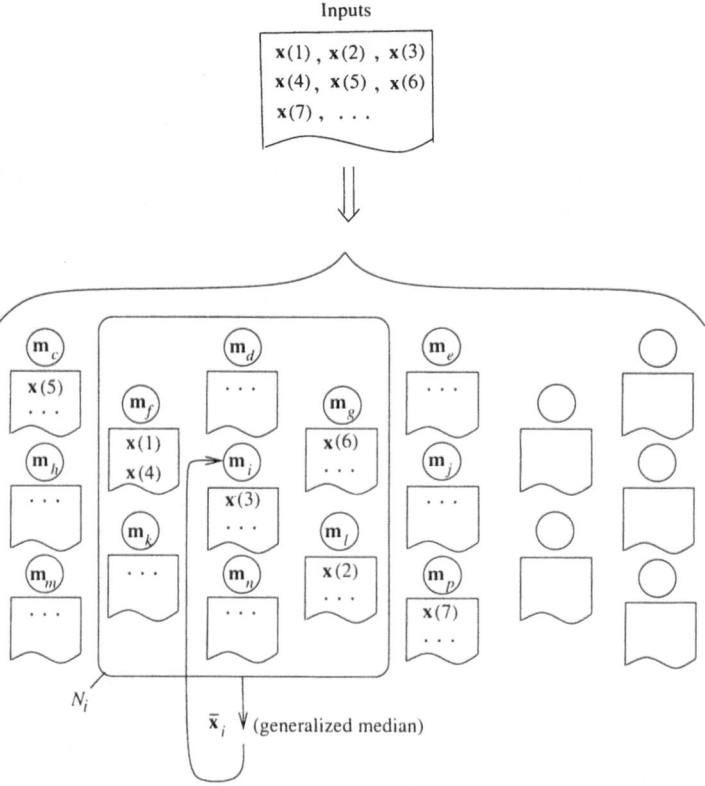

Fig. 1.3. Illustration of the batch process in which the input samples are distributed into sublists under the best-matching models, and then the new models are determined as generalized medians of the sublists over the neighborhoods N_i.

Consider Fig. 1.3 where a two-dimensional ordered array of nodes, each one having a general model $\mathbf{m}_i$ associated with it, is shown. The initial values of the $\mathbf{m}_i$ may be selected as random, preferably from the domain of the input samples, but any kind of rough order in the initial values will speed up convergence significantly. Then consider a list of input samples $\mathbf{x}(t)$. Let us recall that in this scheme, the $\mathbf{x}(t)$ and $\mathbf{m}_i$ may be vectors, strings of symbols, or even more general items; henceforth we shall use the vector notation for them, however. Compare each $\mathbf{x}(t)$ with all the $\mathbf{m}_i$ and copy each $\mathbf{x}(t)$ into a sublist associated with that node, the model of which is most similar to $\mathbf{x}(t)$ relating to the general distance measure. When all the $\mathbf{x}(t)$ have been distributed into the respective sublists in this way, consider the *neighborhood set* N_i around model $\mathbf{m}_i$. Here N_i consists of all nodes up to a certain radius in the grid from node i. In the union of all sublists in N_i, the next task is to find the "middlemost" sample $\bar{\mathbf{x}}_i$, defined as that sample that has the smallest sum of distances from all the samples $\mathbf{x}(t), t \in N_i$. We can also weight the distances within N_i by the neighborhood function h_{ci}. This sample $\bar{\mathbf{x}}_i$ is now called the *generalized median* in the union of the sublists. If $\bar{\mathbf{x}}_i$ is restricted to being one of the samples $\mathbf{x}(t)$, we shall indeed call it the *generalized set median;* on the other hand, since the $\mathbf{x}(t)$ may not cover the whole input domain, it may be possible to find another item $\bar{\mathbf{x}}_i'$ that has an even smaller (weighted) sum of distances from the $\mathbf{x}(t), t \in N_i$. For clarity we shall then call $\bar{\mathbf{x}}_i'$ the *generalized median*. Notice too that for the Euclidean vectors the generalized median is equal to their *arithmetic mean* if we look for an arbitrary Euclidean vector that has the smallest *sum of squares* of the Euclidean distances from all the samples $\mathbf{x}(t)$ in the union of the sublists.

The next phase in the process is to form $\bar{\mathbf{x}}_i$ or $\bar{\mathbf{x}}_i'$ for each node in the above manner, always considering the neighborhood set N_i around each node i, and to replace each old value of $\mathbf{m}_i$ by $\bar{\mathbf{x}}_i$ or $\bar{\mathbf{x}}_i'$, respectively, in a simultaneous operation.

The above procedure shall now be iterated: in other words, the original $\mathbf{x}(t)$ are again distributed into the sublists (which now change, because the $\mathbf{m}_i$ have changed), and the new $\bar{\mathbf{x}}_i$ or $\bar{\mathbf{x}}_i'$ are computed and made to replace the $\mathbf{m}_i$, and so on. This is a kind of *regression process*.

There is an important question that we are not able to answer completely at the moment: even if we keep the $\mathbf{x}(t)$ the same all the time, does this process converge? Do the $\mathbf{m}_i$ finally coincide with the $\bar{\mathbf{x}}_i$? If the $\mathbf{x}(t)$ and $\mathbf{m}_i$ are Euclidean vectors, and the weighted distance measure is used in the winner search (cf. Sec. 1.4.1), the convergence has been proved by *Cheng* [9].

At first glance one might also think that the global order in the "map" reflects some kind of *harmonicity*: it seems as if every model were the average of the neighboring models, like in the theory of the harmonic functions. On a closer look, however, one can find several properties of the SOM that differ from the harmonicity. First of all, *there are no fixed boundary values for the models at the edges of the grid*: the values are determined freely in the regression process, when the neighborhood set N_i of the edge nodes is made to contain nodes from the inside of the grid. The nature of regression is slightly different at the edges and in the inside of the grid, resulting in certain border effects. Another deviation from harmonicity, as demonstrated already in Fig. 1.1, is that there are areas in the map where the models are very similar, but then there are again places where a bigger "jump" between the neighboring models is discernible. If the collection of the models has to approximate the distribution of the inputs, such uneven places must exist in the map.

On the other hand, if one considers the process depicted in Fig. 1.3, one can easily realize how the definition of "harmonicity" must be modified in order to describe the input data: *the collection of models is ordered by definition, if each model is equal to the average of input data mapped to its neighborhood.*

1.3   Component Planes

Note that the models associated with the grid points are usually vectorial, where-upon each component of the vectors may correspond to a different indicator or in-put variable. It is then possible to display the value of a particular component of all models on the two-dimensional groundwork formed by the SOM grid, e.g., using shades of gray. Such diagrams are called *component planes*, and some component planes of the "poverty map" of Fig. 1.2, corresponding to certain socioeconomic indicators, are shown in Fig. 1.4. The scales of the indicators have been normalized before computation of the SOM, and the high values are shown as white, while the low values are shown as black, respectively. The model vector is the combination of values picked up from the same location in all the component planes.

By virtue of the component planes it is possible to gain an insightful view of how the various indicators are globally correlated or otherwise interrelated. For instance, when traversing the map from location to location, the corresponding simultaneous changes in all variables are clearly discernible. Thus, the component planes of the SOM constitute a powerful generalization of the concept of correlations.

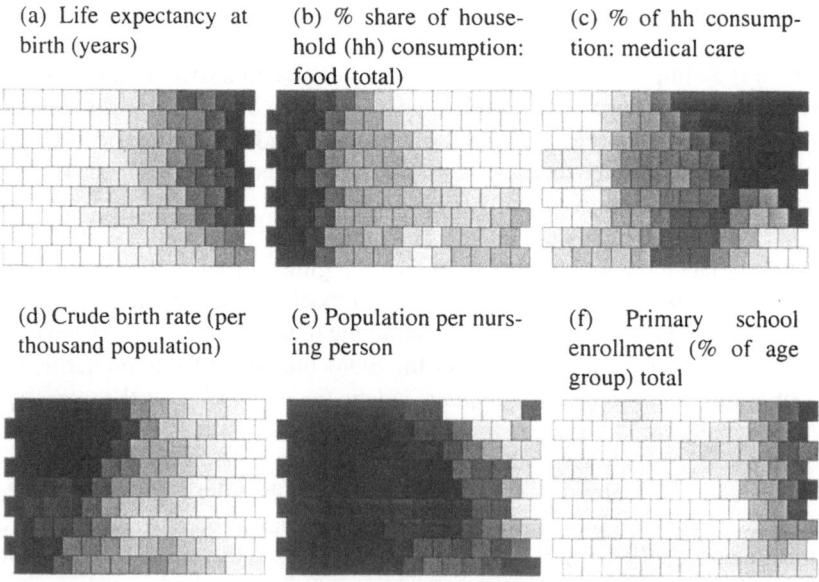

(a) Life expectancy at birth (years)

(b) % share of house-hold (hh) consumption: food (total)

(c) % of hh consump-tion: medical care

(d) Crude birth rate (per thousand population)

(e) Population per nurs-ing person

(f)   Primary   school enrollment (% of age group) total

Fig. 1.4. Component planes of the "poverty map." Each partial diagram here represents one component plane of the array of the $\mathbf{m}_i$ vectors corresponding to the "poverty map" of Fig. 1.2. These values, due to the SOM process, have been defined for all nodes, not only for those labeled by the country symbols in Fig. 1.2. In each component plane, white de-notes the largest value of the corresponding components among the countries, and black the smallest value, respectively. Intermediate values have been encoded by the shades of gray.

1.4   Special Problems Associated with the SOM

1.4.1   Convergence

Satisfactory mathematical convergence proofs of the SOM algorithms have only been presented for the following partial cases:

- one-dimensional (scalar) input and linear (1D) topology of the array [10]
- 2D input and 2D topology [11]
- discrete-valued input data [12]
- variants of the SOM such that when the winner is searched on the basis of weighted matching, the weights are the same as in the neighborhood function $h_{ci}(t)$ [9], [13], [14], [15]

The SOM thus seems to constitute an ill-posed problem; nonetheless much theoretical work is in progress, and we may see new results in near future.

On the other hand, as several criteria for the degree of ordering have been presented [16], [17], [18], [19], one may safely use the SOM algorithms as long as the produced results can be tested.

1.4.2   Point Density of the Models

Strictly speaking, the density of even vectorial models located in the signal space has a meaning only in either of the following cases: 1. Volume elements in the input space can be defined, and the number of models in any reasonably large volume differential used for the approximation of density functions is large, or 2. The input items are stochastic variables, and their probability for falling into a given volume differential can be defined analytically.

Notwithstanding it is a customary practice in vector quantization problems, to which the SOM also belongs, to compare the input probability density function $p = p(\mathbf{x})$ with the asymptotic point density $q = q(\mathbf{x})$ of the model vectors, which only makes a sense if $\mathbf{x}$ and the models are metric vectors. The following explicit results have transpired:

- With scalar input and 1D array we have $q = p^e$, where the exponent e depends on the neighborhood function. For instance, if $h_{ci} = 1$ for the nearest neighbors, i.e., when $i = c$, or when i and c are neighboring grid points, and $h_{ci} = 0$ otherwise, $e = 0.6$ in the usual SOM [20]. This result seems to hold also for arrays of finite but sufficient length [21]. On the other hand, if the winner is defined on the basis of h_{ci} used in weighted matching, we have $e = 1/3$ [15].
- One might argue that the SOM model vectors can be obtained from some objective function like in traditional vector quantization problems [22], [23]. For instance, if one carries out the minimization of the error function $E = \int \sum_i h_{ci} \|\mathbf{x} - \mathbf{m}_i\|^2 p(\mathbf{x}) d\mathbf{x}$ where the h_{ci} have been used to weight the quantization errors, and it is hoped that the SOM algorithm would ensue as the steepest descent of E, one may be able to use the values $\mathbf{m}_i$ that minimize E

as the SOM model vectors. However, in general these $\mathbf{m}_i$ do not agree exactly with the asymptotic values obtained by the incremental SOM algorithm. In the one-dimensional case, the minimization of E produces $q = p^e, e = 1/3$, as in classical vector quantization.

The point-density problem with two-dimensional regular arrays and input vectors of higher dimensionality than two is further complicated by the fact that the neighboring models cannot be located freely in the input space like in classical vector quantization [22], [23], but they tend to lie along an "elastic surface." Therefore, if the $\mathbf{m}_i$ are computed as approximate optima of the error function E, and if the law $q = p^e$ at all holds for SOMs of higher dimensionality, e may be expected to be in the vicinity of the value $1/2$ [24]; notice that in classical vector quantization with squared quantization error one would have $e = n/(n + 2)$, where n is the dimensionality of the vectors. When n grows, in the classical vector quantization $e \to 1$, which does not hold for the SOM, however, as argued above.

1.4.3   Variations and Generalizations of the SOM

The general setting described in Sec. 1.2 allows many variations and generalizations of the basic SOM, for instance, a SOM for string variables. The batch computation scheme has been used in our largest SOM so far constructed: it had over one million models with input dimensionality 500 [25].

A plethora of other SOM variants has been suggested. One special class is worth mentioning here: the structurally growing SOM, introduced by *B. Fritzke*, is a step towards ultrametric approximation of input data.

Then one should point out that if the SOM is meant for a model to be used in computational neuroscience, its information processing functions ought to be relatable to realistic biological components. The winner-take-all function can be implemented by lateral excitatory and inhibitory connections, but it seems that the neighborhood function h_{ci} would operate much more effectively, if the neighborhood interactions during learning were implemented by chemical messengers, which select the neighboring cells for learning [26].

1.4.4   Unsupervised and Supervised SOM

If the training samples fall in a finite number of predetermined classes, one can label the nodes of the SOM by these class labels, by inputting the samples once again after the learning phase and looking what labels there are in the samples that hit a particular node. Since there are usually multiple hits at every node, the latter are labeled according to the majority of labels of the hits.

The class labels of the calibrated SOM are usually clustered into connected regions. When then a new, unknown sample is presented at the input, its most probable class label is found from the winning node.

The general view is that the SOM algorithm defines an unsupervised learning process, similar to clustering. Nonetheless it is easy to modify the learning step such that the correction depends on the tentative classification results. In this way one can apply a reward-punishment scheme that produces improved "effective" model values such that when the classification of inputs is referred to then class separation and statistical classification accuracy are improved. To this end the training samples with known class labels are input reiteratively again, but this time the learning algorithm (1.2) is modified such that $h_{ci}(t)$ is replaced by a factor $\pm\alpha(t)\delta_{ci}$, where $\alpha(t)$ is a small, monotonically decreasing scalar (learning rate), and δ_{ci} is the Kronecker delta ($\delta_{ci} = 1$ for $i = c$, $\delta_{ci} = 0$ otherwise). The plus sign is used if the class labels of the training sample and the winning grid point agree, but the minus sign is used if they disagree. This fine-tuning algorithm of the models is called the LVQ1. Also other fine-tuning algorithms exist. Compared with the usual (labeled) SOM, the classification accuracy of the LVQ1-tuned SOM is usually improved by a factor of two or three ([3], Ch. 6).

1.5 Applications

Over 4000 scientific publications on the SOM have been written. For a documented list of most of them, see [27]. The main application areas are:

- Statistical analysis at large, in particular data mining and knowledge discovery in databases.
- Analysis and control of industrial processes and machines.
- New methods in telecommunications, especially optimization of telephone traffic and demodulation of digital signals.
- Medical and biological applications.

The SOM has been used in Finnish forest industries for control purposes since 1985. Recently, for the continuous casting of steel, an on-line monitor based on the SOM has been developed [28].

Practical applications have been introduced to finance, for instance bankruptcy analysis, profiling of customers, and analysis of macroeconomic systems [29]. A promising area is in real estate business. The new Finnish forest taxation legislation in 1992 was based on segmentation results obtained by the SOM.

Of the numerous applications from very different scientific fields one may mention the analysis of the Hubble Space Telescope data, where a new quantitative classification of thousands of galaxies has been developed [30].

Another novel example is from criminology, where a SOM-based system for computer-aided tracking of homicides and sexual assaults has been developed by the Battelle Pacific Northwest Division, in cooperation with the Attorney General of the State of Washington.

Perhaps the biggest SOM, with over one million models of dimensionality 500, has been developed for document organization, viz. for the similarity diagram of about seven million patent abstracts [25]. This text corpus is over 20 times as big as all the 34 parts of Encyclopaedia Britannica together. Each document is described by the collection of the words it contains. Five hundred statistical indicators of the high-dimensional word histograms, with a vocabulary of about 50 000, are used as real 500-dimensional input vectors to the SOM. Standard browsing tools are used to display and search for the documents of interest. The two-dimensional order in the document maps makes it possible to find additional relevant information, after the starting point in the document collection has been defined (Kohonen et al., 2000).

Several monographies, text books and edited volumes that concentrate on the SOM have appeared: in addition to [3] and [12] one may mention [29], [31], [32], [33], [34]. A special issue of *Neurocomputing* has been dedicated to the SOM [35].

Bibliography on Chapter 1

1. Kohonen, T. (1982) Self-organizing formation of topologically correct feature maps. Biol. Cybern. **43**, 59–69
2. Kohonen, T. (1982) Clustering, taxonomy, and topological maps of patterns. Proc. 6th Int. Conf. Pattern Recognition, Munich, Germany, 114–128
3. Kohonen, T. (2001) Self-Organizing Maps, 3rd ed. Springer, London
4. Young, G., Householder, A. S. (1938) Discussion of a set of points in terms of their mutual distances. Psychometrika **3**, 19–22
5. Kruskal, J. B., Wish, M. (1978) Multidimensional Scaling. Sage University Paper Series on Quantitative Applications in the Social Sciences No. 07-011. Sage Publications, Newbury Park
6. Ultsch, A., Siemon, H. (1989) Exploratory Data Analysis: Using Kohonen's Topology Preserving Maps. Technical Report 329. Univ. of Dortmund, Dortmund, Germany
7. Kraaijveld, M. A., Mao, J., Jain, A. K. (1992) A non-linear projection method based on Kohonen's topology preserving maps. Proc. 11ICPR, Int. Conf. on Pattern Recognition. IEEE Comput. Soc. Press, Los Alamitos, CA, 41–45. Also IEEE Trans. on Neural Networks **6**, 548–559 (1995)
8. Cottrell, M., Fort, J. C., Pagés, G. (1997) Theoretical aspects of the SOM algorithm. Proc. WSOM'97, Workshop on Self-Organizing Maps. Helsinki University of Technology, Neural Networks Research Centre, Espoo, Finland, 246–267
9. Cheng, Y. (1997) Convergence and ordering of Kohonen's batch map. Neural Computation **9**, 1667–1676
10. Cottrell, M., Fort, J.-C. (1987) Étude d'un processus d'auto-organisation. Annales de l'Institut Henri Poincaré **23**, 1–20
11. Flanagan, J. A. (1994) Self-Organizing Neural Networks. Ph.D. Thesis, Swiss Federal Inst. of Tech. Lausanne (EPFL)
12. Ritter, H., Martinetz, T., Schulten, K. (1992) Neural Computation and Self-Organizing Maps: An Introduction. Addison-Wesley, Reading, MA
13. Heskes, T. (1993) Guaranteed convergence of learning rules. Proc. Int. Conf. on Artificial Neural Networks (ICANN'93), Amsterdam, The Netherlands, 533–536
14. Heskes, T. and Kappen, B. (1993) Error potential for self-organization. Proc. Int. Conf. Neural Networks (ICNN'93), vol. III, 1219–1223
15. Luttrell, S. P. (1992) Code Vector Density in Topographic Mappings. Memorandum 4669. Defense Research Agency, Malvern, UK
16. Zrehen, S. (1993) Analyzing Kohonen maps with geometry. Proc. Int. Conf. on Artificial Neural Networks (ICANN'93), Amsterdam, The Netherlands, 609–612
17. Villmann, T., Der, R., Martinetz, T. (1994) A new quantitative measure of topology preservation in Kohonen's feature maps. Proc. IEEE Int. Conf. on Neural Networks (ICNN'94), 645–648
18. Kiviluoto, K. (1996) Topology preservation in self-organizing maps. Proc. IEEE Int. Conf. on Neural Networks (ICNN'96), 294–299
19. Kaski, S., Lagus, K. (1996) Comparing self-organizing maps. Proc. Int. Conf. on Artificial Neural Networks (ICANN'96), 809–814
20. Ritter, H. (1991) Asymptotic level density for a class of vector quantization processes. IEEE Trans. Neural Networks **2**, 173–175
21. Kohonen, T. (1999) Comparison of SOM point densities based on different criteria. Neural Computation **11**, 2081–2095

22. Gersho, A. (1979) Asymptotically optimal block quantization. IEEE Trans. Inform. Theory **IT-25**, 373–380
23. Zador, P. (1982) Asymptotic quantization error of continuous signals and the quantization dimension. IEEE Trans. Inform. Theory **IT-28**, 139–149
24. Kohonen, T. (1998) Computation of VQ and SOM Point Densities Using the Calculus of Variations. Report A52, Helsinki University of Technology, Laboratory of Computer and Information Science, Espoo, Finland.
25. Kohonen, T., Kaski, S., Lagus, K., Salojärvi, J., Honkela., J., Paatero, V., Saarela, A. (2000) Self-organization of a massive document collection. IEEE Trans. Neural Networks **11**, 574–585
26. Kohonen, T. (1993) Physiological interpretation of the self-organizing map algorithm. Neural Networks **6**, 895–905
27. Kaski, S., Kangas, J., Kohonen, T. (1998) Bibliography of self-organizing map (SOM) papers: 1981–1997. Neural Computing Surveys **1**, 1–176 (http://www.icsi.berkeley.edu/˜jagota/NCS/)
28. Alhoniemi, E., Hollmén, J., Simula, O., Vesanto, J. (1999) Process monitoring and modeling using the self-organizing map. Integrated Computer-Aided Engineering **6**, 3–14
29. Deboeck, G., Kohonen T. (Eds.) (1998) Visual Exploration in Finance with Self-Organizing Maps. Springer, London (Japanese translation: Springer, Tokyo, 1999)
30. Naim, A., Ratnatunga, K. U., Griffiths, R. E. (1997) Galaxy morphology without classification: self-organizing maps. Astrophys. J. Suppl. Series **111**, 357–367
31. Miikkulainen, R. (1993) Subsymbolic Natural Language Processing: An Integrated Model of Scripts, Lexicon, and Memory. MIT Press, Cambridge
32. Tokutaka, H., Kishida, S., Fujimura, K. (1999) Applications of Self-Organizing Maps. Kaibundo, Tokyo, Japan (in Japanese)
33. van Hulle, M. M. (2000) Faithful Representations and Topographic Maps – From Distortion- to Information-Based Self-Organization. John Wiley, New York
34. Oja, E., Kaski, S. (Eds.) (1999) Kohonen Maps. Elsevier, Amsterdam
35. Neurocomputing **21**, Special issue on self-organizing maps, Nos. 1–3, October 1998

2 Measures for the Organization of Self-Organizing Maps

Daniel Polani

Abstract. The "self-organizing" dynamics of Self-Organizing Maps (SOMs) is a prominent property of the model that is intuitively very accessible. Nevertheless, a rigorous definition of a measure for the state of organization of a SOM that is also natural, captures the intuitive properties of organization and proves to be useful in practice, is quite difficult to formulate. The goal of this chapter is to give an overview over the relevant problems in and different approaches towards the development of organization measures for SOMs.

2.1 Introduction

2.1.1 Self-Organization

The first use of the notion of "self-organization" can be traced back to 1947 (Ashby 1947; Shalizi 1996). Yet, through the last half-century, a satisfactory universal definition of self-organization has not been found. Instead, the degree of self-organization has always been measured with measures constructed *ad hoc* for the system at hand, requiring some interpretation of the system induced by an observer.

Kohonen Maps represent the phenomenon of self-organization in such a paradigmatic way that they are often simply referred as *Self-Organizing Maps* (SOMs), as in the title of this review. The whole spectrum of questions associated with the phenomenon of self-organization emerges in conjunction with the study of SOMs. The topic of the present review bears close relations to this fundamental question. Because of the limitation of this review's scope, this question will not be studied in its own right, but its connection to the present discussions should always be borne in mind.

2.1.2 Supervised and Unsupervised Learning

An important distinction that is made in the study of neural networks is determined by the learning rule which is applied to the network under consideration. Learning rules are divided into the main classes of *supervised* and *unsupervised* rules as well as into some smaller classes which cannot be associated with the former classes in a definite way and which will not be considered further.

In the supervised case, the teacher provides a set of training examples, each of which consists of input data and corresponding target output data. The usual application of neural nets is then to model an input-output relation implied by this data set. The learning rule is expected to achieve a "suitable" correspondence between the target and the actual output values by adapting the structure and weights of the network. The suitability is measured by an explicit deviation function; e.g., in the case of backpropagation networks, by the squared error $1/2 \sum_\mu \|\mathbf{y}^{(\mu)} - \psi(\mathbf{x}^{(\mu)})\|^2$, where $(\mathbf{x}^{(\mu)}, \mathbf{y}^{(\mu)})$, are the input-output pairs of the training data and $\psi(\mathbf{x}^{(\mu)})$ is the real output of the network when the input $\mathbf{x}^{(\mu)}$ is applied.

In the unsupervised case, however, no target output and no explicit deviation from a learning target is specified. In the case of Kohonen's SOM, however, the dynamics of the system is given *a priori* by the learning rule and not derived from some explicit measure of deviation from a target state. Nevertheless the Kohonen learning algorithm causes a process to take place which human intuition is prone to describe as "self-organizing". The "organization" of the system can be intuitively detected by visual inspection of the graphical representation of SOMs during training. But "intuitive detection" and mathematical quantification are different things. This is in contrast to supervised learning methods with a canonical organization measure. Such a measure would determine how well data are being described by the neural network. For a self-organizing network such a measure would provide some a priori characterization of intrinsic organization.

In the presentation, no explicit distinction will be made between different types of notions, like organization measures, quantization measures, notions or measures of topology preservation. Instead, throughout the chapter the term *organization measure* will be used for functions that assign real values to a SOM in a given state if it fulfills certain conditions (Sec. 2.2.4).

2.1.3 On this Chapter

Before embarking on details of the different models, some remarks are in place to make clear how this chapter should be used. It provides an introduction, overview and categorization of a large selection of different approaches to quantify organization in SOMs and topographic mapping models related to it. The selection of available approaches is vast and cannot be comprehensively treated, so this review will concentrate on most important and influential approaches and give pointers for further information. The idea is to provide the reader with an overview of the most relevant aspects of structure, philosophy and properties of currently existing approaches and with pointers for more extensive studies.

2.2 General Aspects and Definitions

In general, we will assume two spaces, an *input space* V and an *output space* A, to which signals from V are mapped. The input space considered will often be continuous (typically the $\mathbb{R}^d$), the output space will often be discrete, but sometimes we will consider other types of output spaces, too.

2.2.1 A Typology of Organization Measures

Several aspects are important to distinguish the different approaches to measure organization. We will discuss them in the following.

1st-order vs. 2nd-order Measures We do not only consider "pure" organization measures that quantify the organizational structure of the mappings, but also include measures of distortion or quantization and information transmission (entropy). The reason is that the latter measures can be considered *1st-order measures*, whereas the former are *2nd-order measures*. If we speak of 1st-order measures, we intend to say that the measure value is obtained by combination of values (e.g. activation frequency or intensity) obtained for individual neurons, whereas in 2nd-order the final value of the measure results from a combination of values (e.g. similarity or distance) obtained for pairs of neurons into a single number.

Data-orientation The approaches to measure organization can be classified according to whether they include the structure of the data to map in their quantification or not. With data-oriented measures, Structurally equivalent mappings can thus obtain completely different measure values for the same type of measure, depending on the structure of the data.

Dynamics and Structure The construction of many measures is oriented at quantifying the quality of typically the final stage of a mapping. The quantification of the dynamic development of the mapping is usually considered in a separate context since in general. Typical representants for measures of the dynamics are Liapunov functions. This class can sometimes be specialized to consider energy functions, if they exist. There exist mapping scenarios (see (Graepel et al. 1998) and also Sec. 2.3.3), for which an energy function can be formulated. Note that it is not obvious what a Liapunov function has to do with what we consider a good organization. Nevertheless it is important for mathematical analysis of the training process.

Predicates, Measures and Order Parameters There are several notions of topology preservation that just distinguish the cases of a mapping being ordered or not. However, this characterization has a different quality than measures that attribute a numerical value to the disordered case. The first case can be regarded as a logical predicate that defines the notion of a mapping being ordered (or disordered) in a strict mathematical sense. When, say, the ordering predicate is not fulfilled, then the *degree* of disorder, i.e. the deviation from the ordered case can now further be quantified. For the quantification of the deviation there exists a high degree of arbitrariness.

An organization measure could be devised in such a way that it attains some extreme (either minimum or maximum) value for the ordered case and deviates from that value the "farther" in the disordered regime the mapping is. We find types of this kind of measures in (Villmann et al. 1997; Goodhill et al. 1995).

The philosophy is similar to that of *order parameters* known for thermodynamics (Reichl 1980). Order parameters are used in physics to distinguish different types of equilibrium states in physical systems. Usually, they are chosen such that their value vanishes in states with a higher distributional symmetry and deviates from 0 when the symmetry of the state distribution is broken. The order parameter view has been directly used for the study of SOMs and related models (Ritter and Schulten 1989; Der and Herrmann 1993; Graepel et al. 1997). A study that reverses that direction and uses an independently constructed organization measure (Zrehen and Blayo 1992) as a kind of order parameter is (Spitzner and Polani 1998).

Topology and Geometry Mapping organization is often seen as equivalent to topology preservation. In the strictest sense, most measures are not measuring topology preservation. Instead, they use a mixture of topological, similarity, metrical, or even further geometrical properties of the spaces mapped. The measures may be incorporating similarity or metric values themselves or just their relative ordering.

We give a short classification over some of the approaches.

"Pure" topology measures Only very few measures for organization have been formulated in notions of "pure" topology. One of the problems in finding such a formulation that one of the spaces involved in a SOM mapping is usually discrete. For these spaces, the canonical topology is the discrete topology. This is a very uninteresting case since it essentially implies a complete lack of structure..

One approach to explicitly solve the problem has been brought forward in (Villmann et al. 1997) by using a collection of "marked" discrete spaces on which a topology is defined w.r.t. to the respective marked element. This approach requires an extension of the regular notion of topology on the discrete space. In addition, the topological structures considered by Villmann et al. are induced by metric structures.

Driven from considerations of the maximization of information transmission by the mappings, equiprobable mapping methods have been investigated (Hulle 1997; Van Hulle 1997; Hulle 2000). In the present review, we wish to point out that the structure introduced in that papers on the discrete space can be interpreted as a *complex*, a structure known from algebraic topology (Henle 1979). A complex can be seen as a generalization of the notion of a graph. However, no invocation of metric structures is required for its definition, thus the method and the measures derived from it can be regarded as a truly pure topological notions.

Similarity and metrics In the spaces, the pairs of elements have to be compared for the mapping organization to be quantified. This can happen on the graph level by determining whether the elements of the space are adjacent. A second, more general way of doing it is to define a measure on the space that determines how similar the elements of the space are. a similarity measure needs not to have a metric or geometrical interpretation as opposed to a metric which has to fulfil the triangle inequality. The triangle inequality of a metric, on the other hand, can be interpreted as a kind of geometrical generalization of the transitivity property of a linear ordering relation.

It is therefore quite natural to include graph-based spaces at this point. A graph can both be interpreted as a topological and as a metrical structure. However, it is not possible to fully exploit the graph structure as a purely topological construct in the case of more than $k = 1$ dimensions and one has to resort to structures like *topological complexes* (Sec. 2.3.7). However, there is always the possibility to exploit the graph as a metric (and thus, e.g. via some neighborhood function also as similarity quantity). This makes the graph structure to a very natural model for the output space V. Indeed, in many mapping scenarios, output spaces are modeled as graphs.

Geometry Further geometrical properties are used only by very few models to quantify organization. Alder et al. (1991) present an approach that assumes a rectangular grid as output space and is related to curvature measures of Riemannian spaces.

2.2.2   Definitions

We have seen in Sec. 2.2.1 that for the output space graph structures prove to be the most versatile models. Therefore, in the definition of the basic SOM model the output space will be based on a graph.

Definition 1 (Self-Organizing Map). Define the *state of a SOM* as a map $\mathbf{w}{:}A{\rightarrow}V$ from a discrete finite set A of (formal) *neurons* to a convex subset V of $\mathbb{R}^d$, the *input space*, mapping each neuron j to its *weight* $\mathbf{w}_j$. We will often simply say SOM instead of SOM state.

Let $\mathcal{C}_K \subseteq A \times A$ be the adjacency structure of an undirected graph without weights with the set A of neurons as vertex set. This graph will be called the *Kohonen graph*.

The Kohonen graph induces a metric d_A on A, by defining $d_A(i, j)$ as the minimum distance between two neurons $i, j \in A$, where $d_A(i, j) = 1$ for two adjacent neurons. Unless stated otherwise, on V the Euclidean metric is used as d_V.

Given an initial SOM state $\mathbf{w}(0)$, a SOM training sequence $\big(\mathbf{w}(t)\big)_{t=1,2,\dots}$ is defined by the applying the Kohonen learning rule

$$\Delta\mathbf{w}_j(t) := \epsilon(t) \cdot h_t\Big(\mathbf{i}^*_{\mathbf{w}(t)}\big(x(t)\big), j\Big)\big(x(t) - \mathbf{w}_i(t)\big) , \qquad (2.1)$$

to all neurons j with $\Delta\mathbf{w}_j(t) = \mathbf{w}_j(t+1) - \mathbf{w}_j(t)$. Here $x(t)$ is the training input at time t, $\epsilon(t)$ the learning rate, h_t the activation profile and $\mathbf{i}^*_{\mathbf{w}(t)}$ a function where $\mathbf{i}^*_{\mathbf{w}(t)}\big(x(t)\big)$ is a neuron i minimizing the distance $d_V\big(x(t), \mathbf{w}_i(t)\big)$. In the following we will write $\mathbf{i}^*$ instead of $\mathbf{i}^*_{\mathbf{w}(t)}$ for notational convenience.

V is called *input space*. $\mathbf{i}^*$ is the *quantization function*. For $x \in V$, we say that input signal x *activates* the neuron $\mathbf{i}^*(x)$. For a given neuron i, the set $V_i := \mathbf{i}^{*-1}(i)$ of inputs from V that activate this neuron is called its *receptive field*. Its closure, $\overline{V_i}$, is called the *Voronoi cell* of i w.r.t. the set of points $\{\mathbf{w}_k \mid k \in A\}$ if d_V is Euclidean. In that case, for a given SOM state $\mathbf{w}$, the receptive field of a neuron i is uniquely defined on its interior except for a Lebesgue null set[1].

---
[1] If the metrics is non-Euclidean, "pathological" cases can occur where the sets of ambiguous inputs need not be Lebesgue null sets (Polani 1996).

The quantization function $\mathbf{i}^*$ can be regarded as a lossy compression for signals from V into an event represented by a neuron $i \in A$. The inverse direction is information-conserving, as $\mathbf{i}_\mathbf{w}^*$ is a left-inverse map to $\mathbf{w}$.

2.2.3   Metrics and Topology

The restriction to graph-induced metrics or *graph metrics* on A nevertheless enables the realization of important metric structures on standard configurations like an n-dimensional grid. Fig. 2.1 shows the realization of two particular metrics on a 2-dimensional grid of neurons. The $\|.\|_1$- and $\|.\|_\infty$-metric can be realized, however not the (Euclidean) $\|.\|_2$-metric.

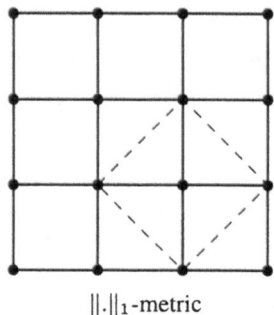
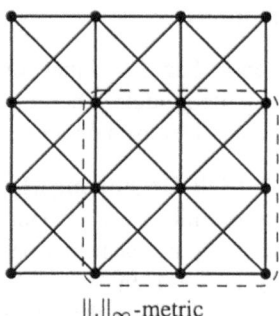

$\|.\|_1$-metric $\|.\|_\infty$-metric

Fig. 2.1. Realization of different 2-dimensional metric structures by Kohonen graphs. The dashed regions indicate the "unit circles" of the corresponding metrics.

The intuitive notion of the output spaces being e.g. one-, two- or higher-dimensional grids (particularly rectangular ones) is widely used in the literature (e.g. (Ritter et al. 1994)). Since our investigations will also consider more general metric structures of A, it is sometimes advantageous to cast this somehow vague notion into a more precise definition (Polani 1995, 1996). Here, when talking about the dimension of a grid, we will appeal to the reader's intuition.

2.2.4   Requirements for an Organization Measure

Due to the lack of a canonical notion or measure of SOM organization, there is a considerable amount of arbitrariness, i.e. of "degrees of freedom", for its choice. It will be therefore necessary to clarify the requirements that should qualify a notion of SOM organization. Let $\mathcal{W}_{A,V} \equiv \mathcal{W}$ be the set of maps with output space A and input space V. We wish to state informally two properties which would qualify a function

$$\mu : \mathcal{W}_{A,V} \to \mathbb{R}$$

to be called an organization measure:

1. It should quantify the process of self-organization during training, i.e. its value should increase (or decrease) monotonously on average.
2. It should measure the quality of the embedding of A into the data manifold, i.e. the embedding shown in Fig. 2.2(a) should obtain a "better" value than, say, the embedding shown in Fig. 2.2(b).

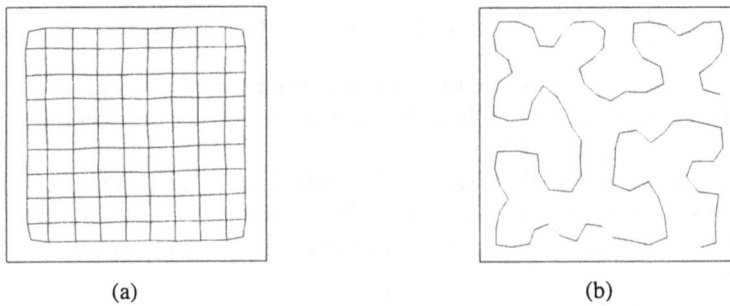

(a)   (b)

Fig. 2.2. Kohonen maps with different embedding quality

The first condition is of interest for mathematical analysis and for the understanding of the organization process. The second one is important for applications which have to estimate the quality of the description of a high-dimensional data manifold by the SOM.

However, an organization measure need not fulfil both properties. Indeed, it is not even clear that a measure μ fulfilling but the first condition could be found. The second property is not a hard constraint as the first, as different measures will most probably yield different estimations for ambiguous cases of embeddings (e.g. when the data manifold consists of several components of different dimension).

Since the first property can be cast into precise form in a fairly straightforward way, one would choose as first candidates for a discussion as organization measures such μ having this property. We can demand a weaker and a stronger version of property 1. The weaker version would only require μ to be a Liapunov-function, i.e. that the μ-measure would decrease on average during the training, or formally

$$E\Big(\mu\big(\mathbf{w}(t+1)\big)\Big) \leq \mu(\mathbf{w}(t))$$

for all $\mathbf{w}(t) \in \mathcal{W}$, $t \in \mathbb{N}$, E denoting the average over a random training signal.

Organization measures can express a deviation, i.e. their value drops when organization becomes better during training or their value may grow during the self-organization process. Yet other sorts of information can be obtained by an organization measure (see Sec. 2.3.6).

2.3   Measures of Organization

2.3.1   Inversion Measures

For the case of a one-dimensional net with $A = \{1 \ldots n\}$ and V a subinterval of $\mathbb{R}$ a Liapunov-function can be given. Following (Cottrell and Fort 1987; Cottrell et al. 1994) and (Kohonen 1989) an organization measure μ_1 can be chosen as the number of inversions, i.e. as

$$\mu_{\text{Inv}}(\mathbf{w}) = \Big| \; \big\{ i \in \{2 \ldots n - 1\} \; | \; \text{sgn}(\mathbf{w}_{i+1} - \mathbf{w}_i) \neq \text{sgn}(\mathbf{w}_i - \mathbf{w}_{i-1}) \big\} \; \Big| \; .$$

In other words, μ_{Inv} counts the times change of directions takes place while i runs from 1 to n. The convergence theorems in above references guarantee that μ_{Inv} thus defined possesses property 1.

As the restriction to dimension 1 is not sufficient for applications, a generalization of this measure to a measure μ_{ZB} on $\mathcal{W}_{A,V}$ for open convex subsets $V \subseteq \mathbb{R}^n$, $n > 1$ has been formulated in (Zrehen and Blayo 1992). It is given by

$$\mu_{\text{ZB}}(\mathbf{w}) = \frac{1}{|\mathcal{C}_K| \cdot (N - 2)} \sum_{(i,j) \in \mathcal{C}_K} D(i,j) \, ,$$

where $\mathcal{C}_K$ is the set of adjacency pairs $(i, j) \in A \times A$ in the Kohonen graph and $D(i, j)$ is the number of receptive fields of neurons $k \neq i, j$ intersecting with the line from $\mathbf{w}_i$ to $\mathbf{w}_j$. The factor $1/|\mathcal{C}_K| \cdot (N - 2)$ normalizes the measure according to the number of possible connections and disturbances.

However, there is no monotonicity theorem guaranteeing a decrease (on average) for μ_{ZB}. Therefore it is not clear, whether it can be considered as a Liapunov function for the SOM.

2.3.2   Entropy

In (Linsker 1988) the information transmission from input to output has been studied for a specialized class of feed-forward neural networks. The investigations showed that under certain conditions the learning process of those networks can be observed to express a maximization of information preservation in the data transmission from the input to the output neurons. This suggests using an information-theoretic principle for the design of networks, *Linsker's infomax principle* and its variants (Haykin 1999). Neural network architectures are not necessarily explicitly designed to obey the infomax principle, but having a look at their information preservation capabilities may promise to yield a better understanding of their information processing properties.

The activation process of a SOM can be considered as a transmission mechanism for information which transfers signals hitting the input space V into signals in the output space A (Hulle 2000). For this information transfer process different kinds of entropies can be calculated; here we concentrate on the *neuron activation entropy* or simply *neuron entropy*.

The neuron entropy attains its maximum when probability of activation is the same for all neurons, i.e. when when all receptive fields have the same probability to be hit under the distribution P. Maximal entropy signifies that the quantization of the input space takes place in a best possible way; of course only the quantization class is then specified as precise as possible with the neurons available and not the position of the incoming signal. The simple neuron entropy does not know anything about the geometrical properties of the embedding of $\mathbf{w}$ in the input space. What neuron entropy can tell us is of purely information-theoretical nature. As long as the receptive fields have the same probability to be activated it does not play a role whether they might be long and narrow, thus leading to a high quantization error μ_Q or close to ideal sphere packing for equidistributed input signals.

Let an input signal distribution be given. If the probability that a neuron i is activated by an input signal is given by p_i, the neuron activation entropy is given by

$$- \sum_{i \in A} p_i \log p_i \; ,$$

The entropy can be seen as that amount of Shannon information that is conveyed by the mapping. Here the activations of the individual neurons are elementary. In that view, the individual activations cannot be compared to each other since it does not assume any semantics structure on the set of transmitted events, and thus also no similarity. In particular, the entropy needs not at all be related to topology preservation.

2.3.3 Energy Function Measures

It would be desirable to have an organization measure with stronger properties than are required for a Liapunov-function (Wiskott and Sejnowski 1997; Heskes 1999). Such a measure could be obtained by using an *energy function* for the training rule.

Definition 2 (Energy Function). Let $\left(T(\epsilon)\right)_{\epsilon \in [0,1]}$ be collection of learning rules with

$$\mathbf{w}_i(t+1) = [T(\epsilon)(x, \mathbf{w})]_i \; ,$$

for $i \in A$ and where ϵ is the learning rate. Assuming fixed $x \in V$ and $\mathbf{w} \in \mathcal{W}$, then $\psi_x(\epsilon) := T(\epsilon)(x, \mathbf{w})$ is a differentiable curve in ϵ with $\psi_x(0) = \mathbf{w}$, which allows us to define

$$\tau_x(\mathbf{w}) = \frac{d\psi_x}{d\epsilon}(0) \; ,$$

which is a vector in the tangent space $T_{\mathbf{w}}\mathcal{W}$ of $\mathcal{W}$ in $\mathbf{w}$ (see Fig. 2.3 and also (Abraham et al. 1983; Forster 1984)).

Since τ_x is defined for every $\mathbf{w} \in \mathcal{W}$, τ_x defines a vector field on $\mathcal{W}$. A differentiable function $U_x : \mathcal{W} \to \mathbb{R}$ is called *energy* or *potential* function for the learning rule T if

$$dU_x = \tau_x^b \; ,$$

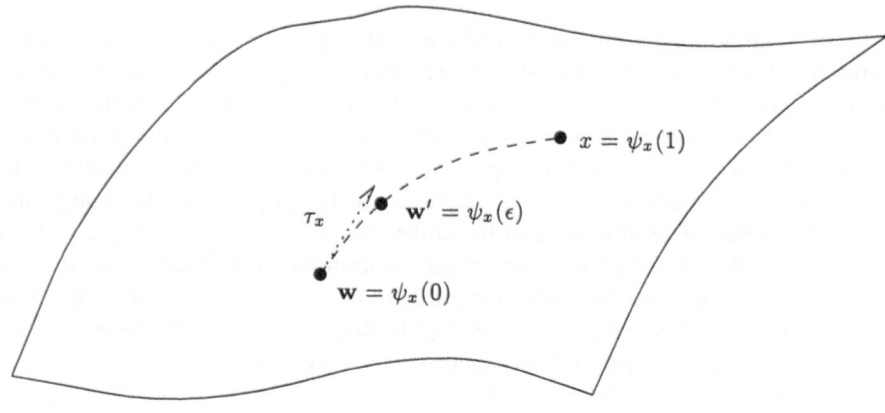

Fig. 2.3. $\psi_x(\epsilon)$

where τ_x^b denotes the adjoint 1-form to the vector field τ_x via the canonical metric[2] on $\mathcal{W}$ (notation from (Abraham et al. 1983)). If x is a random variable then

$$\tau := \mathbf{E}(\tau_x)$$

defines a vector field on $\mathcal{W}$ ($\mathbf{E}$ denoting the expectation value w.r.t. the distribution of x). In analogy to above a function $U : \mathcal{W} \to \mathbb{R}$ is called *energy* function for τ if $dU = \tau^b$.

In certain cases it is possible to specify an energy function explicitly.

Energy Function for Training Signal with Discrete Support If the training signal is concentrated on a set $\{\hat{x}_1, \ldots, \hat{x}_q\}$ which are assumed to have probabilities $p_1, \ldots, p_q$, the function

$$U(\mathbf{w}) = \frac{1}{2} \sum_{i,j} \left(h(i,j) \sum_{\hat{x}_k \in V_j(\mathbf{w})} p_k d_V(\hat{x}_k, \mathbf{w}_i)^2 \right) \tag{2.2}$$

is an energy function for the learning rule (Ritter et al. 1994).

Energy Function for 0-Neighbour Case (Quantization Error) For a SOM with $A = \{1, \ldots, n\}$, $h(i,j) = \delta_{ij}$ (δ_{ij} denoting the Kronecker delta) and a training signal with probability density p a potential function is given by

$$U(\mathbf{w}) = \sum_{i=1}^{n} \int_{V_i(\mathbf{w})} d_V(\mathbf{w}_i, x)^2 \, p(x) dx \quad \text{(Cottrell et al. 1994).} \tag{2.3}$$

---

[2] Assuming $V \subseteq \mathbb{R}^d$, $\mathcal{W}$ can be identified with a subset of $\mathbb{R}^{d \cdot N}$, inheriting the corresponding metric.

This is identical to the squared mean quantization error.

The mean quantization error does not contain any information about topology. However, a connection between the quantization error and the organization of the SOM can be made. Luttrell (1989) shows that by minimizing the quantization error and assuming a suitable error model on the output space, one arrives at a learning rule similar to the SOM. Furthermore, (Ritter et al. 1994) show that a neighborhood improves convergence. This also is observed for other topographic mapping models (Hulle 2000). In (Polani 1996), it is shown that to enhance a fast improvement of the quantization error, a Genetic Algorithm optimization creates a neighborhood, but this neighborhood needs not be topology preserving (Polani and Uthmann 1992, 1993).

Nonexistence of Energy Function for the Continuous Case However, for a case as simple as the SOM with grid dimension 1, the question whether there exists a potential function for more general h or for a continuous distribution of x has been answered negatively. Indeed, by use of an elegant argument, even a stronger result is shown in (Erwin et al. 1992a). As it gives insight into the complex configurations that can arise in the general case, we give a reformulation of the argument in the language of differential forms subsequently yielding a geometric interpretation.

The analysis in (Erwin et al. 1992a) is focused on SOMs with one-dimensional grid topology, with $A = \{1, \ldots, n\}$ and $V = [0, 1]$. As training input an i.i.d. sequence of random variables $(x_t)_{t \in \mathbb{N}}$ is considered, distributed according to the uniform distribution on $[0, 1]$. Analysis of the vector field τ discussed above is restricted to the subset $\mathcal{W}' \subseteq \mathcal{W}$ with

$$\mathcal{W}' = \{\mathbf{w} \in \mathcal{W} | \mathbf{w}_i \neq \mathbf{w}_j \text{ for } i \neq j\},$$

the complement $\mathcal{W} \setminus \mathcal{W}'$ of which is a Lebesgue-null set. To simplify calculation, τ is considered only on the subset

$$\mathcal{W}'' = \{\mathbf{w} \in \mathcal{W} | \mathbf{w}_i < \mathbf{w}_j \text{ for } i < j\}$$

and extended to $\mathcal{W}'$ by application of permutations of A, making use of symmetry properties of $\mathcal{W}'$.

The existence of a function U with $dU = \tau^b$ requires $d\tau^b$ to vanish, which does not happen in general. Indeed, even the weaker condition

$$d\tau^b \wedge \tau^b = 0 \tag{2.4}$$

($\wedge$ denoting the wedge product of differential forms) is not fulfilled in general, as demonstrated in (Erwin et al. 1992a). By the Frobenius theorem (Abraham et al. 1983), this implies that τ^b is not integrable or, equivalently, τ^b cannot be represented as $g\,dU$ with suitable functions $g, U : \mathcal{W} \to \mathbb{R}$. Integrability of the form τ^b would have an immediate geometrical interpretation. It would mean that locally there would exist a system of submanifolds of $\mathcal{W}'$, the tangent bundles of which would be annihilated by τ^b. In other words, if M were such a manifold, $\tau^b(v)$ would

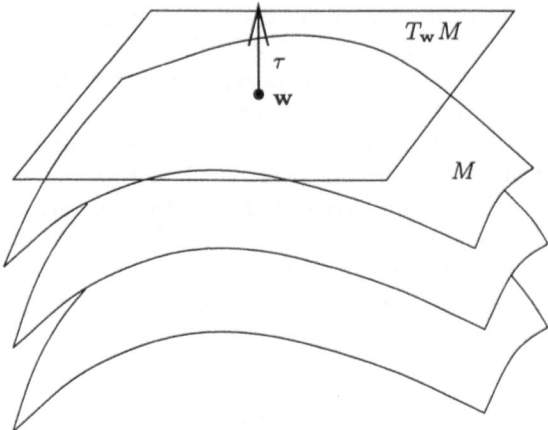

Fig. 2.4. Geometrical interpretation of integrability at point with nonvanishing τ

vanish for a vector v from a tangent space $T_{\mathbf{w}}M$, i.e. $T_{\mathbf{w}}M$ is perpendicular to τ (Fig. 2.4). The importance of the non-integrability of $\tau^{\flat}$ arises from the fact that this implies the nonexistence of local "equipotential" submanifolds M (whose tangent spaces $T_{\mathbf{w}}M$ would be perpendicular to τ at every $\mathbf{w} \in M$).

Modified SOM Algorithms and Energy Functions There exist further modifications of the SOM algorithm and similar models that allow the formulation of an exact energy function for the learning algorithm, e.g. (Wiskott and Sejnowski 1997; Graepel et al. 1997, 1998; Heskes 1999; Hulle 2000). These cannot be discussed here in detail and we refer the readers to the original works.

The question may now be why one would bother to study the original SOM algorithm which poses so much resistance to conventional mathematical analysis. A reason to do so is that the simplicity of the SOM learning rule on the one hand and the complexity of its mathematical analysis on the other hand bring the essential questions of topographic self-organization to the point. Topographic self-organization is *not* just about minimizing an appropriate energy function, even if such energy functions may be devised. It was probably a piece of luck for research motivation that the original rule has not been derived from an energy function. Otherwise the fact that the energy function had be devised to create an organized map would have obscured the fact that topographically organized maps can be created by a much larger variety of dynamics than are defined by an energy function of a certain character.

It will be an interesting question for the future to determine how strong the conditions have to be to guarantee a certain degree of system organization.

2.3.4 Further Approaches for Dynamics Quantification

Energy and Liapunov functions as discussed above are measures based on the dynamics of the learning rule and the SOM learning rule may be modified in such a way that it becomes the gradient of some energy function. However, one could adopt here another view, namely looking at the SOM purely from the dynamics system view. One would then like the energy function to be derived from a given dynamical system and not from additional external interpretation of the system as a topology preserving system. Since the construction of energy functions is impossible for the general SOM, Spitzner and Polani (1998) used a different approach to condense the dynamics of a 1-dimensional SOM into a smaller number of parameters, based on a principle from synergetics (Haken 1983). It separates the dynamics into fast and slow subsystems (Mees 1981; Jetschke 1989). Applied to the 1-dimensional SOM, the method revealed that close to the fixed point of the dynamics a condensation of the essential dynamics into few relevant degrees of freedom or "order parameters" was not possible. However, this might be possible if also states far from the fixed point are taken into account. An indication that this view could be useful to formulate a quantification of (self-)organization not based on a topographic interpretation of the map is the hierarchical structure of the metastable states for the linear SOM (Erwin et al. 1992b). This view will be an interesting research field for the future.

2.3.5 Measures of Curvature

In (Alder et al. 1991) a "crinkleness" measure is defined for SOMs. It is restricted to Kohonen maps with A having rectangular grid topology. For every "inner" neuron of A (every neuron surrounded by neighbours in every dimension of A) a local deformation is calculated. The crinkleness measure then is given by averaging these local deformations.

The Discrete Version The local deformation is calculated in (Alder et al. 1991) as follows: We consider the distortion with respect to the k-th dimension of A. If A has $n_1 \times n_2 \times \cdots \times n_k \times \cdots \times n_r$ grid topology and $i \equiv (i_1, \ldots, i_k, \ldots, i_r)$ is an inner neuron (i.e. $i_k \in \{2, \ldots, n_k - 1\}$), let $i^\pm = (i_1, \ldots, i_k \pm 1, \ldots, i_r)$, further $v_k^- := \mathbf{w}_i - \mathbf{w}_{i^-}$ and $v_k^+ := \mathbf{w}_{i^+} - \mathbf{w}_i$.

The local deformation $c(i)$ of neuron i is then given by

$$
\begin{aligned}
c(i) &= \frac{1}{r} \sum_{k=1}^{r} \left(1 - \frac{\langle v_k^-, v_k^+ \rangle}{|v_k^-| \cdot |v_k^+|} \right) \\
&= \frac{1}{r} \sum_{k=1}^{r} \left(1 - \cos \angle(v_k^-, v_k^+) \right) \\
&= \frac{1}{r} \sum_{k=1}^{r} (1 - \cos \phi_k) ,
\end{aligned}
\tag{2.5}
$$

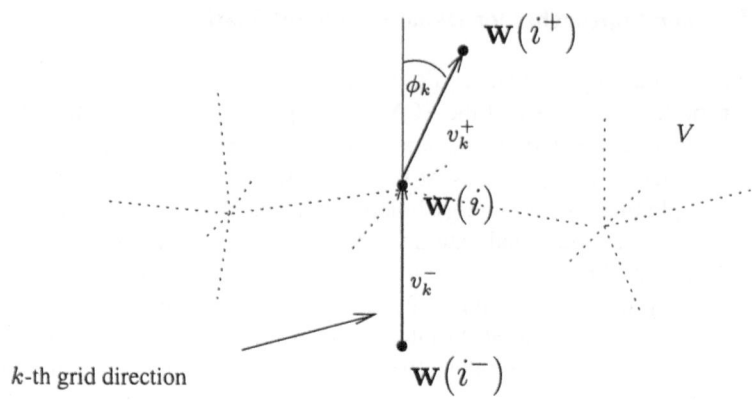

k-th grid direction

Fig. 2.5. Local deformation at a neuron

where ϕ_k is the angle $\angle(v_k^-, v_k^+)$ between v_k^- and v_k^+. The geometric situation can be visualized as in Fig. 2.5.

The average of $c(i)$ on all inner neurons of A will be called μ_{ATA}. As μ_{ATA} assumes a rectangular grid as topology of A, it presupposes significantly more structure than other organization measures only assuming some graph structure on A and one could hope to relate this measure to some of the well known curvature measures of the continuous case. For this purpose certain quantities of the discrete case must be converted to quantities of the continuous model.

To achieve this, define for given $i_1, \ldots, i_{k-1}, i_{k+1}, \ldots, i_r$ the function

$$
\begin{aligned}
s : \{1, \ldots, n_k\} &\to & A \\
t &\mapsto (i_1, \ldots, i_{k-1}, t, i_{k+1}, \ldots, i_r) \, .
\end{aligned}
$$

Then, if $i = s(t)$, obviously

$$
\begin{aligned}
v_k^+ &= [\mathbf{w} \circ s](t+1) - [\mathbf{w} \circ s](t) \\
v_k^- &= [\mathbf{w} \circ s](t) - [\mathbf{w} \circ s](t-1) \, .
\end{aligned}
$$

A Continuous Version To obtain the continuous version of the model, replace A, s and $\mathbf{w}$ by appropriate continuous versions of the corresponding sets and maps, thereby we are able to turn $\mathbf{w} \circ s$ into a smooth function $\psi : [0, 1] \to V$; the small displacement v_k^+ becomes $\dot{\psi}(t)$ in the continuous version.

Now in the crinkleness measure the change of displacement vectors is measured in terms of $1 - \cos \phi_k$. To make the connection between the continuous and the discrete model and to obtain the differential analogy for $1 - \cos \phi_k$, we consider $\angle(\dot{\psi}(t), \dot{\psi}(t + \Delta t))$ instead of ϕ_k and investigate the Taylor expansion of $1 - \cos \angle(\dot{\psi}(t), \dot{\psi}(t + \Delta t))$ with respect to Δt around t. A somehow tedious calculation shows that the first non-vanishing coefficient of the expansion is that of Δt^2, i.e. the 2nd derivative of $1 - \cos \angle(\dot{\psi}(t), \dot{\psi}(t + \Delta t))$ with respect to Δt at $\Delta t = 0$

can be considered as the continuous version of the crinkleness at $\psi(t)$. It is given by

$$c(\psi(t)) = \frac{\langle\dot\psi(t), \dot\psi(t)\rangle\langle\ddot\psi(t), \ddot\psi(t)\rangle - \langle\dot\psi(t), \ddot\psi(t)\rangle^2}{\langle\dot\psi(t), \dot\psi(t)\rangle^2}. \qquad (2.6)$$

Because of the Cauchy-Schwarz-inequality this term is always ≥ 0 (as is its discrete counterpart) and vanishes (this means local crinkleness 0) exactly when $\dot\psi(t)$ is parallel (or antiparallel) to $\ddot\psi(t)$, which means that change of the tangent takes only place in direction parallel (antiparallel) to the tangent.

In exactly this case $c(\psi(t))$ is minimal. Note that in this respect the continuous model differs from the discrete one, since in the latter v_k^- may be antiparallel to v_k^+, yielding a maximal crinkleness value, but in the continuous (smooth) case this cannot happen, since tangent vectors for t and $t + \Delta t$ are always close for small Δt.

This can be reformulated for a general non-vanishing vector field Z on the curve ψ (O'Neill 1983):

$$c_k(\psi(t)) = \left.\frac{\langle Z, Z\rangle\langle D_Z Z, D_Z Z\rangle - \langle Z, D_Z Z\rangle^2}{\langle Z, Z\rangle^2}\right|_{\psi(t)}, \qquad (2.7)$$

where D denotes the canonical connection on V. Thus the differential geometric version of the crinkleness measure μ_{ATA} at location $\mathbf{w}_i \equiv \mathbf{w}(i) \in V$ is obtained by assuming A and V to be smooth manifolds of dimensions r and d, respectively, the latter furnished with a scalar product; then a collection $\{E_j | j = 1 \ldots r\}$ of vector fields on A is chosen forming a basis (orthonormal, if A is equipped with a scalar product) and setting:

$$c(\mathbf{w}(i)) := \sum_{j=1}^{r} c_j(D\mathbf{w}_i(E_j)), \qquad (2.8)$$

$D\mathbf{w}_i$ denoting the Jacobian of $\mathbf{w}$ at $i \in A$ in this context. The global crinkleness measure is then obtained by averaging over A, i.e. by integrating over A and normalizing afterwards.

Unfortunately this measure is not invariant with respect to the choice of the basis $\{E_j | j = 1 \ldots n\}$ (even when restricted to orthonormal ones), as can be shown by a straightforward example. Therefore it cannot be expressed as an abstract property of the embedding of a smooth manifold A into the smooth manifold V. This means that in the transition from the discrete to the smooth model, i.e. from the grid to the manifold the "special" character of certain selected directions of A is not lost. One would, however, prefer a measure taking advantage of the local isotropy of the smooth model and being insensible to the choice of a basis.

Possible Generalizations Through these considerations one is led to consider the possible derivation of a whole collection of crinkleness or – better – curvature measures for application to discrete SOMs from the well-known curvature notions of Riemannian geometry. One may apply the distinction between intrinsic curvature quantities measuring the "inner" distortion of the Kohonen embedding and quantities measuring the distortion of $\mathbf{w}(A)$ with respect to V. For $\mathbf{w}(A)$ having an "inner" distortion at least two dimensions are required, since the intrinsic curvature of a 1-dimensional manifold vanishes (O'Neill 1983). However, notions of the relative curvature of a submanifold embedded into a manifold also exist. Hence measures might be derived enabling to focus on different aspects of the embedding quality.

One should nevertheless keep in mind that these characterizations will be purely geometrical and restricted to evaluate the current map $\mathbf{w}$ without referring to the data manifold it is expected to describe. Thus, one obtains two possible informations by the measures: On condition that a SOM is an appropriate model for the given data manifold, the curvature measures give an information on the geometrical structure of the data manifold. In this case, large curvature values will have to be traced back to the form of the data manifold. But much more often large curvatures will be result of the well-known crinkling effects of nets where dimension of A is inappropriately chosen (as e.g. in Fig. 2.2(b)).

2.3.6   Similarity- and Metric-Based Measures

Goodhill et al. (1995); Goodhill and Sejnowski (1997a) discuss measures that are based essentially on notions of similarity. A measure of similarity can be realized by a metric, but need not fulfil the triangle inequality. In the mentioned papers, the similarity measures are realized as either a (often Euclidean) metric or as some monotonous functions applied to a metric Depending on the orientation of the similarity value, a good match is reflected by a low or a high value.

We cite a proposition from (Goodhill et al. 1995) that is useful for generalizing the notion of a topological homeomorphism to discrete spaces based on similarity measures. First, we first require a definition.

Definition 3. Given metric spaces $\langle X, d_X \rangle$ and $\langle Y, d_Y \rangle$, a map $M : X \to Y$ is called *similarity preserving* if

$$\begin{aligned}
\forall x_1, x_2, x_3, x_4 \in X : d_X(x_1, x_2) < d_X(x_3, x_4) \Rightarrow \\
d_Y(M(x_1), M(x_2)) \leq d_Y(M(x_3), M(x_4))
\end{aligned} \tag{2.9}$$

Proposition 1. *Let $\langle X, d_X \rangle$ and $\langle Y, d_Y \rangle$ be identical metric spaces with countable dense subsets. If M is a bijection such that both M and M^{-1} are similarity preserving, then M is a (topographic) homeomorphism and X and Y are topologically equivalent via M.*

The similarity preservation condition is, as noted by Goodhill et al. (1995), stronger than required to guarantee homeomorphism, therefore the stronger notion of *topographic homeomorphism* is used. A naive definition of topology in discrete spaces is unsatisfactory because the discrete topology is trivial and makes all mappings continuous. The conditions of Proposition 1, however, make sense for discrete spaces, too, and can be used to define a nontrivial notion of topography preservation for mappings between discrete spaces.

A Measure for Topographic Homeomorphism In the following, we use our standard notation for input and output space also for the scenarios of Goodhill et al. Given a measure of similarity F_V on the input space and F_A on the output space, Goodhill et al. (1995); Goodhill and Sejnowski (1997a,b) define a measure C by[3]

$$C := \sum_{\substack{(i,j)\in A\times A \\ i\neq j}} F_A(i,j)F_V(\mathbf{w}_i,\mathbf{w}_j)\ . \tag{2.10}$$

Goodhill et al. (1995) show that minimizing C w.r.t. $\mathbf{w}$ yields a topographic homomorphism if it exists. They also point out that the measure C is related to quadratic assignment problems. Some special cases of the measure C are discussed in (Goodhill and Sejnowski 1997a). A measure very similar to the *inverted minimal distortion* introduced there has already been used in (McInerney and Dhawan 1994).

Further measures of map quality studied in (Goodhill and Sejnowski 1997a) are

Metric Multidimensional Scaling: introduced in (Torgerson 1952), given by

$$\sum_{\substack{(i,j)\in A\times A \\ i\neq j}} \big(d_A(i,j) - d_V(\mathbf{w}_i,\mathbf{w}_j)\big)^2\ , \tag{2.11}$$

a low value denoting a good match;

Sammon Measure: (Sammon 1969) a measure treating input and output space unsymmetrically via

$$\frac{1}{\displaystyle\sum_{\substack{(i,j)\in A\times A \\ i\neq j}} d_A(i,j)} \sum_{\substack{(i,j)\in A\times A \\ i\neq j}} \frac{\big(d_A(i,j) - d_V(\mathbf{w}_i,\mathbf{w}_j)\big)^2}{d_A(i,j)}\ , \tag{2.12}$$

a low value denoting a good match;

Spearman Coefficient: (Bezdek and Pal 1995), based on an order statistics (and not the numerical values) of the similarity values for the different (i,j) pairs. Let R_k and S_k enumerate the respective ranks of the values $d_A(i,j)$ and $d_V(\mathbf{w}_i,\mathbf{w}_j)$ for all the pairs $(i,j) \in A \times A$. Then the Spearman coefficient of this order statistics is given by

$$\frac{\sum_k (R_k - \overline{R})(S_k - \overline{S})}{\sqrt{\sum_k (R_k - \overline{R})}\sqrt{\sum_k (S_k - \overline{S})}} \tag{2.13}$$

its value lies between 0 and 1, a high value denoting a good quality mapping. Its one of the measures mentioned in Sec. 2.2.1 that uses only the ranking information from the metrics of the spaces.

---

[3] The notation used here differs slightly from that of the original papers, but is — apart from a constant factor 2 — essentially equivalent to that given there, assuming that the similarity measures are symmetrical in their arguments.

The Topographic Product The *topographic product* is a measure motivated by the study of dynamical systems that has been modified for use with SOMs.

The topographic product μ_{BP} as defined in (Bauer and Pawelzik 1992) is calculated as follows:

- For all $j \in A$ determine $n_1^A(j), n_2^A(j), \ldots, n_{N-1}^A(j) \in A$ such that

$$d_A(j, n_1^A(j)) \leq d_A(j, n_2^A(j)) \leq \cdots \leq d_A(j, n_{N-1}^A(j))$$

and $n_1^V(j), n_2^V(j), \ldots, n_{N-1}^V(j) \in A$ such that

$$d_V(\mathbf{w}_j, \mathbf{w}_{n_1^V(j)}) \leq d_V(\mathbf{w}_j, \mathbf{w}_{n_2^V(j)}) \leq \cdots \leq d_V(\mathbf{w}_j, \mathbf{w}_{n_{N-1}^V(j)}) .$$

- For $j, k \in A$ set

$$Q_1(j, k) := \frac{d_V(\mathbf{w}_j, \mathbf{w}_{n_k^A(j)})}{d_V(\mathbf{w}_j, \mathbf{w}_{n_k^V(j)})}$$

and

$$Q_2(j, k) := \frac{d_A(j, n_k^A(j))}{d_A(j, n_k^V(j))} .$$

- Set

$$P_3(j, k) := \left(\prod_{l=1}^{k} Q_1(j, l) Q_2(j, l) \right)^{1/2k} .$$

- The final measure is then given by the logarithmic average:

$$\mu_{BP} := \frac{1}{N \cdot (N-1)} \sum_{j=1}^{N} \sum_{k=1}^{N-1} \log \big(P_3(j, k) \big) .$$

A value near 0 signifies good adaptation, negative values of μ_{BP} are expected to indicate folding of A into V (Bauer and Pawelzik 1992) (typically when "dimension" of A – considered as grid – is chosen too small), positive values are expected to indicate a topology mismatch in the other direction.

In the first step, the choice of $n_1^A(j), n_2^A(j), \ldots, n_{N-1}^A(j)$ need not be unique, when there are different pairings of neurons, such that e.g. $d_A(j, k) = d_A(j, k')$ with $k \neq k'$. No standard procedure is given in (Bauer and Pawelzik 1992) to resolve this ambiguity. Therefore different implementations of the measure may yield different values for the intermediate values Q_1 and Q_2, while – empirically – the global result seems to be largely independent of these details[4].

---

[4] The author's implementation of μ_{BP} has been compared to the implementation from (Speckmann et al. 1994).

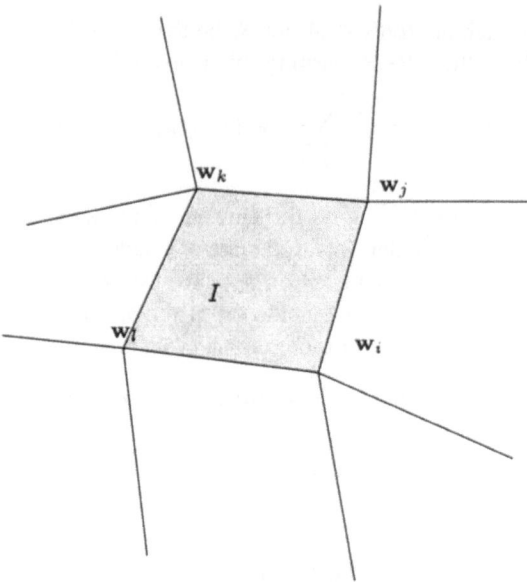

Fig. 2.6. A VBAR quadrilateral defined by four neurons in two-dimensional space

2.3.7   A "Topological" Learning Rule

With (Hulle and Martinez 1993) and (Van Hulle 1997), the topological learning rules BAR and VBAR are introduced. They carry both the structural advantage of achieving or approximating an equiprobabilistic representation of the data manifold, thus maximizing entropy and, even more, being, in a sense, true topological learning rules.

The VBAR Algorithm   Here we give a short sketch of VBAR and refer the reader to (Hulle 1997, 2000) for more details. In VBAR, the output space is considered a rectangular grid of some given dimension d. Similar to the SOM, a neuron i has a corresponding "weight" vector $\mathbf{w}_i$ in the input space. Unlike in the SOM, however, the neurons do not just denote individual points to be mapped to the input space, but define the corners, and thus the borders of quadrilaterals H_k (d-dimensional rectangular intervals). Fig. 2.6 shows this for $d = 2$.

In VBAR, an input signal is not considered to activate an individual neuron, but a complete quadrilateral. Exactly the neurons defining the corners of the respective quadrilateral will be updated. Let I define the image of a quadrilateral (multidimensional interval) under $\mathbf{w}$ in input space (Fig. 2.6). Let χ_I be the characteristic function of the set I, i.e.

$$\chi_I(x) := \begin{cases} 1 & \text{if } x \in I \\ 0 & \text{else}. \end{cases} \tag{2.14}$$

for all $x \in V$. For each neuron $i \in A$, let S_i be the set of those quadrilaterals that have i as corner. Then the VBAR learning rule is given by

$$\Delta \mathbf{w}_i(t) := \eta \sum_{I \in \mathbf{w}(S_i)} \chi_I(x) \, \mathrm{sgn}(x - \mathbf{w}_i(t)) \qquad (2.15)$$

for every neuron i, where $x \in V$ is the current input, $\mathbf{w}(S_i)$ denotes the image of the quadrilateral set S_i under $\mathbf{w}$, i.e. the set of quadrilateral images under $\mathbf{w}$, η is the learning rate and sgn operates component-wise. For further details and the activation of quadrilaterals when x outside the grid map, see e.g. (Hulle 2000). For the VBAR algorithm, the notion of topographic order is well-defined.

Definition 4. A VBAR mapping is *topographically ordered* if for all inputs $x \in V$ the condition

$$\sum_I \chi_I(x) = 1 \qquad (2.16)$$

holds where the sum is over all quadrilaterals I.

Condition (2.16) guarantees that, for a topologically ordered VBAR map, all points in V are covered by exactly one quadrilateral image $\mathbf{w}(I)$. For a point x that would not be covered, the term in (2.16) would become 0. A point x that would be covered multiply would return a number larger than 1. Thus, an organization measure could be defined based on the deviation of the term from (2.16) from unity for a given distribution of inputs x.

In addition to that, the VBAR rule also guarantees that, on convergence, the probability of activation of quadrilaterals is equidistributed, maximizing the entropy of the activation. Thus, in this context, the entropy could be used as (indirect) indicator of organization.

A "Topological" View of VBAR There is one property that makes the VBAR algorithm particularly interesting from the viewpoint of a topological mapping. As becomes clear throughout the present review, most measures of organization of topographic mappings are, in fact, not pure topological quantities. Most of them require a metric or at least a similarity structure on the input and output spaces. This is a stronger structural requirement than topology preservation (Sec. 2.3.6).

Even the predicate notion of Delaunay-topology preservation as in (Villmann et al. 1997) (see Sec. 2.3.8) requires a metric structure on the spaces, in particular a metric on V to construct the Delaunay graph, and two different metrics on A, depending on whether one considers the topology preservation of $\mathbf{w}$ or of its left inverse $\mathbf{i}^*$.

Here, this review wishes to direct the attention to a property particular to the VBAR model and the models directly related to it which, as far as known to the author, has not been pointed out before. Namely, the fact that the output space used in VBAR can be seen as a special case of using the topological notion of a *complex*

(see e.g. (Henle 1979) or any other standard reference on algebraic topology). This review cannot give a full technical definition of a complex and only outlines it to make its point.

In algebraic topology, important notions are based on the study of d-dimensional simplices. For those, the notion of orientation and of boundary is defined. A d-dimensional complex is a structure that is constructed from d-dimensional simplices by topological identification (*gluing together*) of simplex boundaries. One of the simplest examples for a complex is a graph. It is constructed by 1-dimensional simplices (its edges), the boundaries of the edges are the vertices. An orientation is given by directing the edges. Higher-dimensional complexes can be seen as a generalization of that view by not only gluing edges together, but polygons, polyhedra and hyperpolyhedra to obtain the relevant structures.

The quadrangles used in VBAR can be constructed from simplices. They carry the full topological information about the structure of the output space, since they are truly d-dimensional, unlike the approaches restricted to graph or metric constructions to model higher dimensional spaces. It is therefore quite satisfying to see that in this model the topology preservation notion is simple to formulate. The VBAR model and its siblings can be seen as natural generalization of the discrete graph models of output space. It will be interesting to explore this type of generalization in the future.

2.3.8 Data-Oriented Measures

Intrinsic Distance Measure Kaski and Lagus (1996) introduce as measure the shortest mapped distance between the neurons whose weights are closest and second closest to a given data point in input space. More formally, let $x \in V$ be an input vector, $\mathbf{i}^*(x)$ be the neuron closest and $\mathbf{j}^*(x)$ the neuron second closest to the input (for simplicity, we assume uniqueness). Be further $\mathcal{S}_{i^* j^*}$ the set of paths starting at $\mathbf{i}^*(x)$ and ending at $\mathbf{j}^*(x)$ with edges from the Kohonen graph $\mathcal{C}_K$. Then define the intrinsic distance as

$$\hat{d}(x) := d_V(x, \mathbf{w}_{\mathbf{i}^*(x)}) + \min_{S \in \mathcal{S}_{i^* j^*}} \sum_{(s_k, s_{k+1}) \in S} d_V(\mathbf{w}_{s_k}, \mathbf{w}_{s_{k+1}}) , \qquad (2.17)$$

where the paths $S \in \mathcal{S}_{i^* j^*}$ are represented as set of edges (s_k, s_{k+1}). The measure is then given by the expectation value $\mathbf{E}\big(\hat{d}(x)\big)$ of the intrinsic distance w.r.t. the data distribution.

$\mathbf{i}^*(x)$ and $\mathbf{j}^*(x)$ can be regarded as neurons that represent a given data point $x \in V$ "similarly" well in input space. In output space, however, they need not be close-by. One could have therefore taken the distance between $\mathbf{i}^*(x)$ and $\mathbf{j}^*(x)$ in output space as measure. Instead, in the approach chosen by Kaski and Lagus, the paths are projected back into input space. This can be interpreted as weighting the distance between both "representants" according to how large a region in data (input) space is in fact covered by the SOM. Thus, even if $\mathbf{i}^*(x)$ and $\mathbf{j}^*(x)$ seem far away in output space, they can still be close by in input space. The authors report the method to be more robust than that from (Villmann et al. 1994a). At the present time, a direct comparison of the measure to further organization measures is still an open task.

Measures Based on Delaunay Graphs The measure described in this section is based on the *Delaunay triangulation* of the set

$$\{\mathbf{w}_i | i \in A\} \equiv \mathbf{w}(A) \tag{2.18}$$

induced by the signal probability distribution on V. The mechanism applied is a Hebbian one as it is based on the simultaneous activation of neurons. The particular algorithm that enables determination not only of the Delaunay triangulation, but also of the structure of *2nd order Voronoi cells* of $\mathbf{w}(A)$ has been introduced in (Martinetz and Schulten 1993) and we will refer to it as the *Hebb-Martinetz-Schulten-* or *HMS-algorithm*[5].

A whole class of measures can be defined by making use of such a mechanism, which we will therefore call *Hebbian measures*. It is applied by construction of graphs as discrete models of data manifolds.

Voronoi Tessellation and Delaunay Triangulation Before we turn to the construction of the 2nd Voronoi triangulation by the Hebb algorithm of Martinetz and Schulten, we will redefine the notions of Voronoi cells and tessellations (consistently with Sec. 2.2.2) as well as the Delaunay triangulation.

Definition 5 (Voronoi Cell). Consider a SOM $\mathbf{w}$. The *Voronoi cell* (also *1st order Voronoi cell*) of a neuron $i \in A$ or the weight $\mathbf{w}_i \in \mathbf{w}(A)$ is the set

$$V_i := \{x \in V | \forall k \in A : d_V(x, \mathbf{w}_i) \leq d_V(x, \mathbf{w}_k)\} .$$

The *2nd order Voronoi cell* of neurons i and j is the set

$$V_{ij} :=$$
$$\{x \in V | \forall k \in A \setminus \{i, j\} : d_V(x, \mathbf{w}_i) \leq d_V(x, \mathbf{w}_k) \wedge d_V(x, \mathbf{w}_j) \leq d_V(x, \mathbf{w}_k)\} .$$

Definition 6 (Voronoi Tessellation). Given a SOM $\mathbf{w}$, its *Voronoi tessellation* is the set

$$\{V_i | i \in A\}$$

of Voronoi cells V_i.

---

[5] We use this term also to distinguish it from the full *Self-Organizing Discrete Manifold* algorithm presented in (Martinetz and Schulten 1993), since the latter also involves a process distributing the $\mathbf{w}_i$, whereas in for our purposes it is applied to a given fixed $\mathbf{w}$.

Definition 7 (Delaunay Triangulation, Hebb Graph). The *Delaunay triangulation* or *Delaunay graph* of a SOM **w** is a graph with vertex set A, i and j being adjacent if V is an open subset of $\mathbb{R}^d$ and $\overline{V}_i$ and $\overline{V}_j$ have a common $d-1$-dimensional section. If i and j are adjacent for $\overline{V}_i \cap \overline{V}_j \neq \emptyset$, we speak of a *pseudo-Delaunay graph*.

The *weighted Delaunay* or *Hebb graph* of **w** induced by a probability distribution P (on V) is obtained by furnishing the edges (i, j) of the (simple) Delaunay graph with the weights $P(V_{ij})$, where $P(V_{ij})$ is the probability that an input signal lies in the 2nd order Voronoi cell V_{ij}. The edges of the Hebb graph are also called *Hebb connections*.

If V is a subset of $\mathbb{R}^d$, d_V being the Euclidean metric, then the Delaunay triangulation defines a triangulation of the space in the usual sense, except for degenerate choices of **w** (see (Okabe et al. 1992)).

The following general assumption will put us in the position to ignore pathological effects occurring at the boundaries of the Voronoi cells: From now on we will assume that $P(V_i \cap V_{i'})$ and $P(V_{ij} \cap V_{i'j'})$ always vanish for $i \neq i'$, $j \neq j'$, P being the probability distribution of input signals on V.

This assumption holds automatically if P has a density and d_V is the Euclidean metric e.g. on a subset of $\mathbb{R}^d$. However, it is not true in general if one of these conditions does not hold (Polani 1996). Our assumption ensures that we need not take such pathologies into consideration.

The Hebb Algorithm for the Self-Organizing Map The algorithm from (Martinetz and Schulten 1993, 1994) is recapitulated in the following. Note that the original algorithm has been modified insofar as to count the signals activating a connection and thus include a weight information about the Hebb connections. Let a SOM **w** be given. Then the HMS-algorithm is given by:

- For every adjacency pair $(i, j) \subset C_K \subseteq \Lambda \times \Lambda$ let the *connection strength* be $c_{ij} := 0$.
- Choose randomly a finite set $S := \{x_1 \ldots x_q\}$ of *Hebb signals* $x_1 \ldots x_q \in V$ according to the probability distribution P.
- For every Hebb signal $x_l \in S$:
  - Determine a neuron $i^* \subset \Lambda$ minimizing $d_V(x_l, \mathbf{w}_{i^*})$ (a "best matching" neuron, typically $\mathbf{i}^*(x_l)$).
  - Determine a neuron $j^* \in A \setminus \{i^*\}$ minimizing $d_V(x_l, \mathbf{w}_{j^*})$ (a "second best matching" neuron).
  - Increment the connection strength $c_{i^* j^*}$ by 1.

If c_{ij}-values have been obtained by the above Hebb algorithm, c_{ij}/q then provide estimates for the $P(V_{ij})$-values and hence for the weights of the Hebb graph of **w** induced by P. The larger q and hence the set S is chosen, the better the estimate. At this stage there are several possibilities to calculate a quality measure for the Kohonen graph from the c_{ij}-values. Since they share the same vertex set A, the

Kohonen graph can be compared directly to the Hebb graph. The Hebb measure μ_H, to be defined in Section 2.3.8, will be based on such a comparison. Villmann et al. also make use of the original (nonweighted) HMS-algorithm to calculate their organization measure.

A Notion of Topology Preservation Based on Delaunay Triangulation Villmann et al. (1997) introduce a notion of topology preservation from the graph structure on A and the neighborhood structure of the Voronoi cells in V. Based on the notion of the pseudo-Delaunay graph, a notion of topology preservation has been defined by Villmann et al.. To distinguish it from other definitions, we call it Delaunay-topology preservation.

Definition 8 (Delaunay-Topology Preservation). Assume A is a rectangular lattice embedded into $\mathbb{R}^d$. Let d_A^1 represent the $\|.\|_1$ and d_A^∞ represent the $\|.\|_\infty$ norm (Sec. 2.2.3) on A. The map $\mathbf{w}$ is then called *Delaunay topology-preserving* iff $i, j \in A$ adjacent w.r.t. d_A^1 (i.e., for which $d_A^1(i, j) = 1$) are also adjacent as vertices of the pseudo-Delaunay graph and $\mathbf{i}^*$ is called *Delaunay topology-preserving* iff $i, j \in A$ adjacent as vertices of the pseudo-Delaunay graph are also adjacent w.r.t. d_A^∞, i.e. $d_A^\infty(i, j) = 1$.

Until now, the definition brought forward does not incorporate the structure of the data. However, the authors point out that it is essential to incorporate the structure of the data manifold, i.e. the support of the probability distribution in the calculation of the measures to obtain a quantity that adequately describes the quality of the mapping (this point has also been emphasized e.g. in (Kaski and Lagus 1996; Polani 1996)). In their model, Villmann et al. solve this problem by operating on an induced or *masked* Delaunay graph that depends on the data manifold. If M is the data manifold, the *masked* Voronoi cells are defined by $\bar{V}_i = V_i \cap M$, where V_i are the standard Voronoi cells. The masked Delaunay graph is constructed from the masked Voronoi cells $\bar{V}_i$ analogously to the nonmasked version by defining two neurons i and j as adjacent in the masked Delaunay graph iff $\overline{V}_i \cap \overline{V}_j \neq \emptyset$. The notion of topology preservation resulting from the masked Delaunay graph is the original version of the notion introduced in their paper.

Topographic Function In the categories from Sec. 2.2.1, the notion of Delaunay topology-preservation from Def. 8 acts as a predicate by determining whether or whether not a mapping is topology preserving. As is the case for many measures described here, e.g. for the "similarity match" measure C from Sec. 2.3.6, when the predicate is not fulfilled, one would like quantify the degree to which the observed map does not comply with the predicate. Instead of restricting to a single numerical value, Villmann et al. (1997) introduce a function that quantifies the structure of the mismatch.

Definition 9 (Topographic Function). Denote the metric defined by the masked Delaunay graph by d_D. Define for all $i \in A, k \in \mathbb{Z} \backslash \{0\}$ the functions $f_i : \mathbb{Z} \backslash \{0\} \rightarrow$

$\mathbb{N}$ by:

$$f_i(k) := \begin{cases} |\{j \mid d_D(i,j) = 1, d_A^\infty(i,j) > |k|\}| & \text{for } k > 0 \\ |\{j \mid d_A^1(i,j) = 1, d_D(i,j) > |k|\}| & \text{for } k < 0 . \end{cases} \tag{2.19}$$

It gives the distribution of mismatches of a size beyond k for $\mathbf{w}$ and its left inverse $\mathbf{i}^*$. For $k > 0$, f_i gives the number of mismatches for the map $\mathbf{i}^*$ from V to A, for $k < 0$ the mismatches for the map $\mathbf{w}$ from A to V. Note that analogously to the definition of the topology preservation the required variant of d_A depends on which direction is considered, whether $\mathbf{i}^*$ or $\mathbf{w}$. The topographic function is then given by

$$\Phi(k) := \begin{cases} \dfrac{1}{|A|} \displaystyle\sum_{j \in A} f_j(k) & \text{for } k \neq 0 \\ \Phi(1) + \Phi(-1) & \text{for } k = 0 , \end{cases} \tag{2.20}$$

i.e. for $k \neq 0$ it is the value of the f_i averaged over all neurons.

The topographic function contains more information about the type of mismatch than the measures giving only a single number. First, it gives a direction of mismatch. If $\Phi(k) > 0$ for positive k, then this means that two neurons whose weights are close together in V (who are connected by a Delaunay edge) have a larger distance in A, i.e. that $\mathbf{i}^*$ is not Delaunay topology-preserving. This is typically the case if A has a too low dimension to accurately map V, as in Fig. 2.2. The largest k with nonzero $\Phi(k)$ indicates the scale size of the deviations. If k is large, then this indicates a deviation up to large size scales. Analogous statements hold for $k < 0$, where $\Phi(k) > 0$ indicates that $\mathbf{w}$ is not Delaunay topology-preserving and that the dimension of A is too high.

Note that to calculate the masked Delaunay graph to determine the topographic function in a concrete case, the authors use the original HMS-algorithm where it is only determined whether $P(V_{ij}) \neq 0$, i.e. whether the c_{ij} determined by the HMS-algorithm from Sec. 2.3.8 does not vanish.

A Hebbian Measure In (Böhme 1994; Polani 1996, 1997b,a), another measure is introduced and discussed that uses the HMS-principle to determine the Hebb (weighted Delaunay) graph with which the Kohonen graph may be compared. A comparison could be done by directly comparing the edges, e.g. counting the spare Kohonen edges (those Kohonen edges having no corresponding one in the Hebb graph) and vice versa. This would ignore the weights of the Hebb edges. However, such an indiscriminate comparison of the edges leads to sensitive discontinuities which are undesirable (Böhme 1994). This is illustrated by the standard example of a SOM with square net topology trained by an equidistribution on $[0,1]^2$ whose weights and receptive fields are shown in Fig. 2.7.

Note that since P is an equidistribution in our example, $P(V_{ij})$ is given by the 2-dimensional volume of the 2nd order Voronoi cells V_{ij}; this volume in turn is nonzero only when the 1st order Voronoi cells V_i and V_j have a common edge (just

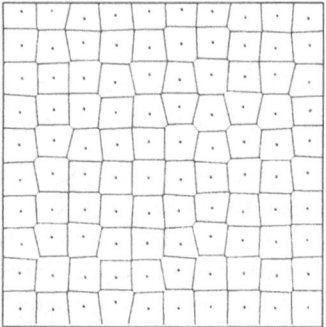

Fig. 2.7. Receptive fields of a SOM with square net topology, $V = [0,1]^2$ and P an equidistribution on V (Kohonen connections are not drawn).

fulfilling $V_i \cap V_j \neq \emptyset$ is not sufficient). The final state reached after a training is the average equilibrium state, i.e. of a state in which the weights $\mathbf{w}_i$ lie on slightly perturbed grid positions.

By inspection of Fig. 2.7 one can observe that even slight deviations from the symmetric equilibrium state of the SOM lead to "diagonal" edges in the Hebb graph for which no equivalent edges in the Kohonen graph are present. If comparison of Hebb and Kohonen graph were done ignoring the Hebb edges' weights, mismatch of a diagonal edge would be counted with the same significance as that of a rectangular edge. However, one would like the rectangular Kohonen still to be characterized as "well organized" when the $\mathbf{w}_i$ display some small deviations from the equilibrium grid positions. Moreover, even small perturbations may flip a diagonal edge (Fig. 2.8). Therefore even if at first correct diagonal Kohonen edges matching the Hebb edges were present, a small perturbation of $\mathbf{w}$ could transform those edges into spare ones, i.e. the resulting measure counting the matches would not be continuous.

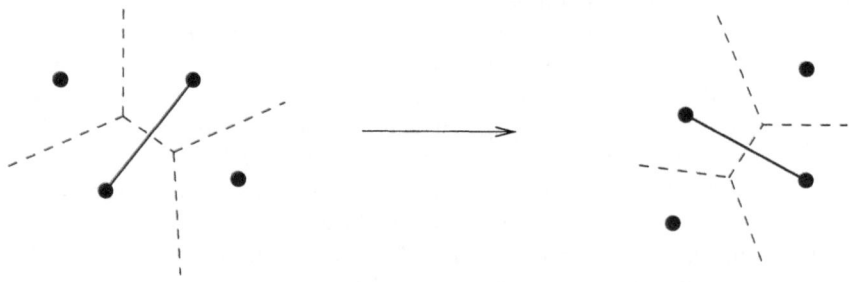

Fig. 2.8. Hebb edge flip induced by perturbation of $\mathbf{w}$ (deviation from square grid equilibrium positions is exaggerated for illustration)

Thus the demand for continuity leads us to take into account the weights of the Hebb edges since the diagonal edges have much smaller weights than the rectangular ones. A comparison of the graphs can be realized by furnishing the Kohonen edges with weights. This can be done in different ways, leading to similar results except for pathological cases. We will follow the conventions of (Böhme 1994), where the weight $\hat{c}_{ij}$ of a Kohonen graph edge $(i,j) \in \mathcal{C}_K$ is set to

$$\hat{c}_{ij} := \hat{c} = \frac{\sum\limits_{(i,j)\in\mathcal{C}_H} c_{ij}}{|\mathcal{C}_H|},$$

i.e. to the average weight of the Hebb graph edges for all edges. The measure is now obtained by summing up the weights of Kohonen edges not matching any Hebb edges and vice versa and normalizing. Finally, the measure μ_H is calculated by subtracting the result from 1; a high degree of organization is therefore represented by a high value of $\mu_H(\mathbf{w})$ since in this case there are only few non-matching edges. The formula for $\mu_H(\mathbf{w})$ reads then[6]:

$$\mu_H(\mathbf{w}) := 1 - \frac{\sum\limits_{(i,j)\in\mathcal{C}_K\setminus\mathcal{C}_H} \hat{c}_{ij} + \sum\limits_{(i,j)\in\mathcal{C}_H\setminus\mathcal{C}_K} c_{ij}}{\sum\limits_{(i,j)\in\mathcal{C}_K} \hat{c}_{ij} + \sum\limits_{(i,j)\in\mathcal{C}_H} c_{ij}}$$

$$= 1 - \frac{\hat{c}\cdot|\mathcal{C}\setminus\mathcal{C}_H| + \sum\limits_{(i,j)\in\mathcal{C}_H\setminus\mathcal{C}} c_{ij}}{\hat{c}\cdot|\mathcal{C}| + \sum\limits_{(i,j)\in\mathcal{C}_H} c_{ij}}. \tag{2.21}$$

2.4   Further Reading

This section will give a couple of pointers for further reading. The measures (Minamimoto et al. 1993) and (Demartines and Hérault 1995) are discussed in (Bauer et al. 1999). An organization measure related to C from Sec. 2.3.6 is introduced and used in (Mehler 1994). Demartines and Blayo (1992) use the variance of the connection lengths between weights in input space as a type of organization measure. This quantity is discussed e.g. in (Polani 1997b).

Different types of measures have been compared to each other in (Villmann et al. 1994a,b; Villmann 1996; Villmann et al. 1997; Villmann 1999; Bauer et al. 1999; Goodhill et al. 1995; Goodhill and Sejnowski 1997a,b; Polani 1997b, 1995, 1996). The last work, together with (Polani and Uthmann 1992, 1993; Polani 1997a, 1999) study the optimization of SOM topologies w.r.t. organization measures using GAs; it is found that this type of analysis can reveal much about the property of a given measure. In particular, for being able to claim that a given measure detects certain topological defects or favors a certain type of organization, that type of analysis is very helpful.

---

[6] For sake of simplicity in our notation we identify the values of $P(V_{ij})$ with their estimates c_{ij} obtained by the Hebb algorithm. Note that in those terms q is cancelled – it only plays a role for the accuracy of the c_{ij} as estimates for $p_{ij}\cdot q$.

2.5 Summary

The present paper gave an overview over existing approaches to quantify the organization of SOMs and related topographic mapping models. The organization measures were discussed according to conceptual, structural and dynamical properties. In particular, the necessary properties relevant for the definition of organization measures were discussed. The overview shows clearly that organization measures can be defined from many conceptually different points of view, like information-theory, dynamical systems, topology, similarity, metrics or curvature. It also shows that the study of organization measures is generally regarded as central for the understanding of SOMs. And perhaps, in future, the study of this paradigmatic system will help to better understand the phenomena of self-organization in general.

Acknowledgments

I wish to thank Thomas Martinetz, Jürgen Perl, Thomas Uthmann, Thomas Villmann, Dirk Böhme, Frank Mehler and Stéphane Zrehen for lots of helpful discussions and valuable comments and Klaus Schulten for drawing my attention to the application of the Hebbian learning mechanism to Self-Organizing Maps.

Bibliography on Chapter 2

Abraham, R., Marsden, J. E., and Ratiu, T., (1983). *Manifolds, Tensor Analysis, and Applications*. Global Analysis, Pure and Applied. Addison-Wesley.

Alder, M., Togneri, R., and Attikiouzel, Y., (1991). Dimension of the speech space. *Proceedings of the IEE-1*, 138(3):207–214.

Ashby, W. R., (1947). Principles of the self-organizing dynamic system. *J. Gen. Psychol.*, 37:125–128.

Bauer, H.-U., Herrmann, M., and Villmann, T., (1999). Neural maps and topographic vector quantization. *Neural Networks*, 12(4-5):659–676.

Bauer, H.-U., and Pawelzik, K. R., (1992). Quantifying the Neighbourhood Preservation of Self-Organizing Feature Maps. *IEEE Transactions on Neural Networks*, 3(4):570–579.

Bezdek, J. C., and Pal, N. R., (1995). Index of Topological Preservation for Feature Extraction. *Pattern Recognition*, 28(3):381–91.

Böhme, D., (1994). *Entwicklung eines Stetigkeitsmaßes für die Einbettung von Kohonen-Netzen*. Diplomarbeit, Universität Mainz, Institut für Informatik.

Cottrell, M., Fort, J., and Pagès, G., (1994). Two or three things that we know about the Kohonen algorithm. In Verleysen, M., editor, *Proceedings of the European Symposium on Artificial Neural Networks (ESANN)*, 235–244. Brussels.

Cottrell, M., and Fort, J.-C., (1987). Étude d'un processus d'auto-organisation. *Ann. Inst. Henri Poincaré*, 23(1):1–20.

Demartines, P., and Blayo, F., (1992). Kohonen self-organizing maps: Is the normalization necessary? *Complex Systems*, 6(2):105–123.

Demartines, P., and Hérault, J., (1995). Curvilinear Component Analysis. In *Quinzième Coloque Gretsi*. Juan-les-Pins.

Der, R., and Herrmann, M., (1993). Phase Transitions in Self-Organized Maps. In Gielen, S., and Kappen, B., editors, *Proc. ICANN'93, International Conference on Artificial Neural Networks*, 597–600. London, UK: Springer.

Erwin, E., Obermayer, K., and Schulten, K., (1992a). Self-Organizing Maps: ordering, convergence properties and energy functions. *Biol. Cybern.*, 67:47–55.

Erwin, E., Obermayer, K., and Schulten, K., (1992b). Self-organizing maps: stationary states, metastability and convergence rate. *Biol. Cybern.*, 67:35–45.

Forster, O., (1984). *Analysis*, vol. 3. Braunschweig/Wiesbaden: Vieweg.

Goodhill, G., Finch, S., and Sejnowski, T., (1995). Quantifying neighborhood preservation in topographic mappings. Technical Report Series INC-9509, Institute for Neural Computation.
http://www.giccs.georgetown.edu/ geoff/pubs.html, January 2001

Goodhill, G., and Sejnowski, T., (1997a). Objective functions for topography: a comparison of optimal maps. In Bullinaria, J. A., Glasspool, D. G., and Houghton, G., editors, *Proceedings of the Fourth Neural Computation and Psychology Workshop: Connectionist Representations*.

Goodhill, G. J., and Sejnowski, T. J., (1997b). A unifying objective function for topographic mappings. *Neural Computation*, 9:1291–1304.

Graepel, T., Burger, M., and Obermayer, K., (1997). Deterministic annealing for topographic vector quantization and self-organizing maps. In *Proceedings of WSOM'97, Workshop on Self-Organizing Maps, Espoo, Finland, June 4–6*, 345–350. Espoo, Finland: Helsinki University of Technology, Neural Networks Research Centre.

Graepel, T., Burger, M., and Obermayer, K., (1998). Self-organizing maps: generalizations and new optimization techniques. *Neurocomputing*, 21(1–3):173–90.

Haken, H., (1983). *Advanced synergetics*. Berlin: Springer-Verlag.

Haykin, S., (1999). *Neural networks: a comprehensive foundation*. Prentice Hall.

Henle, M., (1979). *A Combinatorial Introduction to Topology*. W. H. Freeman.

Heskes, T., (1999). Energy functions for self-organizing maps. In Oja, E., and Kaski, S., editors, *Kohonen Maps*, 303–316. Amsterdam: Elsevier. Keywords: self-organising maps, energy functions, soft assignments.

Hulle, M. M. V., (2000). *Fathful Representations and Topographic Maps*. Wiley.

Hulle, M. M. V., and Martinez, D., (1993). On an unsupervised learning rule for scalar quantization following the maximum entropy principle. *Neural Computation*, 5:939–953.

Hulle, V., (1997). Topology-preserving map formation achieved with a purely local unsupervised competitive learning rule. *Neural Networks*, 10(3):431–446.

Jetschke, G., (1989). *Mathematik der Selbstorganisation*. Braunschweig: Vieweg.

Kaski, S., and Lagus, K., (1996). Comparing Self-Organizing Maps. In von der Malsburg, C., von Seelen, W., Vorbrüggen, J. C., and Sendhoff, B., editors, *Proceedings of ICANN96*, vol. 1112 of *Lecture Notes in Computer Science*, 809–814.

Kohonen, T., (1989). *Self-Organization and Associative Memory*, vol. 8 of *Springer Series in Information Sciences*. Berlin, Heidelberg, New York: Springer-Verlag. Third edition.

Linsker, R., (1988). Self-Organization in a Perceptual Network. *Computer*, 21(3):105–117.

Luttrell, S., (1989). Self-Organization: a derivation from first principles of a class of learning algorithms. In *Proceedings 3rd IEEE Int. Joint Conf. on Neural Networks*, vol. 2, 495–498. IEEE Neural Networks Council, Washington.

Martinetz, T., and Schulten, K., (1993). Competitive Hebbian Rule Forms Delaunay Triangulations, Perfect Topology Preserving Maps and Discrete Models of Manifolds. Technical Report UIUC-BI-TB-93-04, Beckman Institute for Advanced Science and Technology, University of Illinois at Urbana-Champaign.

Martinetz, T., and Schulten, K., (1994). Topology Representing Networks. *Neural Networks*, 7(2).

McInerney, M., and Dhawan, A., (1994). Training the Self-Organizing Feature Map using Hybrids of Genetic and Kohonen Methods. In *Proc. ICNN'94, Int. Conf. on Neural Networks*, 641–644. Piscataway, NJ: IEEE Service Center.

Mees, A. I., (1981). *Dynamics of feedback systems*. John Wiley & sons, Ltd.

Mehler, F., (1994). *Selbstorganisierende Karten in Spracherkennungssystemen*. Dissertation, Institut für Informatik, Johannes Gutenberg-Universität Mainz.

Minamimoto, K., Ikeda, K., and Nakayama, K., (1993). Topology Analysis of Data Space Using Self-Organizing Feature Maps. In *Proc. ICNN'95, IEEE Intl. Conf. on Neural Networks*, vol. II, 789–794.

Okabe, A., Boots, B., and Sugihara, K., (1992). *Spatial Tesselations, Concepts and Applications of Voronoi Diagrams*. Wiley.

O'Neill, B., (1983). *Semi-Riemannian Geometry*. Pure and applied mathematics. San Diego: Academic Press.

Polani, D., (1995). On the Choice of Organization Measures for Self-Organizing Feature Maps. Technical Report 1/95, Institut für Informatik, Universität Mainz.

Polani, D., (1996). *Adaption der Topologie von Kohonen-Karten durch Genetische Algorithmen*, vol. 143 of *Dissertationen zur Künstlichen Intelligenz*. Infix. (In German).

Polani, D., (1997a). Fitness Functions for the Optimization of Self-Organizing Maps. In Bäck, T., editor, *Proceedings of the Seventh International Conference on Genetic Algorithms*, 776–783. Morgan Kaufmann.

Polani, D., (1997b). Organization Measures for Self-Organizing Maps. In Kohonen, T., editor, *Proceedings of the Workshop on Self-Organizing Maps (WSOM '97)*, 280–285. Helsinki University of Technology.

Polani, D., (1999). On the Optimization of Self-Organizing Maps by Genetic Algorithms. In Oja, E., and Kaski, S., editors, *Kohonen Maps*. Elsevier.

Polani, D., and Uthmann, T., (1992). Adaptation of Kohonen Feature Map Topologies by Genetic Algorithms. In R. Männer, and Manderick, B., editors, *Parallel Problem Solving from Nature, 2*, 421–429. Elsevier Science Publishers B.V.

Polani, D., and Uthmann, T., (1993). Training Kohonen Feature Maps in different Topologies: an Analysis using Genetic Algorithms. In Forrest, S., editor, *Proceedings of the Fifth International Conference on Genetic Algorithms*, 326–333. San Mateo, CA: Morgan Kaufmann.

Reichl, L., (1980). *A Modern Course in Statistical Physics*. Austin: University of Texas Press.

Ritter, H., Martinetz, T., and Schulten, K., (1994). *Neuronale Netze*. Addison-Wesley.

Ritter, H., and Schulten, K., (1989). Convergence Properties of Kohonen's Topology Conserving Maps: Fluctuations, Stability and Dimension Selection. *Biological Cybernetics 60*, 59–71.

Sammon, Jr., J. W., (1969). A nonlinear mapping for data structure analysis. *IEEE Trans. Comput.*, C-18:401–409.

Shalizi, C. R., (1996). Is the Primordial Soup Done Yet?.
http://www.santafe.edu/~shalizi/Self-organization/soup-done/,
Jan 23, 2001

Speckmann, H., Raddatz, G., and Rosenstiel, W., (1994). Improvement of learning results of the self-organizing map by calculating fractal dimensions. In Verleysen, M., editor, *ESANN '94 – Proceedings of the European Symposium on Artificial Neural Networks*, 251–255. Brussels, Belgium: D facto.

Spitzner, A., and Polani, D., (1998). Order Parameters for Self-Organizing Maps. In Niklasson, L., Bodén, M., and Ziemke, T., editors, *Proc. of the 8th Int. Conf. on Artificial Neural Networks (ICANN 98), Skövde, Sweden*, vol. 2, 517–522. Springer.

Torgerson, W. S., (1952). Multidimensional scaling I. Theory and method. *Psychometrika*, 17:401–419.

Van Hulle, M. M., (1997). Nonparametric Density Estimation and Regression Achieved with Topographic Maps Maximizing the Information-Theoretic Entropy of Their Outputs. *Biological Cybernetics*, 77:49–61.

Villmann, T., (1996). *Topologieerhaltung in selbstorganisierenden neuronalen Merkmalskarten.* PhD thesis, Universität Leipzig.

Villmann, T., (1999). Topology Preservation in Self-Organizing Maps. In Oja, E., and Kaski, S., editors, *Kohonen Maps*, 279–292. Amsterdam: Elsevier. Keywords: self-organising map, topology preservation, growing self-organizing map.

Villmann, T., Der, R., Herrmann, M., and Martinetz, T., (1997). Topology Preservation in Self-Organizing Feature Maps: Exact Definition and Measurement. *IEEE Trans. Neural Networks*, 8(2):256–266.

Villmann, T., Der, R., and Martinetz, T., (1994a). A New Quantitative Measure of Topology Preservation in Kohonen's Feature Maps. In *Proc. ICNN'94, Int. Conf. on Neural Networks*, 645–648. Piscataway, NJ: IEEE Service Center.

Villmann, T., Der, R., and Martinetz, T., (1994b). A Novel Approach to Measure the Topology Preservation of Feature Maps. In Marinaro, M., and Morasso, P. G., editors, *Proc. ICANN'94, Int. Conf. on Artificial Neural Networks*, vol. I, 298–301. London, UK: Springer.

Wiskott, L., and Sejnowski, T. J., (1997). Objective Functions for Neural Map Formation. *Lecture Notes in Computer Science*, 1327.

Zrehen, S., and Blayo, F., (1992). A geometric organization measure for Kohonen's map. In *Proc. of Neuro-Nîmes*, 603–610.

3 Unsupervised Learning and Self-Organization in Networks of Spiking Neurons

Thomas Natschläger, Berthold Ruf and Michael Schmitt

Abstract. One of the most prominent features of biological neural systems is that individual neurons communicate via short electrical pulses, the so-called action potentials or spikes. In this chapter we investigate possible mechanisms of unsupervised learning and self-organization in networks of spiking neurons. After giving a brief introduction to spiking neuron networks we describe a biologically plausible algorithm for these networks to find clusters in a high dimensional input space or a subspace of it. The algorithm is shown to work even in a dynamically changing environment. Furthermore, we study self-organizing maps of spiking neurons showing that networks of spiking neurons using temporal coding can achieve a topology preserving behavior quite similar to that of Kohonen's self-organizing map. For these networks a mechanism of competitive computation is proposed that is based on action potential timing. Thus, the winner in a population of competing neurons can be determined locally and in generally faster than in approaches which use rate coding. The models and algorithms presented in this chapter establish further steps toward more realistic descriptions of unsupervised learning in biological neural systems.

3.1 Introduction

Undoubtedly, unsupervised learning and self-organization belong to the most impressive capabilities of biological neural systems. Since the beginning of research in artificial neural networks, numerous models have been proposed that have lead to successful applications in machine learning and pattern recognition tasks. When asking how these models are related to their biological counterparts, it turns out that computations in these networks almost exclusively use a particular type of coding that is known as rate coding. Biological neurons communicate in terms of short electrical pulses, so-called action potentials or spikes. One possible scheme of coding information in a sequence of such spikes is to consider the temporal average and interpret the frequency of the spikes as the value that is encoded. In recent years, however, neurobiological experiments have exhibited that rate coding might not be the only mechanism on which computations in biological nervous systems are based. A complementary coding scheme has appeared as crucial in particular for fast information processing: temporal coding, i.e. coding in terms of single spikes. There it is assumed that information is represented in the temporal pattern of neural activity and in the timing of single spikes.

When considering temporal coding, however, the question arises how learning can be based on this coding scheme and how networks that use spike timing for computing can learn so successfully. Two of the most widespread unsupervised learning mechanisms for artificial neural networks are radial basis function (RBF) neural networks and Kohonen's self-organizing map (SOM). In the investigations presented in this chapter we consider spiking neurons for the computation of radial basis functions and the realization of topology preserving mappings, and propose and study possible learning mechanisms for networks of these model neurons. We emphasize that this contribution is not about biology, but about possibilities of computing and learning with models of spiking neurons inspired by biology. Networks of spiking neurons have been shown to be powerful computing devices [23,24], but there is still not much known about how learning can be performed in these model networks. Although they are still rather simplified compared to biological networks, a more thorough analysis might contribute to the design of neuromorphic hardware circuits, e.g. in pulse-coded VLSI (very large scale integration), and to the description and understanding of the mechanisms in biological systems.

This chapter is organized as follows: In Sec. 3.2 we give a brief introduction to models of spiking neurons and discuss the two major coding mechanisms they use: rate coding and temporal coding. In Sec. 3.3 we propose a model for spatial and temporal pattern analysis with spiking neurons based on temporal coding. In particular, we introduce spiking neurons as radial basis function units and suggest an unsupervised learning mechanism for networks of these model neurons. The detailed simulation studies presented show that these networks find clusters in the input space, and even in subspaces, and recognize temporal sequences which may be distorted in time and form. In Sec. 3.4 we consider self-organizing maps of spiking neurons. First, we introduce a mechanism of competitive computing and learning using temporal coding. Then we extend this approach to self-organizing maps of spiking neurons. Computer experiments show that these networks indeed exhibit the desired topology preserving behavior known from the Kohonen map. Finally, in Sec. 3.5 a discussion of the results and concluding remarks can be found.

3.2   Networks of Spiking Neurons

A model for computations in biological neural systems that captures the essential aspects of biological neural systems has to take into account the way how biological neurons transmit and process information. The output of a biological neuron consists of a sequence of almost identical electrical pulses, or *spikes* (see Fig. 3.1). These so-called *spike trains* are the time series which are processed by a biological neural system. In the following section we discuss a formal model of such networks of

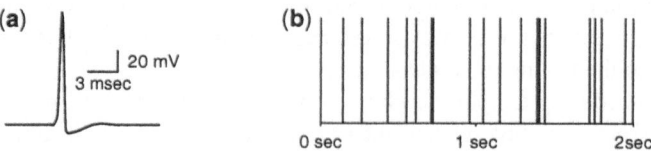

Fig. 3.1. (a) Typical action potential (spike). (b) A typical spike train produced by a neuron (each firing time marked by a bar).

spiking neurons.[1] For an excellent survey about modeling neural systems we refer to [8].

3.2.1   A Formal Model for a Network of Spiking Neurons

If one ignores all dynamic aspects, then a spiking neuron has some similarity to the familiar threshold gate in computer science (see for example [35]). A threshold gate outputs 1 if and only if the weighted sum of its inputs reaches some threshold. Similarly a spiking neuron i "fires", i.e. generates a short electrical pulse, which is called *action potential*, or *spike* (see Fig. 3.1(a)), if the current input at time t drives the membrane potential $h_i(t)$ above some threshold θ_i. All such spikes have an almost identical shape. Hence the output of a spiking neuron is a sequence of spikes at certain points in time, informally called *spike train* (see Fig. 3.1(b)). Formally the spike train generated by neuron i is simply the set of firing times $F_i \subset \mathbb{R}^+$ ($\mathbb{R}^+ = \{x \in \mathbb{R} : x \geq 0\}$).

In the simplest (deterministic) model of a spiking neuron one assumes that a neuron i fires whenever the membrane potential h_i (which models the electric membrane potential at the *trigger zone* of neuron i) reaches the threshold θ_i. h_i is the sum of so-called excitatory postsynaptic potentials (EPSPs) and inhibitory postsynaptic potentials (IPSPs), which result from the firing of presynaptic neurons j that are connected through a synapse to neuron i (see Fig. 3.2).

The firing of a neuron j at time $\hat{t}$ contributes to the potential $h_i(t)$ at time t an amount that is modeled by the term $w_{ij}(\hat{t}) \cdot \varepsilon_{ij}(t-\hat{t})$, which consists of the *synaptic strength* $w_{ij}(\hat{t}) \geq 0$ and a *response function* $\varepsilon_{ij}(t - \hat{t})$. Biologically realistic shapes of such response functions are indicated in Fig. 3.2(b). If Γ_i is the set of all neurons presynaptic to neuron i, then the membrane potential $h_i(t)$ at the trigger zone of neuron i at time t is given in terms of the sets F_j of firing times of these presynaptic neurons j by

$$h_i(t) := \sum_{j \in \Gamma_i} \sum_{\hat{t} \in F_j : \hat{t} < t} w_{ij}(\hat{t}) \cdot \varepsilon_{ij}(t - \hat{t}) . \tag{3.1}$$

The membrane potential $h_i(t)$ does not really correspond to the weighted sum of a threshold gate since it varies over time. Unfortunately not even the threshold θ_i is static . If a neuron i has fired at time $\hat{t}$, it will not fire again for a few milliseconds (ms) after $\hat{t}$, no matter how large its current potential $h_i(t)$ is (*absolute refractory period*). Then for a few further ms it is still "reluctant" to fire, i.e. a firing requires a larger value of $h_i(t)$ than usual (*relative refractory period*). Both of these refractory effects are modeled by a suitable *threshold function* $\theta_i(t - \hat{t})$, where $\hat{t}$ is the time of the most recent firing of i. A typical shape of the function $\theta_i(t - \hat{t})$ for a biological neuron is indicated in Fig. 3.2(c).

---

[1] The "spike trains" demo software which illustrates information processing with spikes can be downloaded from http://www.cis.TUGraz.at/igi/tnatschl/ spike_trains_eng.html.

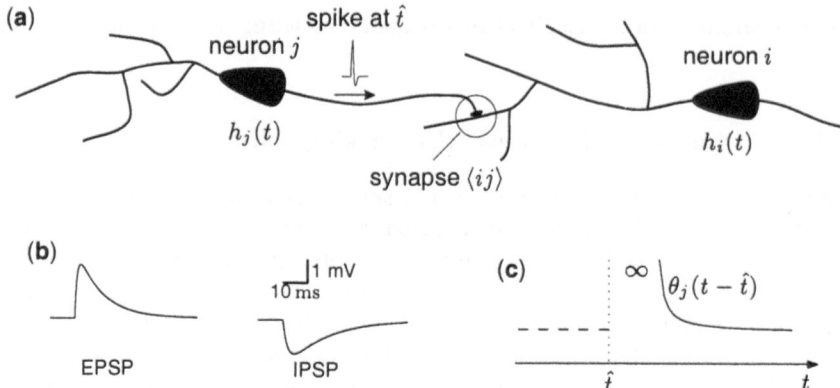

Fig. 3.2. Information processing with spikes. (a) Spikes are generated by a threshold process whenever the membrane potential $h_j(t)$ crosses the threshold $\theta_j(t - \hat{t})$. The spike travels down the axon of neuron j. Via the synaptic connection $\langle ij \rangle$ the spike is transformed into a postsynaptic response at neuron i. (b) Typical shape of a postsynaptic response, which is either positive (EPSP) or negative (IPSP), of a biological neuron. (c) Typical shape of the threshold function $\theta_j(t - \hat{t})$ of a biological neuron ($\hat{t}$ is the time of its most recent firing).

The synaptic strength $w_{ij}(\hat{t})$ can be interpreted as the amplitude of the postsynaptic response triggered by the firing of neuron j at time $\hat{t}$, whereas the time course and the sign (EPSP or IPSP) of the response are determined by the response function $\varepsilon_{ij}(t - \hat{t})$. The restriction of $w_{ij}(\cdot)$ to non-negative values — in combination with positive (EPSP) or negative (IPSP) response functions $\varepsilon_{ij}(\cdot)$ — is motivated by the empirical result that a biological synapse is either excitatory or inhibitory, and that it does not change its sign in the course of a learning process. In addition, for most biological neurons j, either all response functions $\varepsilon_{ij}(\cdot)$ for postsynaptic neurons i are excitatory (i.e. positive), or all of them are inhibitory (i.e. negative). As we will see in the following sections $w_{ij}(\cdot)$ is changed by suitable learning algorithms on rather short time scales.[2]

We assume that for some specified subset of *input neurons* their firing times (spike trains) are given from the outside as *input* to the network. The firing times for all other neurons are determined by the previously described rules, and the output of the network is given in the form of the spike trains for the neurons in a specified set of *output neurons*.

We would like to point out that the formal model for a spiking neuron that we have outlined so far is a coarse simplification. In particular the membrane potential

---

[2] Note however, that a large number of experimental studies have shown that biological synapses have an inherent short term dynamics, which controls how the pattern of amplitudes of postsynaptic responses depends on the temporal pattern of the incoming spike train, see e.g. [38,26]. This dynamics is independent of potential learning processes. In this chapter we do not take into account this effect an assume that $w_{ij}(\cdot)$ changes only due to some learning process.

$h_i(t)$ at the trigger zone of a neuron i is in general not a linear sum of incoming pulses. Both *sublinear* and *superlinear* summation occur in biological neurons. Also the threshold function θ_i (see Fig. 3.2(c)) varies from neuron to neuron. For example, in the case of periodically firing neurons (pacemaker neurons) the threshold function θ_i may also rise again after its initial decline. With regard to further details about biological neural systems we refer to [15–17,28].

3.2.2   Neural Codes

When one thinks about computations in neural systems the first question is: *How is relevant information encoded by a set of spike trains?* At present, a definite answer to this question of neural coding is not known. In this section we give a review of some potential coding schemes, see [8] for a survey.

Rate Code

The firing rate is usually defined by a temporal average.[3] The experimentalist sets a time window of, let us say, $\Delta = 100$ ms and counts the number of spikes $n(\Delta)$ that occur in this time window. Division by the length of the time window gives the mean firing rate $r = n(\Delta)/\Delta$ usually reported in units of s^{-1} or Hz. Traditionally it has been thought that most, if not all, of the relevant information was contained in the mean firing rate of the neuron. It is clear, however, that an approach based on a temporal average neglects all the information possibly contained in the exact timing of the spikes. It is therefore no surprise that the firing rate concept has been repeatedly criticized and is subject of an ongoing debate [3,14,25,28].

Temporal Coding

During recent years, more and more experimental evidence has been accumulated which suggests that a straightforward firing rate concept based on temporal averaging may be too simple for describing brain activity. One of the main arguments is that reaction times in behavioral experiments are often too short to allow slow temporal averaging [28,37]. Moreover, in experiments on a visual neuron in the fly, it was possible to 'read the neural code' and reconstruct the time-dependent stimulus based on the neurons' firing times [28]. There is evidence of precise temporal correlations between pulses of different neurons [3] and stimulus dependent synchronization of the activity in populations of neurons [31]. Most of these data are inconsistent with a naive concept of coding by mean firing rates where the exact timing of spikes should play no role.

---

[3] A quick glance at the experimental literature reveals that there is no unique and well-defined concept of a rate code. In fact, there are at least three different notions of rate which are often confused and used simultaneously: rate as an *average over time*, rate as an *average over several trails*, rate as an *average over a population* of neurons. An excellent discussions of rate codes can be found in [8,28].

Other recent experimental results indicate that it is in fact questionable whether biological neural systems are *able* to carry out analog computation with analog variables represented as firing rates. Due to "synaptic depression" the amplitude of postsynaptic potentials tends to scale like $1/f$ where f is the firing rate of the presynaptic neuron (see e.g. [2]). Therefore both slowly firing neurons and rapidly firing neurons inject roughly the same amount of current into a postsynaptic neuron during a given time window. This suggests that both a McCulloch-Pitts neuron and a sigmoidal neuron model overestimate the computational capability of a biological neuron for rate coding.

In this chapter we will explore some of the possibilities of encoding relevant information by the timing of single spikes. We will refer to such encodings as *temporal coding*. In temporal coding the relevant information may be represented by the firing times of neurons relative to the stimulus onset [7] or relative to the firing times of other neurons [27,37].

3.3   Spatial and Temporal Pattern Analysis via Spiking Neurons

Radial basis function (RBF) networks are among the most powerful artificial neural network types, e.g. for the purposes of function approximation, pattern classification, and data clustering. There is also growing interest in the relevance of this approach for biological neural networks. The question if biological neurons can realize one of the main advantages of RBF neurons, namely their ability to discover clusters in the input space, is not yet resolved. In the following we present a learning algorithm for spiking neurons realizing RBFs which is not based on rate coding but on the timing of single spikes and which is able to find centers of clusters in an unsupervised fashion.

In this context, Hopfield presented a model for computing RBFs with spiking neurons [14]. The basic idea is that an "RBF neuron" encodes a particular input spike pattern in the delays available across its synapses. If an input pattern is close to the encoded spike pattern of an RBF neuron (called center of the RBF neuron in the following), the delays even out the differences of the firing times of the input neurons such that the RBF neuron fires. This approach is motivated by many neurobiological findings: In [27] evidence has been given that the relative timing of spikes in the rat hippocampus might carry information about the relative location in an environment which is common to the rat. In [12] it was shown that in the rat piriform cortex a wide range of time delays (up to 20 ms) can occur, which would allow a large interval for the input firing times.

In [9] a learning mechanism was introduced for a *single* RBF neuron. The proper delays, which encode the center of an RBF neuron, are chosen by using a neuronal learning rule which increases the efficacy of synapses with corresponding proper delays and decreases the efficacy of other synapses. A basic assumption underlying this approach is that there are several paths with different delays between each input and each RBF neuron.

Here we extend this approach by considering not only the firing/non-firing of a neuron but also its firing *time*. We show how on the basis of these ideas networks of such RBF neurons can be constructed and trained to divide the input space into several clusters, where the selection of the proper delays uses exclusively local information (essentially the difference between pre- and postsynaptic firing time). We performed computer simulations with the neuron models described in Sec. 3.2.1 and achieved promising results:

- The RBF neurons converged very reliably to the centers of the clusters even in the presence of noise.
- The RBF neurons were able to reconfigure themselves dynamically, if during the learning process clusters were added or removed. This adaptation to a changing environment seems to be of particular importance for biological neural systems.
- In the case where each cluster is formed in a certain subspace of the input space we show that with our learning mechanism RBF neurons can find the proper delays for the coordinates of the subspace and "deactivate" the remaining inputs. This can be considered as some kind of feature extraction, where the coordinates of the subspace carry the relevant information.
- Especially in the context of biological neural networks temporal sequences and their analysis seem to be of great importance. It turns out that by employing postsynaptic potentials of variable width such RBF neurons can detect temporal sequences even if those sequences are distorted in various ways.

3.3.1   A Computational Model for RBFs Using Spiking Neurons

The model for a spiking neuron which we use in the following is described in Sec. 3.2.1. Here we consider a network of such spiking neurons with input neurons $u_1, \ldots, u_m$ and output neurons $v_1, \ldots v_n$, the latter ones will be denoted in the following as RBF neurons (see Fig. 3.3).

In the simplest case each input neuron u_j forms one synaptic connection to each RBF neuron v_i with weight w_{ij} and delay d_{ij}, where the delay is given by the difference between the presynaptic firing time and the time the synapse between the two neurons is activated. We assume that the postsynaptic delay is always constant and neglect it here for the sake of simplicity. The center of an RBF neuron v_i is given by the vector $\mathbf{c}_i = \langle c_{i1}, \ldots, c_{im} \rangle$ with $c_{ij} = d_{ij} - \min\{d_{ij} | 1 \leq j \leq m\}$. For our basic construction we consider the simple coding scheme, where each input neuron u_j fires exactly once at time t_j within a certain time interval $[0, T]$, to which we will refer as the *coding interval*, in the following. The firing times represent the input vector $\mathbf{x} = \langle x_1, \ldots, x_m \rangle$ with $x_j = \max\{t_i | 1 \leq j \leq m\} - t_j$. No reference spike is needed for this type of competitive temporal coding. The input $\mathbf{x}$ is close to the center $\mathbf{c}_i$ of an RBF neuron v_i if the spikes of the input neurons reach the soma of v_i due to the corresponding delays at similar times, i.e. if $\|\mathbf{x} - \mathbf{c}_i\|$ is small. This is basically the approach presented in [14]. There the case is considered where the input vector is close enough to the center of an RBF neuron to make v_i fire.

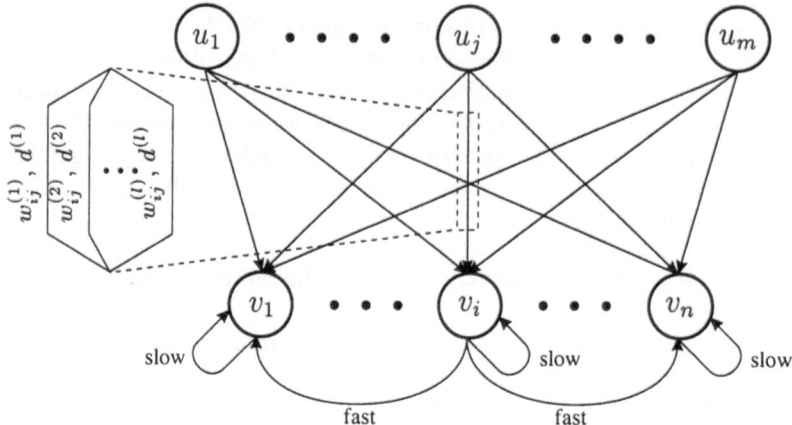

Fig. 3.3. Our basic architecture for an RBF network of spiking neurons. Each connection from an input neuron u_j to an RBF neuron v_i consists actually of a set of l excitatory synapses with different weights $w_{ij}^{(k)}$ and corresponding delays $d^{(k)}$, as indicated by the magnification on the left. Each RBF neuron forms a slow inhibitory connection to itself and fast inhibitory connections to all other RBF neurons.

However, in addition the firing time of v_i indicates, how close an input vector $\mathbf{x}$ is to c_i (see Fig. 3.4). If the distance between $\mathbf{x}$ and v_i is too large, v_i does not fire at all.

Let us consider now a set of RBF neurons $\{v_1, \ldots, v_n\}$. If for some input vector $\mathbf{x}$ the difference $\|\mathbf{x} - \mathbf{c}_i\|$ is small enough for various i to make v_i fire then the RBF neuron whose center is closest to $\mathbf{x}$ fires first.[4] Hence a set of such RBF neurons can be used to separate inputs into various clusters.

3.3.2 Learning RBFs and Clustering with Spiking Neurons

If one uses RBF neurons for clustering as described above, then the centers of the clusters can be found by the RBF neurons in an unsupervised way: we assume similar as in [9] that each u_j forms to each v_i instead of one synapse a set of synapses with varying delays (see Fig. 3.3). Throughout this chapter we use the set $D = \{d^{(1)}, \ldots, d^{(l)}\}$ of delays which is the same for every input with

---

[4] Strictly speaking, there are some cases where a small deviation in the input spikes from the center of an RBF neuron v_i may cause an *earlier* firing. Consider e.g. the case where the input vector equals the center of v_i except one coordinate j, such that neuron u_j fires slightly earlier. This makes v_i fire earlier if the EPSP caused by u_j is still in its rising segment when v_i fires. However, since we choose a threshold being high enough to make v_i only fire when it received synchronous spikes from nearly all input neurons, v_i tends to fire more toward the end of the rising segment of the EPSPs. This shift in the firing time of v_i requires very special constellations of input spikes, which are especially for large numbers of inputs very unlikely to occur. Hence this undesired effect can be neglected.

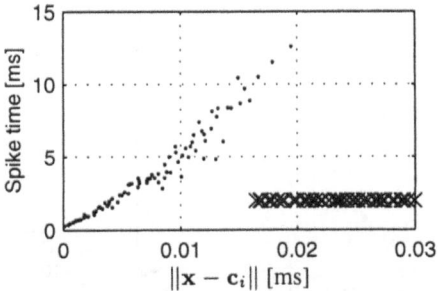

Fig. 3.4. Dependence of the firing time of an RBF neuron v_i on the distance of the input vector $\mathbf{x}$ to the center $\mathbf{c}_i$. For this simulation 120 randomly chosen inputs $\mathbf{x} \in [0, 20 \text{ ms}]^{20}$ are presented to v_i, with equal weights and randomly chosen delays from $[0, 20 \text{ ms}]$. Crosses indicate the case that the RBF neuron has not fired.

$d^{(k)} - d^{(k-1)} > 0$ for $2 \leq k \leq l$. The length T of the coding interval where the input neurons are allowed to fire is assumed to be smaller than $d^{(l)} - d^{(1)}$. We say the delay of length $d^{(k)}$ for the connection from u_j to v_i is *active* if its corresponding weight $w_{ij}^{(k)}$ is significantly greater than zero.

The goal of our learning algorithm can be formulated as follows: given a cluster C of input vectors $\mathbf{x}$ in the input space, an RBF neuron v_i should activate for each input coordinate one delay, such that the resulting center $\mathbf{c}_i$ minimizes $\sum_{\mathbf{x} \in C} \|\mathbf{x} - \mathbf{c}_i\|$. In addition we also want to allow that several neighboring delays become activated because the center of a cluster may not be representable due to a finite number of available delays. As it will turn out (see Sec. 3.3.2), the activation of a larger number of neighboring delays may also reflect a larger variance of the input cluster.

For such a constellation of several active neighboring delays we define the center $\mathbf{c}_i$ of an RBF neuron v_i as follows: we compute for every input coordinate j the mean delay $\bar{d}_{ij} = \sum_{k=1}^{l} w_{ij}^{(k)} \cdot d^{(k)} / \sum_{k=1}^{l} w_{ij}^{(k)}$. The center of v_i is then the vector $\mathbf{c}_i = \langle c_{i1}, \ldots, c_{im} \rangle$ with $c_{ij} = \bar{d}_{ij} - \min\{\bar{d}_{ij} | 1 \leq j \leq m\}$.

Each RBF neuron should converge to the center of some cluster, which can be achieved in the following way: initially all weights are set to random values such that no RBF neuron can fire until it has received at least one spike from every $u_j, 1 \leq j \leq m$. As soon as an RBF neuron fires, the spike is propagated backwards to its synapses, where a weight is changed according to some "learning function" $L(\Delta t)$, with $\Delta t = t_{pre} - t_{post}$ denoting the difference between the arrival times t_{pre} and t_{post} of the pre- and postsynaptic spike at the synapse (see Fig. 3.5(a)). See [1] for a survey about synaptic plasticity; in particular about such spike-timing dependent synaptic plasticity.

In the following, we will use the learning function as shown in Fig. 3.5(b). $L(\Delta t)$ is chosen here such that the weights of those synapses, which received shortly before the postsynaptic firing a presynaptic spike, are increased (i.e. the peak of L

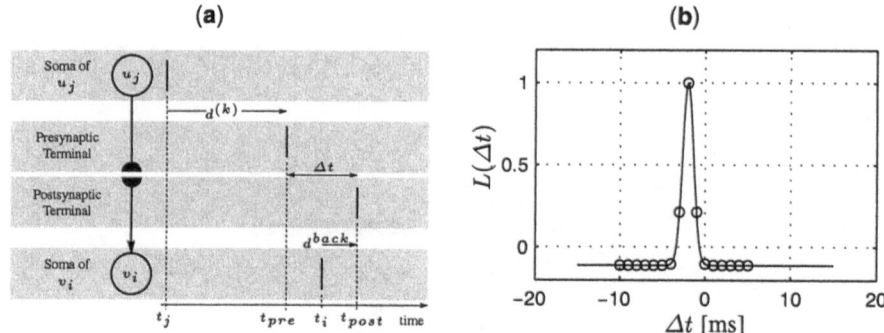

Fig. 3.5. (a) A spike emitted at u_j (v_i) at time t_j (t_i) reaches the synapse between u_j and v_i at time $t_{pre} = t_j + d^{(k)}$ ($t_{post} = t_i + d^{back}$). (b) Learning function $L(\Delta t)$, describing how the weight change is influenced by the difference $\Delta t = t_{pre} - t_{post}$ between pre- and postsynaptic firing times at the synapse. The circles denote the weight changes for various delays occurring when u_j fires 10 ms before v_i. The leftmost (rightmost) circle corresponds to a delay of 1 ms (16 ms). We chose here $d^{back} = 1$ ms and $L(\Delta t) = (1-b) \cdot \exp\left(-(\Delta t - c)^2/\beta^2\right) + b$, with $b = -0.11$, $\beta = 1.11$ ms and $c = -2$ ms.

is shifted by 2 ms from 0), whereas synapses, which received a presynaptic spike much earlier or later are decreased.

More precisely, the weight $w_{ij}^{(k)}$ of the synapse between input neuron u_j and RBF neuron v_i with delay $d^{(k)}$ is updated during the whole learning process due to the following learning rule:

$$\Delta w_{ij}^{(k)} = \eta L\left((t_j + d^{(k)}) - (t_i + d^{back})\right) \tag{3.2}$$

for every pair of input firing time t_j and firing time t_i of the RBF neuron (see Fig. 3.5(b) for an example). $\eta > 0$ denotes the learning rate. The delay d^{back} describes the time the postsynaptic spike needs to propagate backwards to the synapses. For all these synapses d^{back} has the same value. Furthermore, we assume that the weights saturate at 0 respectively at a certain $w_{max} > 0$ (see Sec. 3.3.5 for details).

By adding fast lateral inhibition among the RBF neurons, one can implement a winner-take-all mechanism where only the RBF neuron fires that is closest to the current input vector. Hence, only the weights of the winning RBF neuron are modified, such that its center moves toward the current input vector. This is a similar approach as in competitive learning (see Sec. 3.4.1), and it enables RBF neurons to learn clusters in the input space.

Since we are interested in the firing *time* of an RBF neuron, we have to make sure that the sum of the weights of all incoming synapses remains approximately constant in order to guarantee that the firing time does not shift due to a simple weight scaling but only because of peaks in the weight distribution. Such peaks can be seen in Fig. 3.7, where there are only a few neighboring delays with large corresponding weights after learning. To achieve such a constant sum, one would usually have to require some non-local interaction. However, if one chooses a proper learning function (where $\int_{-\infty}^{\infty} L(t)dt \approx 0$), this effect can also be achieved using only locally available information. Fig. 3.6(b) shows that the peaks of the membrane potential caused by one input vector indeed stabilize during the learning process at a certain level.

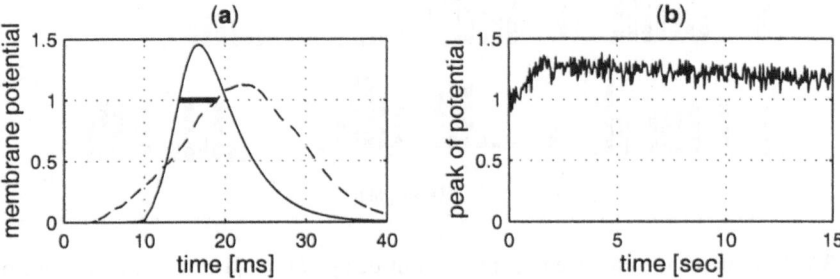

Fig. 3.6. (a) Membrane potential of an RBF neuron for one input vector before (*dashed line*) and after (*solid line*) learning, causing a shift of the firing time from 19 ms to 14 ms, denoted by the horizontal bar at the height of the threshold. (b) Development of the peaks of the membrane potential for an RBF neuron during the learning process. The jitter results from the varying input vectors chosen from one cluster. For the threshold we chose for both figures $\Theta = 1$.

A general problem with clustering, which also occurs here, is that it is very likely that some RBF neurons specialize on sets of clusters (thus computing their mean) and other RBF neurons are not used at all. We solved this problem by adding slow self-inhibition to every RBF neuron, decreasing the probability that an RBF neuron fires twice within a few iterations of the learning rule. A similar "conscience" mechanism is often used in competitive learning [11]. However, with a surplus of RBF neurons and a proper choice for the learning function and the initialization of the weights, such that each RBF neuron needs at least one spike from every input neuron, it is also possible to achieve good results without self-inhibition.

Simulation Results

We performed computer simulations in order to investigate how these RBF neurons behave on a high-dimensional input space. In the following we use RBF neurons with 40 inputs. The length of the coding-interval was 10 ms, the delay-interval 15 ms such that the available delays were 1 ms, 2 ms, ..., 16 ms. Every 50 ms a new learning cycle was started by presenting a new input vector $\mathbf{x}$ to the network. The input vectors were randomly chosen from an input space which has p clusters C_μ for $1 \leq \mu \leq p$. For each component of an input vector $\mathbf{x} \in C_\mu$ the standard deviation was chosen between 2 ms and 4 ms. There was no overlap between the clusters. We set $L(\Delta t) = 0$ for $|\Delta t| > 15$ ms to make sure that weights are changed only due to the current input vector.

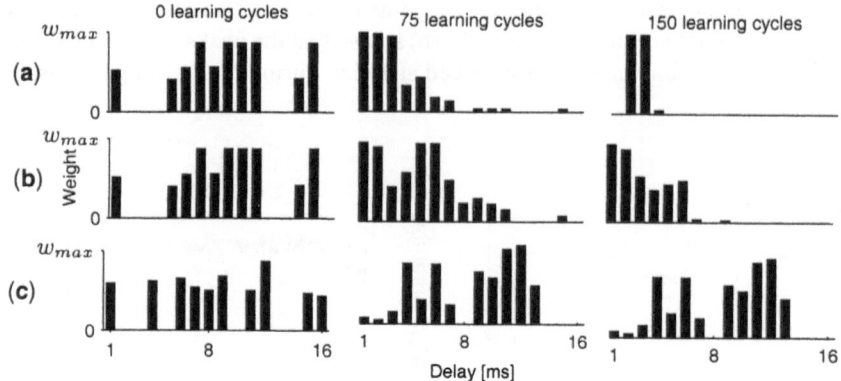

Fig. 3.7. Development of the weights for a set of delays $D = \{1\,\text{ms}, \ldots, 16\,\text{ms}\}$ for one input neuron and one RBF neuron. (a) Input patterns chosen from a cluster C_μ with small standard deviation. (b) Same as (a), but with large standard deviation. (c) Input patterns chosen from two different clusters.

First we considered the case $n = p$, i.e. there were as many RBF neurons as clusters. After about 50 examples from each cluster each RBF neuron had converged to one cluster. As Fig. 3.7 indicates, the variance of the clusters is reflected in the distribution of the weights of the delays. A small standard deviation of a cluster results in a sharp peak, whereas larger values cause several similar delays to have strong weights nearby the saturation value w_{max}, such that the RBF neuron will respond stronger, i.e. fires earlier, for a wider range of input vectors.

Furthermore, we could observe that an RBF neuron fires considerably earlier after learning a cluster, when a pattern from that cluster is presented. This results from the fact that the membrane potential increases more steeply after learning (see Fig. 3.6(a)).

If one considers the definition of a center c_i given in Sec. 3.3.1, there can be several values of $\bar{d}_{ij}$ which yield the same center c_i. Observe that our model implies that $\min\{\bar{d}_{ij} | 1 \leq j \leq m\}$ almost always is nearly equal to the smallest available delay $d^{(1)} = 1$ ms. This is not obvious since we use no reference spike to define our input interval. It has the advantage that the firing times of several RBF neurons are comparable on an absolute time scale as suggested in [37].

In the case of more clusters than RBF neurons, some RBF neurons will converge to the mean of a subset of the clusters. This may result in one wide peak or several smaller peaks in the distribution of the strengths of the delays, which is the best one can expect in this case (cf. Fig. 3.7(c)). The latter case occurs if an RBF neuron has specialized on several clusters, i.e. it has more than one center.

If there are initially more RBF neurons than clusters, it is obvious that several RBF neurons can specialize on the same cluster or that, depending on the initial conditions, some RBF neurons do not learn anything. Furthermore, we observed that a dynamic reconfiguration of the RBF neurons in a changing environment is possible. We assume that some of the RBF neurons have already learned to represent some clusters. If one adds during the learning process another cluster, one of the "free" RBF neurons will quickly specialize on this cluster (see Fig. 3.8). If one continues adding clusters, such that the number of clusters exceeds the number of RBF neurons, a dynamic reconfiguration occurs in the sense, that an RBF neuron switches from an old to a new cluster or learns the mean of a set of clusters including new ones.

Fig. 3.8. Spike trains for four RBF neurons. At the beginning, two clusters were presented, after 8 s a third and after 20 s a fourth cluster was added.

We also investigated the influence of noise by making each input neuron fire in addition at random time points. This means that each input neuron generates a Poisson spike train (of low frequency) and fires in addition corresponding to the sequence of input patterns. Fig. 3.9 shows that despite of these underlying Poisson spike trains the RBF neurons are still able to detect the patterns in the noisy environment and assign them to the proper cluster. We were able to show that such noise may be even present during the learning phase without deteriorating the learning quality and speed considerably.

3.3.3   Finding Clusters in Subspaces

If one interprets the input coordinates as a representation of certain features, then the clusters represent typical constellations of frequently occurring combinations of input values. A cluster may then be formed out of some features, whereas other features are irrelevant. In our context, this means that only certain input neurons describe a cluster whereas the remaining input neurons simply produce noise. The task of the RBF neuron is to extract the "relevant" inputs for every cluster.

More precisely, we consider an input space of dimension n with basis $E = \{e_1, \ldots, e_n\}$, where e_j is the jth unit vector. Let us assume that for an input $x = \sum_{j=1}^{n} \lambda_j e_j$ the input neuron u_j receives the value $\lambda_j \in [0, T]$, with $[0, T]$ being the coding interval. We consider only subspaces where the basis E_{sub} is a subset of E.

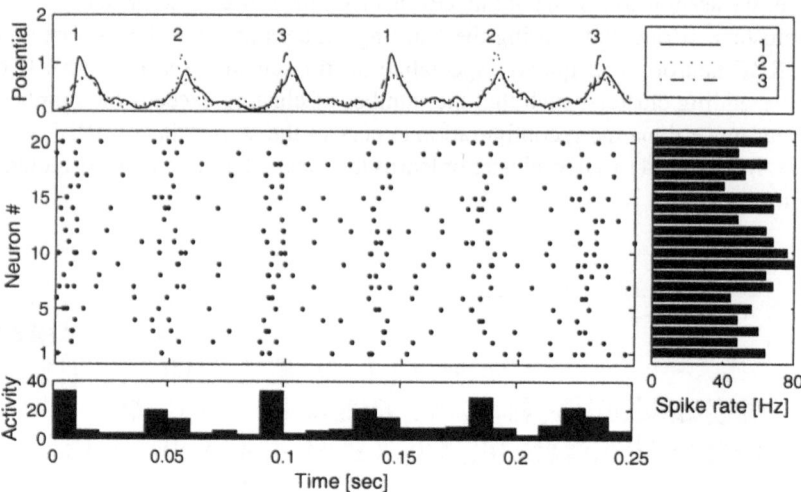

Fig. 3.9. Behavior of 3 RBF neurons on spike trains including Poisson distributed spikes of 25 spikes/sec after 270 learning cycles. The vectors presented to the network (every 50 ms) are chosen cyclically from 3 clusters. The lower box shows the number of spikes within time bins of size 10 ms. Although the input neurons fire all approximately with the same frequency (right box) the RBF neurons can distinguish the pattern in the noise (upper box). The threshold was set to 1.0.

Samples from a cluster C in such a subspace are presented to the network as

$$\mathbf{x} = \sum_{\mathbf{e}_j \in E_{sub}} \lambda_j \mathbf{e}_j + \sum_{\mathbf{e}_j \in E \setminus E_{sub}} n_j \mathbf{e}_j$$

where the n_j are random variables, uniformly distributed over $[0, T]$ (see Fig. 3.10 for an example). Assume that RBF neuron v_j is to converge to the center $\mathbf{c}$ of C. The goal is that besides finding the proper delays for the input neurons receiving the inputs for E_{sub}, the remaining "noisy" coordinates should have no impact on v_j, i.e. the corresponding weights should become zero.

Our construction described in the previous section can handle this situation if one slightly changes the learning function, such that larger values of $|\Delta t|$ cause a larger weight decay. This results on the long run in a weight decay for all delays of these noisy coordinates and in sharp peaks of the weight distribution for the other inputs u_j with $\mathbf{e}_i \in E_{sub}$. Fig. 3.11 shows an example of a delay distribution after a successful learning process. It also indicates that RBF neurons which have finally not specialized on some cluster may have quite similar weight distributions as "successful" RBF neurons. This results from a rather long period of competition.

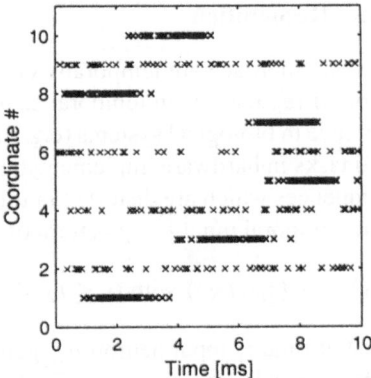

Fig. 3.10. Distribution of 100 input vectors for a cluster of dimension 5 in the subspace $E_{sub} = \{e_1, e_3, e_5, e_7, e_8, e_{10}\}$. The coding interval is of length 10 ms.

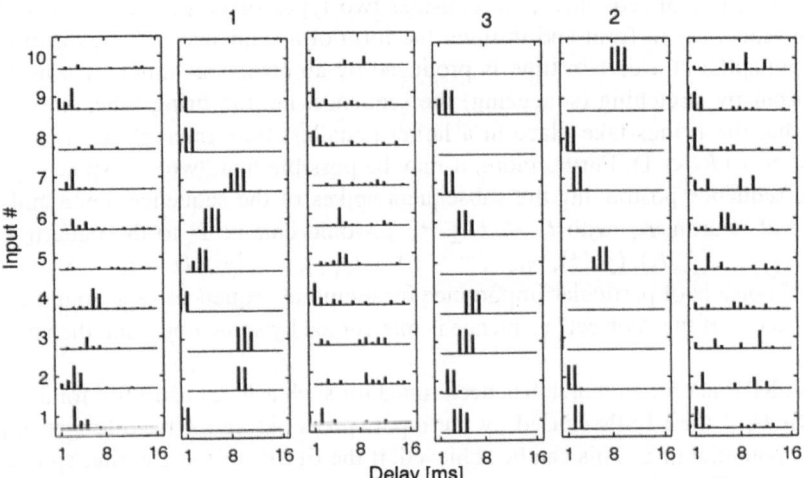

Fig. 3.11. Weight distribution for a set of delays $D = \{1\,\text{ms}, \ldots, 16\,\text{ms}\}$ after a successful learning process of 660 learning cycles with three clusters. Each column represents one RBF neuron with 40 inputs, of which 10 are shown. A number above a column indicates that the corresponding RBF neuron has specialized on the cluster with that number. Cluster 1 had 5, cluster 2 had 10 and cluster 3 had 15 noisy coordinates. RBF neuron 1 has a similar weigh distribution as RBF neuron 4, since it has competed for a rather long period with neuron 4 for cluster 3. The parameters for $L(\Delta t)$ (see Fig. 3.5(b)) were chosen as $c = -1.5\,\text{ms}$, $\beta = 1.67ms$ and $b = -0.2$.

3.3.4   Temporal Sequence Recognition

Since the input patterns in our approach are temporally encoded, it is quite natural
to use it for the recognition of regularities in temporal sequences of input stimuli.
This is of particular importance in biological systems (e.g. in the auditory and visual
cortex) but also for related tasks in hardware implementations. We will show that it
is possible to recognize sequences which are distorted in time and form.

Let us consider a spatio-temporal input firing pattern of the form

$$P = \langle (j_1, t_1), \ldots, (j_N, t_N) \rangle \text{ with } t_1 \leq t_2 \leq \cdots \leq t_N ,$$

where (j_i, t_i) indicates that at time t_i input neuron u_{j_i} generated a spike.[5] In contrast to our previous assumption an input neuron may fire here several times, i.e. we
allow $j_i = j_k$ for $i \neq k$. The detection of such patterns corresponds to the clustering problem described above, except that we allow here that an input neuron may
fire more than once within one sequence. In Sec. 3.3.2 we generated the clusters
by adding Gaussian noise to the centers of the clusters. Since we are now dealing with temporal sequences, we consider two types of noise, namely distortion
of the sequences in form and in time: the form of a sequence may be distorted if
an input spike at a certain time is produced by an erroneous input. A time warp
can occur by stretching (squeezing) the sequence, i.e. the firing times are scaled
such that the firings take place in a larger (smaller) time interval $K \cdot (t_N - t_1)$
for $K > 1$ ($K < 1$). Furthermore, it may be possible that "wrong" spikes appear
in the sequence postponing the subsequent spikes of the sequence, i.e. a spike at
time t of neuron u_{j_k} with $t_i \leq t \leq t_{i+1}$ would change P to the pattern $\tilde{P} =
\langle (j_1, t_1), \ldots, (j_i, t_i), (j_k, t), (j_{i+1}, (t - t_i) + t_{i+1}), \ldots, (j_N, (t - t_i) + t_N) \rangle$. This
type of noise is of particular importance for temporal sequences (e.g. sound waves
like speech), if the "correct" sequence is interrupted by some noise and then contin-
ued.

An RBF neuron v, which has been tuned for such a P, can still fire for an input
pattern $\tilde{P}$, if the EPSPs caused by the input spikes before t have a longer impact
on the potential of v. This can be achieved, if the EPSPs of these earlier spikes last
longer (see Fig. 3.12), such that a shift of the later EPSPs of the amount $t - t_i$,
caused by the input spikes after t, still suffices to make v fire. Generally we assume
that the earlier an input spike occurs the wider is the resulting EPSP.

This approach is similar to an idea presented in [36], where it is shown how an
artificial neural network can recognize such sequences using certain "delay filters".
We were able to realize this concentration in time very naturally by spiking neurons
employing EPSPs of variable width which roughly correspond to the delay filters.

Our construction is able to deal with all these above-mentioned types of noise.
Fig. 3.13 shows as an example that a stretching or squeezing of an input sequence
up to a factor of $K = 1.4$ respectively $K = 0.5$ still allows a proper detection of
each sequence. If a sequence is too strongly squeezed or stretched, the RBF unit,

---

[5] We assume that P describes a valid spike train, where a neuron does not fire within its
absolute refractory period.

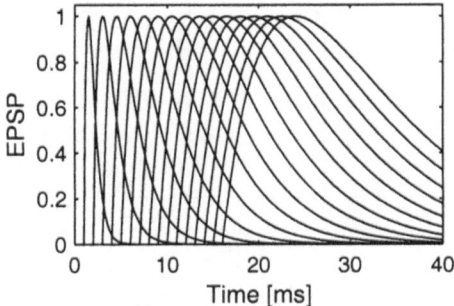

Fig. 3.12. EPSPs used for our simulations described in Sec. 3.3.4. The EPSPs shown here were α-functions, given by $\varepsilon(t) = t/t_p \exp(1 - t/t_p)$ with $t_p \in \{0.5 \text{ ms}, 1.0 \text{ ms}, \ldots, 8 \text{ ms}\}$. The delays (i.e. the time until the onset of the EPSP) were chosen $2 \cdot t_p$ such that a wider EPSP corresponds to a larger delay.

which was tuned to the corresponding sequence, does not fire at all, i.e. the network does either respond correctly or does not respond at all.

3.3.5   Stability

It is not obvious that the RBF neurons stabilize at a certain delay constellation during a long learning process. Indeed, it could be the case that the center of an RBF neuron becomes smaller during the learning process such that for certain inputs only the smallest available delays are active (i.e. have weights greater zero) whereas for other inputs the active delays vanish (see Fig. 3.14).

This undesirable behavior occurs if some weights become too large during the learning process. Large weights make the neuron fire earlier, such that the learning rule (3.2) increases the weights corresponding to shorter delays. Our learning rule tends to increase weights until they have reached the saturation value w_{max} (i.e. the largest possible weight). Hence this shift of delays can be avoided if w_{max} is not too large. The opposite effect, where larger delays are activated, may occur if w_{max} is too small. Thus the choice of w_{max} is crucial for the stability of the learning process.[6]

To overcome this problem one can use a continuous saturation function $S(w)$ such that a weight which is close to saturation is less strongly modified than a weight being more in the middle of $[0, w_{max}]$ (see Fig. 3.14(c) for an example). Hence, such a saturation function supports the stabilization of possible peaks in the weight distribution. One may use the following modification of the learning rule (3.2):

$$w_{ij}^{(k),new} = w_{ij}^{(k),old} + \eta \cdot S(w_{ij}^{(k),old}) \cdot L((t_j + d^{(k)}) - (t_i + d^{back})) \qquad (3.3)$$

---

[6] In our simulations in Sec. 3.3.2 we used $w_{max} = \Theta/(m \cdot h)$ where m is the number of inputs, Θ the threshold and h is the length of the interval $[\Delta t_1, \Delta t_2]$ for which the learning function $L(\Delta t)$ is greater zero.

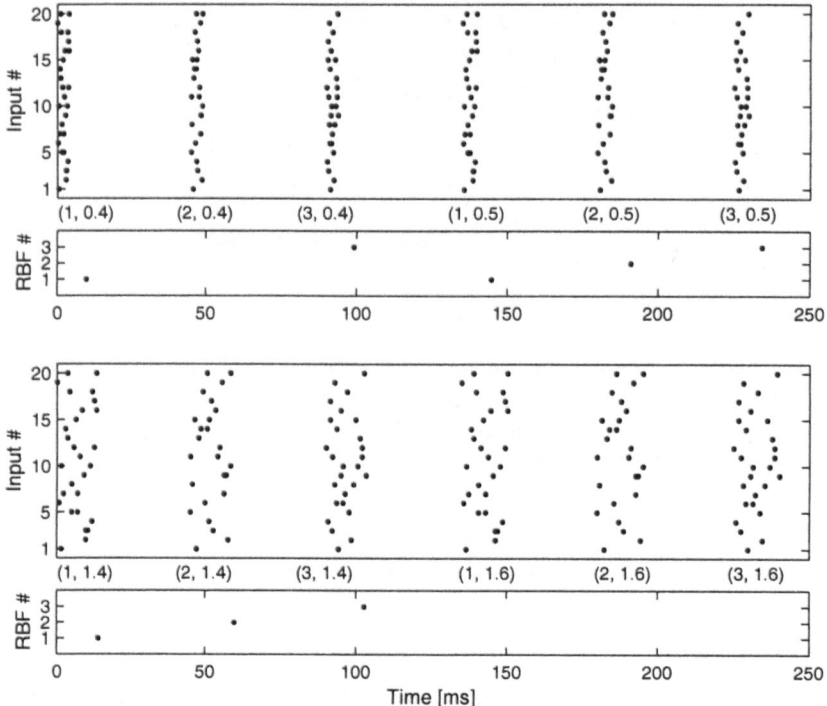

Fig. 3.13. Input- and output spike trains of an RBF network with 20 inputs and 3 RBF neurons. The network was trained on three clusters. After 270 learning cycles RBF neuron u had converged to the center of cluster i. A tuple (i, K) indicates that an input vector from the ith cluster with a stretch/squeeze factor of K was presented.

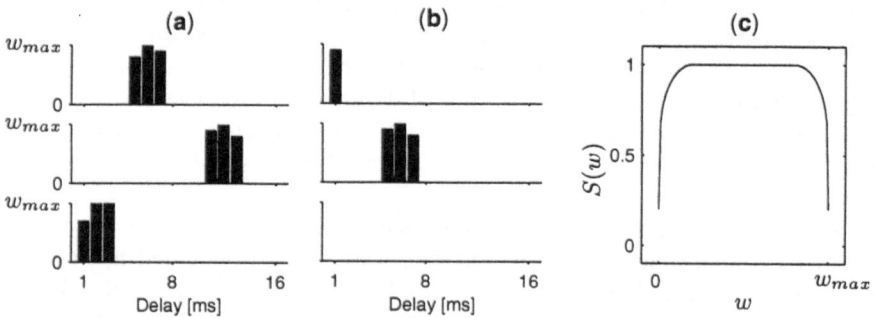

Fig. 3.14. (a) and (b) Shift of the weight distribution for a set of delays $D = \{1\,ms, \ldots, 16\,ms\}$. (a) 120 learning cycles, (b) 800 learning cycles. The maximum weight w_{max} was chosen too large, such that an initial learning success after 120 learning cycles shifted toward smaller delays, where for certain inputs even no delay had a weight greater zero. (c) One possible choice for a saturation function.

If the weights are outside of the interval $[0, w_{max}]$ after applying this rule, an additional weight clipping is made, such that a weight can never exceed its boundaries. In order to allow dynamic reconfigurations (like adding/removing clusters during the learning process), one has to assume that $S(0) > 0$ and $S(w_{max}) > 0$.

A reasonable choice of w_{max} is important especially in the context of finding clusters in subspaces (Sec. 3.3.3): One basic assumption in our construction is that one spike of each input neuron within the coding interval suffices to make an RBF neuron fire. If there are some noisy coordinates, the corresponding input neurons have no impact on the RBF neuron after a successful learning process. Hence the weights for the remaining input neurons must be large enough to make the RBF neuron fire. The straightforward solution to this problem, namely to have various w_{max} for different RBF neurons depending on the number of noisy coordinates is not very desirable in the framework of unsupervised learning since one must put some knowledge about the training set into the parameters of the network. In our simulations we were able to show that it suffices to use a saturation function as shown in Fig. 3.14(c) and one *global* value for w_{max}, which is large enough that an RBF neuron with many noisy coordinates can still fire but also can stabilize at a certain delay constellation.

3.4   Self-Organizing Maps of Spiking Neurons

Having dealt thus far with networks of spiking RBF neurons and the task of clustering, we shall now be concerned with the question how topology preserving mappings emerge in neural systems from unsupervised learning processes. On the basis of the model for networks of spiking neurons introduced in Sec. 3.2.1 we first provide a mechanism for competitive learning in networks of spiking neurons that is based on temporal coding schemes as discussed in Sec. 3.2.2. We then extend this idea to establish learning capabilities in networks of spiking neurons that are closely related to one of the most successful paradigms of unsupervised learning: the self-organizing map (SOM) by Kohonen [19].

Several regions of the brain are known to be topographic representations of some sensory area, e.g. in the visual, auditory, or somatosensory cortex [4]. Researchers have been able to reproduce some of the main qualitative features of these maps by the SOM (see, e.g., [29,30]). In particular, feature maps in area V1 of the visual cortex including retinal position, orientation, and ocular dominance could remarkably well be reproduced by the SOM [5]. Because of numerous studies with similar successful outcomes, the SOM is in general believed to provide a possible explanation how such maps arise from learning processes.

Most previous implementations of the SOM assume that input and output values of neurons are represented by their firing rates and not by the timing of single firing events. If neural network implementations for detecting the so-called winner neuron are considered then a recurrent network is used in most cases. The recurrence in these networks is then introduced in terms of lateral connections. This construction implies, however, that the winner neuron cannot be detected before the network has settled down to an equilibrium state. Since the computation relies on the convergence of the network, which may last for a considerable number of cycles, this approach disregards the benefits that temporal coding offers to fast information processing in networks of spiking neurons.

In addition to these conventional implementations there has also been some research on biologically more realistic models of self-organizing map algorithms, see e.g. [6,18,34]. Also in these approaches the output of a neuron is assumed to correspond to its firing rate, and learning takes place in terms of this rate after the network has reached a stable state of firing. There has also been the idea to implement the neighborhood relations between neurons through varying extracellular concentrations of the molecule nitric oxide [18]. This approach, however, has the disadvantage that the diffusion and change in concentration of nitric oxide are rather slow (see also [20]).

In the following we propose a mechanism for unsupervised learning in networks of spiking neurons where computing and learning are based on the timing of single firing events. In contrast to the standard formulation of the SOM, our construction has the additional advantage that the winner among competing neurons can be determined fast and locally by using lateral excitation and inhibition. These lateral connections also constitute the neighborhood relation among the neurons. We assume that initially neurons which are topologically close together have strong excitatory lateral connections whereas remote neurons have strong inhibitory connections. During the learning process the lateral weights are decreased, thus reducing the size of the neighborhood.

In a series of computer simulations we investigated the capability of the model to form topology preserving mappings. For the evaluation of these maps, instead of relying on visual inspection, we used a measure for quantifying the neighborhood preservation [10]. Our results show that the model exhibits the same main characteristics as the SOM. The typical emergence of topology preserving behavior could indeed be observed for a wide range of parameters.

We also studied the effect of weight normalization. This operation is used in several implementations of the SOM (see e.g. [6,19,34]) but its biological relevance is controversial. We performed simulations with and without weight normalization after each learning cycle. When comparing the results we found that after a certain number of learning cycles approximately the same degree of topology preservation has been achieved regardless whether the weights were normalized or not. The model of competitive learning in terms of action potential timing that we propose here is therefore a candidate for a more realistic description of self-organization in biological neural systems.

3.4.1 Competitive Computing and Learning

The mechanism of neural computation employed here is similar to that of Sec. 3.3 and based on the fact that the models of spiking neurons introduced in Sec. 3.2.1 can compute weighted sums in temporal coding, where the value encoded by the action potential of a neuron is larger the earlier the firing takes place [22,24] (see also

[32]). More precisely, consider a neuron v which receives excitatory input from m neurons $u_1, \ldots, u_m$, where the corresponding weights are denoted by $w_1, \ldots, w_m$. Each u_j fires exactly once within a sufficiently small time interval at a time t_j with $t_j = T_{in} - s_j$, where s_j is the jth input to v and T_{in} some constant. Usually T_{in} is given by the time when a reference spike is fired by some additional input neuron of v. Under certain weak assumptions (basically the initial segments of the excitatory postsynaptic potentials have to be linear) it is guaranteed that v fires at a time determined by $T_{out} - \sum w_j s_j$ with T_{out} being some constant. This computation relies on reference times T_{in} and T_{out} provided by the firing of specific neurons, but it can also be performed on the basis of *competitive temporal coding*, such that these explicit reference times are not required (see Sec. 3.3.1 and [22,24,32]).

Using this construction we can implement mechanisms of unsupervised learning in networks of spiking neurons as follows: Suppose S is a set of m-dimensional input vectors $\mathbf{s}^l = (s_1^l, \ldots, s_m^l) \in [0, \gamma]^m$ and consider a network of spiking neurons with m input neurons and n competitive neurons, where each competitive neuron v_i receives synaptic *feedforward* input from each input neuron u_j with weight w_{ij} and *lateral* synaptic input from each competitive neuron $v_k, k \neq i$, with weight $\tilde{w}_{ik}$. At each cycle of the learning procedure one input vector $\mathbf{s}^l \in S$ is randomly chosen and the input neurons are made fire such that they temporally encode $\mathbf{s}^l$. Each v_i then starts to compute $\sum_j w_{ij} s_j^l$ as described in Sec. 3.2.1. There is an intuitive explanation of what these neurons compute: If we assume that the input vector and the weight vector for each neuron are normalized, then the weighted sum represents the similarity between the two vectors with respect to the Euclidean distance. Hence the earlier v_i fires, the more similar is its weight vector to the input vector, i.e. the winner among the layer of competitive neurons fires first. Note that the firing time of the winner is not influenced by the firing of the other competitive neurons (see Fig. 3.15).

If the lateral connections are strongly inhibitory, such that the firing of the winner neuron, say v_k, prevents all other neurons in the competitive layer from firing, competitive learning can be implemented in a straightforward way: The standard competitive learning rule is applied to the winner neuron v_k (see, e.g., [19]). This rule is given for v_k by

$$\Delta w_{kj} = \eta(s_j^l - w_{kj}) , \qquad j \in \{1, \ldots, m\}$$

where $\mathbf{s}^l$ is the current input vector and η the learning rate. Thus, one learning cycle consists of the presentation of an input vector $\mathbf{s}^l \in S$ making one of the competitive neurons fire. The spikes representing the input vector cause weight changes at the synapses of the winner neuron due to the above learning rule.

We note that the learning rule does not ensure that the weights are always positive. A weight vector that becomes similar to some input vector during learning, however, will naturally have all weights nonnegative. A change of the sign of a weight, which is biologally highly implausible, can also formally be avoided by introducing a saturation function that guarantees the values of the weights to be from some restricted interval (see also Sec. 3.3.5).

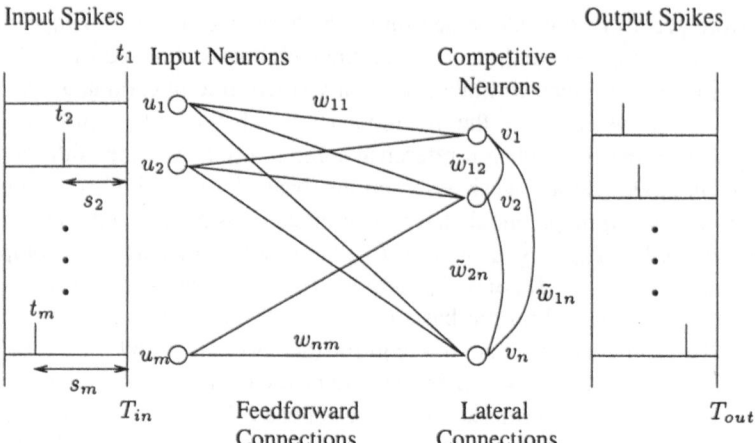

Fig. 3.15. The basic architecture of a network of spiking neurons that is capable of learning topology preserving mappings.

3.4.2   Self-Organization

In order to extend this type of competitive learning to a mechanism that results in topology preserving behavior, a way has to be found for representing a given neighborhood matrix $(m_{kj})_{1 \leq k,j \leq n}$ on the basis of locally available information, where m_{kj} describes the distance between the kth and jth competitive neuron. We use a monotonically increasing function $m_{kj} \mapsto \tilde{w}_{kj}$ such that the lateral connections $\tilde{w}_{kj}$ among the competitive neurons reflect the structure of the neighborhood: Initially, neurons which are topologically close together have strong excitatory lateral connections whereas remote neurons have strong inhibitory connections. This means that the firing of the winner neuron, say v_k, at time t_k drives the firing times of neurons in the neighborhood of v_k toward t_k, thus increasing the values they encode. The firing of remote neurons is postponed through the lateral inhibitions. Hence v_k encodes with its firing the weighted sum $\sum s_i^l \cdot w_{ik}$ for an input pattern $s^l \in S$. In the example of Fig. 3.15, the firing of the winner v_1 shifts the firing time of a topologically close neuron v_2 (which would compute a similar weighted sum if its firing were not influenced by v_1) toward the firing time of v_1 and delays the firing of a topologically remote neuron v_n. For the modification of weights we propose the following learning rule:

$$\Delta w_{ji} = \eta \frac{T_{out} - t_j}{T_{out}} (s_i^l - w_{ji}) , \qquad (3.4)$$

where t_j is the firing time of the jth competitive neuron. The learning rule applies only to neurons that have fired before a certain time T_{out} (which has to be chosen sufficiently large). The factor $(T_{out} - t_j)/T_{out}$ implements the neighborhood function, which is largest for the winner neuron and decreases for neurons which fire at later times.

During the learning process the lateral weights $\tilde{w}_{ik}$ are decreased, thus reducing the size of the neighborhood.[7] As in the standard formulation of the SOM, the learning rate η is slowly reduced approaching zero toward the end of the learning process.

3.4.3 Simulations

We performed computer experiments with one- and two-dimensional input patterns. In the one-dimensional case, 10 input vectors, uniformly distributed over the coding interval, were presented to a layer of 10 competitive neurons, which had initially random weights of values around the midpoint of the patterns. The lateral weights for the immediate neighbors were chosen slightly positive, for the second neighbors zero and for all other neurons negative. The goal was the formation of a linear map. Fig. 3.16 shows that initially nearly all neurons react strongly, i.e. fire early on each input, whereas after learning only few neurons, being topologically close, react on a certain input pattern.

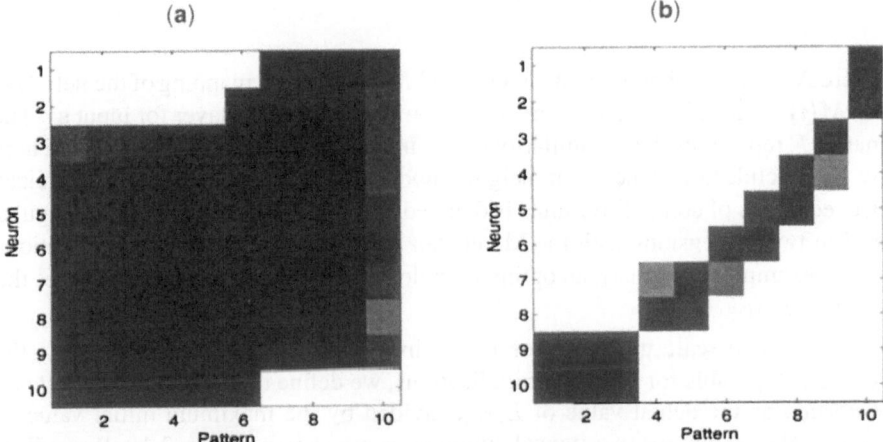

Fig. 3.16. Learning process for one-dimensional input patterns after (a) 10 cycles and (b) 4000 cycles. Each column represents the response of the network to a particular input pattern. The grey tones of the squares indicate the firing times: the darker a square, the earlier the corresponding firing time. The non-firing of a neuron is represented by a white square.

We furthermore investigated the behavior of the network on one of the standard examples for the SOM, where two-dimensional input patterns are chosen randomly

---

[7] This may require that the lateral weights change their sign during learning, which is known to be not very realistic for biological networks. This can be avoided replacing a synapse by two connections, one excitatory, which decreases, and one inhibitory, which increases during learning.

from a square and the competitive neurons are expected to self-organize in a topology preserving grid. For the tests we used an array of 5×5 competitive neurons. We normalized the input vectors by adding a third input component, which was chosen such that all input vectors have equal norm. The weights were initialized in the same fashion as in the previous type of experiment. In order to examine the effect of normalization we performed two series of experiments. In the first one the feed-forward weights were normalized after each application of the learning rule (3.4), in the second they were kept unnormalized. Fig. 3.17 shows the typical result of an experiment run with weight normalization.

The topology preserving quality of self-organizing maps is usually assessed by visual inspection. This may be appropriate when observing the development of a single mapping but it is hardly possible to compare the degree of topology preservation of two different mappings. In our second experiment we therefore used a measure for quantifying the neighborhood preservation known as *metric multidimensional scaling* (see e.g. [10]). This measure is based on the objective function

$$E_{\mathrm{MDS}} = \sum_{i=1}^{N} \sum_{j<i} (F(i,j) - G(M(i), M(j)))^2 \,, \tag{3.5}$$

where N is the number of input patterns and M denotes the mapping of the network, i.e. $M(i)$ is the index of the winner neuron in the competitive layer for input $\mathbf{s}^i$. The matrix F represents the dissimilarity of a pair of input patterns $\mathbf{s}^i, \mathbf{s}^j$ expressed here by the Euclidean distance. The neighborhood matrix which defines the distances between pairs of competitive units is denoted by G. We use the familiar rectangular grid in two dimensions with the Manhattan metric for the neighborhood relations in this example. Obviously, an optimal topology preserving mapping minimizes the value of E_{MDS}.

In order to scale the values of E_{MDS} into the interval $[0, 1]$ and to make the results comparable for different initializations, we define the *relative neighborhood distortion* as the actual value of E_{MDS} divided by the maximum initial value of E_{MDS}. The results for two typical simulations are shown in Fig. 3.18. Regardless whether the weights are normalized or not—the relative neighborhood distortions indeed approach approximately the same small value.

We have found that a standard implementation of the SOM, including various heuristics, e.g. for the decrementation of the learning rate and the reduction of the neighborhood size, converges about 10 times faster. This is mainly due to the usage of a look-up table providing precise values for the neighborhood function and the above-mentioned heuristics. In the construction provided here, distances between neurons are computed each learning cycle anew and may represent only approximate values of the exact distances. On the other hand, the distances are computed in a way that is biologically more realistic than in any of the conventional SOM implementations. Thus, the observed increase of time required for learning may be interpreted as having traded speed of learning for biological plausibility.

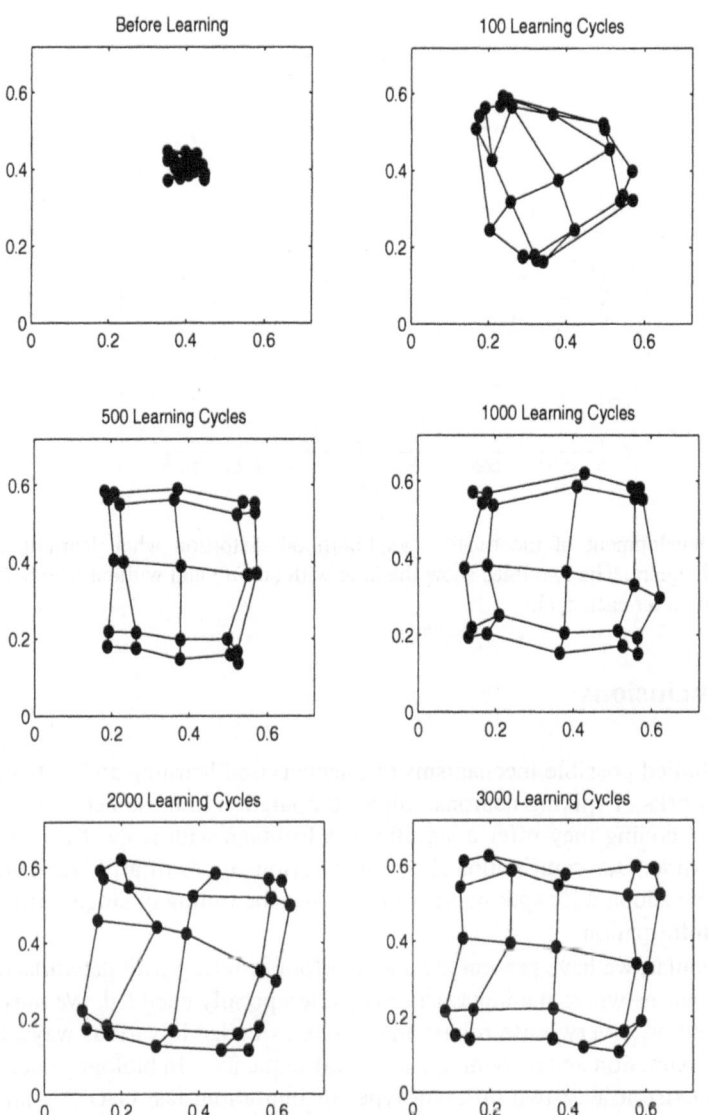

Fig. 3.17. Learning a square region of the plane with 5×5 competitive spiking neurons. Each dot represents the location of a neuron in the plane specified by its weight vector. The lines connect neurons that are immediate neighbors.

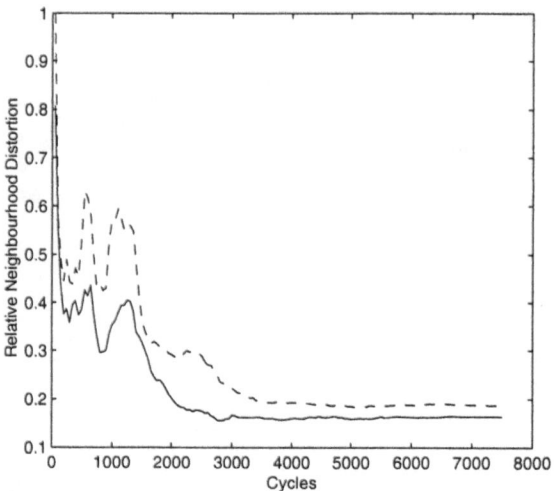

Fig. 3.18. Development of the relative neighborhood distortion when learning the two-dimensional square. The two lines show the case with (*solid*) and without (*dashed*) weight normalization after each cycle.

3.5   Conclusions

We have studied possible mechanisms of unsupervised learning and self-organization in networks of spiking neurons. Since computation in these networks is based on temporal coding they offer a significant advantage with respect to computing power and time. Our contribution shows that various types of artificial neural networks can be adopted for spiking neurons that use the timing of single firing events to encode information.

In particular, we have presented a method for clustering with networks of spiking RBF neurons where the input patterns are temporally encoded. We have found that this method is very noise-robust and can be extended in various ways, such as for feature extraction and recognizing temporal sequences. In biological neural systems there exist at least two different types of inhibition: fast $GABA_A$ and slow $GABA_B$ mediated inhibitory synapses (see e.g. [33]). Our model combines them in a new way: Fast lateral inhibition is used to implement the winner-take-all mechanism and slow self-inhibition to allow a competition among all RBF neurons, such that no RBF neuron dominates the competition over a longer period of time. In artificial RBF networks, which are used for classification and function approximation tasks, there is an additional linear gate following the RBF layer. If the inputs are temporally encoded as described in this chapter, spiking neurons can compute such linear gates in a very simple and straightforward way. This idea can be incorporated in our construction: If one omits the fast lateral inhibition among the RBF neurons, the firing time of each RBF neuron relative to the others expresses its similarity to the input vector. A spiking neuron receiving input from all these RBF neurons can then perform the task of the linear gate.

We have also proposed a mechanism for networks of spiking neurons to learn topology preserving mappings. Our results show that temporal coding in these networks leads to a simple and straightforward way for determining the winner among competing neurons. Furthermore, a topology preserving behavior emerges from learning processes solely based on locally available information. We emphasize that the learning rule used in this process is not an exact replication of Kohonen's learning algorithm. However, since our implementation is inspired by the latter, it constitutes a further step toward showing that biological neurons can indeed achieve a topology preserving behavior using learning methods like those suggested by Kohonen.

The constructions presented in this contribution seem to be suitable for hardware implementations in pulse stream VLSI, where analog computational devices are built for processing information encoded in pulses. The method of pulse phase modulation, where values are represented by the temporal delay of discrete pulses with fixed amplitude, is closely related to the scheme of temporal coding using action potentials of spiking neurons. Approaches for VLSI implementations of local learning rules with spiking neurons as we have considered exist already [13,21].

Acknowledgments

We would like to thank Oswin Aichholzer and Peter Auer for helpful discussions about the topics of Sec. 3.3. Peter Auer also pointed us to the problem of finding clusters in subspaces. Part of this work has been supported by the ESPRIT Working Group in Neural and Computational Learning II, NeuroCOLT2, No. 27150. T.N. acknowledges support by the "Fonds zur Förderung der wissenschaftlichen Forschung (FWF), Austrian Science Fund", project number P12153.

72   Thomas Natschläger et al.

References

1. L. F. Abbott and S. B. Nelson. Synaptic plasticity: taming the beast. *Nature Neuroscience*, 3(Supp):1178–1183, 2000.
2. L. F. Abbott, J. A. Varela, K. Sen, and S. B. Nelson. Synaptic depression and gain control. *Science*, 275:220–224, 1997.
3. M. Abeles, H. Bergman, E. Margalit, and E. Vaadia. Spatiotemporal firing patterns in the frontal cortex of behaving monkeys. *Journal of Neurophysiology*, 70(4):1629–1638, 1993.
4. M. A. Arbib, editor. *The Handbook of Brain Theory and Neural Networks*. MIT Press, Cambridge, Mass., 1995.
5. G. Blasel and K. Obermayer. Putative strategies of scene segmentation in monkey visual cortex. *Neural Networks*, 7:865–881, 1994.
6. Y. Choe and R. Miikkulainen. Self-organization and segmentation in a laterally connected orientation map of spiking neurons. *Neurocomputing*, 21:139–157, 1998.
7. T. J. Gawne, T.Kjaer, and B. Richmond. Latency: Another potential code for feature binding in striate cortex. *Journal of Neurophysiology*, 76(2):1356 – 1360, 1996.
8. W. Gerstner. Spiking neurons. In W. Maass and C. M. Bishop, editors, *Pulsed Neural Networks*, pages 3–53. MIT Press, Cambridge, Mass., 1999.
9. W. Gerstner, R. Kempter, L. van Hemmen, and H. Wagner. A neuronal learning rule for sub-millisecond temporal coding. *Nature*, 383:76–78, 1996.
10. G. J. Goodhill and T. J. Sejnowski. A unifying objective function for topographic mappings. *Neural Computation*, 9:1291–1303, 1997.
11. S. Grossberg. Adaptive pattern classification and universal recording: II. Feedback, expectation, olfaction, illusions. *Biological Cybernetics*, 23:187–202, 1976.
12. L. Haberly. Neuronal circuitry in olfactory cortex: Anatomy and functional implications. *Chemical Senses*, 10(2):219–238, 1985.
13. P. Häfliger, M. Mahowald, and L. Watts. A spike based learning neuron in analog VLSI. In *Advances in Neural Information Processing Systems 9*, pages 692–698, MIT Press, Cambridge, Mass., 1997.
14. J. J. Hopfield. Pattern recognition computation using action potential timing for stimulus representation. *Nature*, 367:33–36, 1995.
15. D. Johnston and S. M. S. Wu. *Foundations of Cellular Neurophysiology*. MIT Press, Cambridge, Mass., 1995.
16. C. Koch. *Biophysics of Computation: Information Processing in Single Neurons*. Oxford University Press, 1999.
17. C. Koch and I. Segev. *Methods in Neural Modeling: From Ions to Networks*. MIT Press, Cambridge, Mass., 1998.
18. T. Kohonen. Physiological interpretation of the self-organizing map algorithm. *Neural Networks*, 6:895–905, 1993.
19. T. Kohonen. *Self-Organizing Maps*. Springer, Berlin, 1995.
20. B. Krekelberg and J. G. Taylor. Nitric oxide: What can it compute? *Network: Computation in Neural Systems*, 8:1–16, 1997.
21. T. Lehmann and R. Woodburn. Biologically-inspired on-chip learning in pulsed neural networks. *Analog Integrated Circuits and Signal Processing*, 18:117–131, 1999.
22. W. Maass. Fast sigmoidal networks via spiking neurons. *Neural Computation*, 9:279–304, 1997.
23. W. Maass. Networks of spiking neurons: The third generation of neural network models. *Neural Networks*, 10:1659–1671, 1997.

24. W. Maass. Computing with spiking neurons. In W. Maass and C. M. Bishop, editors, *Pulsed Neural Networks*, chapter 2, pages 55–85. MIT Press, Cambridge, Mass., 1999.
25. W. Maass and C. M. Bishop, editors. *Pulsed Neural Networks*. MIT Press, Cambridge, Mass., 1999.
26. H. Markram, Y. Wang, and M. Tsodyks. Differential signaling via the same axon of neocortical pyramidal neurons. *Proc. Nat. Acad. Sci. USA*, 95:5323–8, 1998.
27. J. O'Keefe and M. L. Reece. Phase relationship between hippocampal place units and the EEG theta rhythm. *Hippocampus*, 3(3):3317–30, 1993.
28. F. Rieke, D. Warland, W. Bialek, and R. de Ruyter van Steveninck. *SPIKES: Exploring the Neural Code*. MIT Press, Cambridge, Mass., 1999.
29. H. Ritter. Self-organizing feature maps: Kohonen maps. In M. A. Arbib, editor, *The Handbook of Brain Theory and Neural Networks*, pages 846–851. MIT Press, Cambridge, Mass., 1995.
30. H. Ritter, T. Martinetz, and K. Schulten. *Neural Computation and Self-Organizing Maps*. Addison-Wesley, Reading, Mass., 1992.
31. R. Ritz and T. J. Sejnowski. Synchronous oscillatory activity in sensory systems: new vistas on mechanisms. *Current Opinion in Neurobiology*, 7:536–546, 1997.
32. B. Ruf. *Computing and Learning with Spiking Neurons—Theory and Simulations*. PhD thesis, Institute for Theoretical Computer Science, Technische Universität Graz, Austria, 1998.
33. I. Segev. Dendritic processing. In M. A. Arbib, editor, *The Handbook of Brain Theory and Neural Networks*, pages 282–289. MIT Press, Cambridge, Mass., 1995.
34. J. Sirosh and R. Miikkulainen. Topographic receptive fields and patterned lateral interaction in a self-organizing model of the primary visual cortex. *Neural Computation*, 9:577–594, 1997.
35. K.-Y. Siu, V. Roychowdhury, and T. Kailath. *Discrete Neural Computation: A Theoretical Foundation*. Information and System Sciences Series. Prentice-Hall, Englewood Cliffs, NJ, 1995.
36. D. W. Tank and J. J. Hopfield. Neural computation by concentrating information in time. *Proc. Nat. Acad. Sci. USA*, 84:1896–1900, Apr. 1987.
37. S. Thorpe, D. Fize, and C. Marlot. Speed of processing in the human visual system. *Nature*, 381:520–522, 1996.
38. J. A. Varela, K. Sen, J. Gibson, J. Fost, L. F. Abbott, and S. B. Nelson. A quantitative description of short-term plasticity at excitatory synapses in layer 2/3 of rat primary visual cortex. *J. Neurosci*, 17:220–4, 1997.

4 Generative Probability Density Model in the Self-Organizing Map

Jouko Lampinen and Timo Kostiainen

Abstract. The Self-Organizing Map, SOM, is a widely used tool in exploratory data analysis. A theoretical and practical challenge in the SOM has been the difficulty to treat the method as a statistical model fitting procedure. In this chapter we give a short review of statistical approaches for the SOM. Then we present the probability density model for which the SOM training gives the maximum likelihood estimate. The density model can be used to choose the neighborhood width of the SOM so as to avoid overfitting and to improve the reliability of the results. The density model also gives tools for systematic analysis of the SOM. A major application of the SOM is the analysis of dependencies between variables. We discuss some difficulties in the visual analysis of the SOM and demonstrate how quantitative analysis of the dependencies can be carried out by calculating conditional distributions from the density model.

4.1 Introduction

The self-organizing map, SOM, is a widely used tool in data mining, visualization of high-dimensional data, and analysis of relations between variables. For a review of SOM applications see other chapters in this volume, and [12].

The most characteristic property of the SOM algorithm [8] is the preservation of topology, or the fact that the neighborhood relationships of the input data are maintained in the mapping. A large part of theoretical work on SOM has been focused on the definition and quantification of topology preservation, but mathematically rigorous treatment is not yet complete. See [16] for up-to-date discussion of the topology preservation in the SOM.

The roots of the SOM are in simplified models for the self-organization process in biological neural networks [7]. In related engineering problems the SOM offers considerable potential, such as in automatic formation of categories in larger artificial neural systems.

Currently, a very active application domain of the SOM is *exploratory data analysis*, where a database is searched for any phenomena that are important in the studied application. In normal statistical data analysis there are usually a set of hypotheses that are validated in the analysis, while in exploratory data analysis the hypotheses are generated from the data in a data-driven *exploratory* phase and validated in a *confirmatory* phase. The SOM is mainly used in the exploratory phase, by visually searching for potentially dependent variables. There may be some problems where the exploratory phase alone might be sufficient, such as visualization of data without more quantitative statistical inference upon it. However, in practical data analysis problems the found hypotheses need to be validated with well understood methods, in order to assess the confidence of the conclusions and to reject those that are not statistically significant.

When using the SOM in data analysis, an obvious criterion for model selection should be generalization of the conclusions to new data, just as it is in the case of any other statistical method. The preservation of topology is also important, to facilitate the visual analysis by grouping the similar states to neighboring map units, but if the positions of the map units are not statistically reliable the map is useless for any generalizing inference.

In this chapter we present the SOM as a probability density estimation method, in contrast to the standard view of the SOM as a method for mapping high dimensional data vectors to a lower dimensional space. There are several benefits in associating a probability density, or a generative model, with a mapping method (see [14] for discussion of a generative model for the PCA mapping):

- The density model enables computation of the likelihood of any data sample (training data or test data), facilitating statistical testing and comparison with other density estimation techniques.
- The selection of hyperparameters of the model (eg., the width of the neighborhood) can be chosen with standard methods, such as cross-validation, to avoid overfitting, in the same way as with other statistical methods.
- The density model facilitates quantitative analysis of the model, for example, by computing conditional densities to test the visually found hypotheses.
- In principle, Bayesian methods could be used for model complexity control and model comparison (see [10] for a review of Bayesian approach for neural networks). However, as shown later, the normalization of the probability density in the original SOM requires a numerical procedure that seems to render the Bayesian approach impractical.

The organization of this chapter is the following:

In Sec. 4.2 we discuss the problem of finding dependencies between variables using visual inspection of the SOM, to demonstrate the need for quantitative analysis tools with the SOM.

In Sec. 4.3 we shortly review some results related to the existence of the error function in the SOM. The SOM algorithm is not defined in terms of an error function, but directly via the training rule, and unfortunately the training rule is not a gradient of any global error function [5]. This makes the exact mathematical analysis of the SOM algorithm fairly difficult. For a discrete data sample the algorithm may converge to a local minimum of an error function [13], which may exist only in a small volume in the parameter space (the error function changes if the best-matching unit of any data sample changes). In Sec. 4.3 we shortly review the results about the existence of the error functions in the SOM and some modifications that make the error function to exist more generally.

The probability density model in the SOM, derived in this chapter, consists of kernels of non-regular shape, whose positions are weighted averages over the neighboring units receptive fields, and thus the model is close to many mixture models where the kernels are confined to a low dimensional latent space. In Sec. 4.4 we review some constraint mixture models that are similar to the SOM.

In Sec. 4.5 we derive the exact probability density model, for which the converged state of the SOM training gives the Maximum Likelihood estimate.

In Sec. 4.6 we discuss the selection of the SOM hyperparameters to avoid overfitting, and demonstrate how quantitative analysis can be carried out with the aid of the probability density model.

In Sec. 4.7 we present conclusions and point some directions for further study.

4.2   SOM and Dependence Between Variables

In practical data analysis problems a common task is to search for dependencies between variables. Statistical dependence means that the conditional distribution of a variable is dependent on the values of other (explanatory) variables, and thus the analysis of dependencies is closely related to the estimation of probability density or conditional probability densities. In regression analysis the goal is to estimate the dependence of the conditional mean of the target variable on the explanatory variables, using, for example, the standard least squares fitting of neural network outputs to the targets. In real data analysis problems, the shape of the conditional distribution needs to be considered also in the regression models, by means of, e.g., error bars, or confidence intervals, in order to assess the statistical significance of the dependence of conditional mean on the explanatory variables.

The most simple goal is to look for pairwise dependencies, where a variable is assumed to depend only on one other variable. For such a problem the advantage of the SOM is rather marginal, as simple correlation analysis is sufficient for the linear case, and in the non-linear case there exist plenty of methods for directly estimating the conditional density and thus the dependencies in such a low dimensional case (see e.g. [1] for review).

The tough problem in exploratory data analysis is to search for non-linear dependencies between multiple variables. With the SOM, the analysis of dependencies is based on visual inspection of the SOM structure. Several visualization methods have been developed for interpreting the SOM, see, e.g., [4]. The basic procedure is to visually search for regions on the map where the values of two or more variables coincide, e.g., have large or small values in the same units. Such a region is interpreted as a hypothesis that the variables are dependent in a way that, for example, low value for one variable is indication of low value for the other variable, given that rest of the variables are close to the corresponding values in the reference vectors. Clearly, efficient visualization methods are necessary, as the number of variable pairs is proportional to the square of the number of variables, and there may be dozens of distinguishable regions in the SOM.

It is very important to notice that any conclusions drawn from models overfitted to the data sample are not guaranteed to generalize to any other situation. In the case of the SOM, overfitting means that reference vectors of some units have been determined by too few data points, so that the reference vectors are not representative of the underlying probability density. Any conclusions based on such units are prone to fail for new data, and thus analysis of statistical dependencies requires some way, heuristic or more disciplined, to avoid overfitting, as in all statistical modeling. On the other hand, it should be noted that when the SOM is used for analyzing the whole population, and measurement errors are considered negligible, there is no need to generalize the conclusions to other data, and overfitting is not an issue. This important distinction between analyzing the population, and analyzing a sample from the population and generalizing the conclusions to the population, often seem to be ignored in the SOM framework.

The main problem in visual inspection of the SOM is that in general *the lack of dependence between variables is difficult to observe visually from the SOM*. That is, even if variables, say, x^1 and x^2 both have high values at map unit M_{ij}, that alone does not show that the variables have any mutual dependence. As a simple example, consider two-dimensional uniform distribution $x_1, x_2 \sim U(-1, 1)$. A 2×2 SOM with zero neighborhood would have component planes (in any order of the columns and rows)

$$M^1 = \begin{bmatrix} -0.5 & -0.5 \\ 0.5 & 0.5 \end{bmatrix} \qquad M^2 = \begin{bmatrix} -0.5 & 0.5 \\ -0.5 & 0.5 \end{bmatrix}$$

The coincidence of high values in unit M_{22} and low values in M_{11} are only a result of the vector quantization. To see that high values in M_{22} do not indicate dependence between x^1 and x^2, one must observe that high value for x^1 occurs also in M_{12} with low value for x^2 (i.e., tallied over the map, high value for x^1 occurs with both high and low value for x^2).

In high dimensional space the visual inspection of the dependencies becomes more difficult, as the map folds into the data space, and the range of values for each variable is distributed around the map. Fig. 4.1 illustrates this for random data with no dependencies between the variables, and Fig. 4.2 shows an example from a real data analysis project, where all hypotheses were later rejected in careful analysis.

4.3   Error Functions in the SOM

The converged state of the SOM is a local minimum of the error function which is given by [13]

$$E(X) = \sum_{n=1}^{N} \sum_{j=1}^{M} H_{bj} \|x^n - m^j\|^2, \tag{4.1}$$

where $X = \{x^n\}, n = 1, \ldots, N$ is the discrete data sample, j is the index (or position) of a unit in the SOM, with reference vector m^j, and $H_{bj} = H(b(x) - j)$ is the neighborhood function, $b(x)$ being the index of the best matching unit for x.

The error function is not defined at the boundaries of the receptive fields, or Voronoi cells, so the function does not exist in the continuous case. In the case of a discrete data set, the probability of any sample lying at any boundary is zero, so

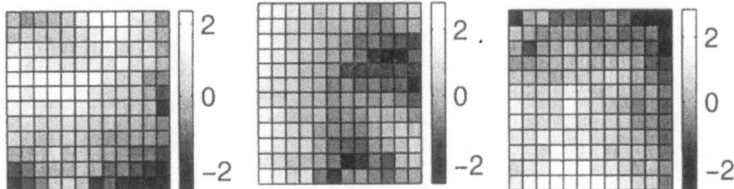

Fig. 4.1. Example of a SOM trained on purely random data. The independence of the variables in the component level display is not trivial to observe. One might, for example, erroneously conclude that high values of x_3 would indicate low values of x_2. Here the neighborhood is trained down to zero.

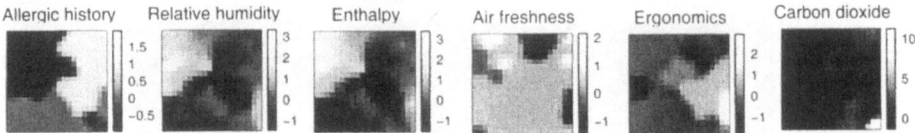

Fig. 4.2. Example of real data analysis. In the case study, the dependence of *Air freshness* on the other variables was investigated. In the final analysis all hypotheses were rejected using methods like RBF models, Bayesian neural networks, and hierarchical generalized linear models using Bayesian inference, etc. [17]. One evident conclusion, from the lower right corner of the map, is that high value for variable *Ergonomics* appears only with low value for *Air freshness*, but careful analysis showed that this was just an effect of vector quantization.

in practice the error function can always be computed for any data set. The error function changes if any sample changes its best matching unit. That is why the error function is only consistent with the SOM training rule when the algorithm has converged. In practical data analysis the data set is always discrete and the algorithm is allowed to converge, so analysis of the error function is thereby justified.

The SOM training rule involves assigning to each data sample one reference vector, the best matching unit $b(x)$. The matching criterion is Euclidian distance, as follows:

$$b(x) = \operatorname*{argmin}_{i} \|x - m_i\|. \tag{4.2}$$

When a sample is near a boundary of two or more receptive fields, a small change in the position of one reference vector can change the best matching unit of that sample. Hence the gradient of the error function with respect to the reference vectors is infinite.

The error function (4.1) can be thought of as the sum of the *distortion function*

$$D(x) = \sum_j H_{bj} \|x - m_j\|^2. \tag{4.3}$$

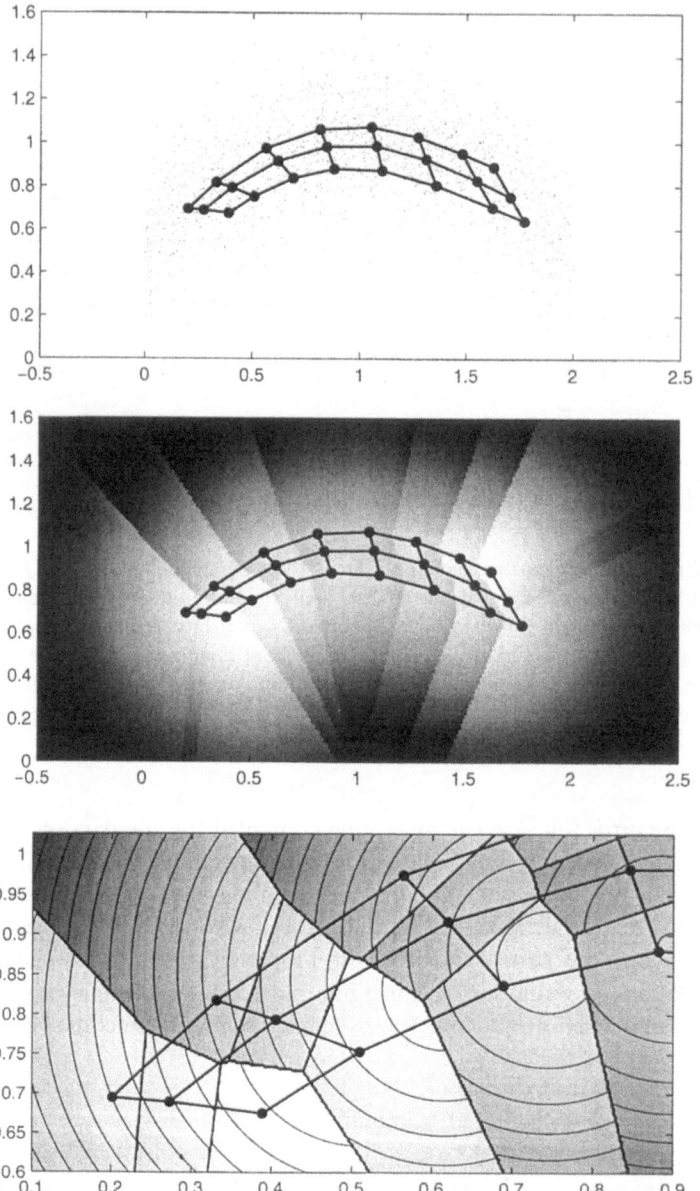

Fig. 4.3. Example of the density model in a 3 × 8 SOM. Top: training data and the resulting SOM lattice using Gaussian neighborhood with $\sigma = 1$. Middle: the density model of the SOM. Bottom: zoomed part of the density model above, with contours and Voronoi cell boundaries added. From the figure it is clear how the Gaussian kernels are located at the neighborhood weighted averages of the reference vectors.

over the data set. The distortion function is also discontinuous at the Voronoi cell boundaries. The discontinuity is a result of the winner selection rule of the training algorithm.

Luttrell [11] has shown that exact minimization of Eq. 4.1 leads to an approximation to the original training rule, where, instead of the nearest neighbor winner rule, the best matching unit is taken to be the one that minimizes the value of the distortion function (4.3), as follows:

$$b(x) = \operatorname*{argmin}_{i} \sum_{j} H_{ij} \|x - m_j\|^2. \tag{4.4}$$

The minimum distortion rule avoids many theoretical problems associated with the original rule, without compromising any desirable properties of the SOM except for an increase in the computational burden [6]. The gradient of the error function becomes continuous at the boundaries of receptive fields (which are no longer the same as the Voronoi tessellation). The distortion function with the modified winner selection rule is also continuous across the unit boundaries.

4.4   Constrained Mixture Models

The main aspects of the SOM algorithm, that make the analysis difficult are 1) the hard assignment of the input samples to the nearest units, which makes the receptive fields irregularly shaped according to the Voronoi tessellation, and 2) the regularizing effect defined as updating the parameters of the neighboring units towards each input, instead of regularization applied directly to the positions of the reference vectors.

It is worth noticing that the first issue is due to the shortcut algorithm [7] devised to speed up the computation and to enchance the organization of the map from a completely random initial state, and is not a characteristic of the assumed model for the self-organization in biological networks. The original on-center off-surround lateral feedback mechanism in [7] produces a possibly multimodal pattern of activity on the map, that was approximated by a single activity bubble around the best-matching unit, BMU, with the shape of the neighborhood function. Then equal Hebbian learning rule in each unit produces the concerted updating towards the input in the neighborhood of the BMU. An interpretation of the mapping of input data point on the SOM lattice, that would be consistent with issues above, and the minimum distortion rule in Eq. 4.4 would thus be, that a point in the input space is mapped to a activity bubble H_{ib} around the BMU b, rather than to a single unit.

By ignoring the winner-take-all mechanism the SOM can be approximated by a kernel density estimator, where the activity of a unit is dependent only on the match of the data point and the unit reference vector. This is often called soft assignment of data samples to the map units, in contrast to the hard assignment of a data point to only the best-matching unit.

The second characteristic of the SOM training, the way of constraining the unit positions, is dictated by the biological origin of the method. A regularizer directly on the reference vector positions would require a means for the neurons to update their weights towards the weights of the neighboring neurons, while in the SOM rule all learning is towards the input data. The biological plausibility is obviously non-relevant in data analysis applications, even though it may have a role in building larger neural system with the SOM as a building block.

In the approach taken by Utsugi [15], small approximations are made to render the model more easily analyzable: the winner-take-all rule is replaced by soft assignment, and the neighborhood effect is approximated by a smoothing prior directly on the reference vector positions. The model is then a Gaussian mixture model with kernels constrained by the smoothing prior. This approach yields a very efficient way to set the hyperparameters of the model, that is, the widths of the kernels and the weighting coefficient of the smoothing prior, by empirical Bayesian approach. For any values of the hyperparameters, the evidence, or conditional marginal probability of the values given the data and the priors, can be computed by integrating over the posterior probability of the model parameters (kernel positions). The values with the maximum evidence are then chosen as the most likely values. Actually, a proper Bayesian approach would be to integrate over the posterior distribution of the hyperparameters (see [10] for a discussion), but clearly the empirical Bayes approach is a notable advance in the SOM theory.

Another model close to the SOM is the Generative Topographic Mapping [2]. In that approach, the Gaussian mixture density model is constrained by a nonlinear mapping from a regularly organized distribution in a latent space to the component centroids in data space. Hyperparameters of the model, which control noise variance, stiffness of the nonlinear mapping and the prior distribution of mapping parameters, can be optimized using Bayesian evidence approximation, similar to the one used by Utsugi [3].

4.5   Probability Density Model in the Self-Organizing Map

In this section we derive the probability density model for the original SOM, with no approximations in the effect of the neighborhood or the posterior probability of units given an input sample (the activity of the units).

The density model is based on the mean square type error function (4.1), discussed in Sec. 4.3. The error function is specific to the given neighborhood parameters, so that it cannot be directly used to compare maps which have different neighborhoods. The maximum likelihood (ML) estimate is based on maximizing the likelihood of data given the model. We wish to find a likelihood function which is consistent with the error function. This can be achieved by making the error function proportional to the negative logarithm of the likelihood of data. Assuming the training samples x^n independent, the likelihood of the training set $X = \{x^n\}, n = 1, \ldots, N$ is the product of probabilities of each sample,

$$p(X|m, H) = \prod_n p(x^n|m, H), \qquad (4.5)$$

where m denotes the codebook (set of reference vectors) and H is the neighborhood. The negative log-likelihood is $L = -\log p(X|m, H)$ and setting it proportional to Eq. 4.1 yields

$$p(X|m, H) = Z' \exp(-\beta E) = Z' \exp(-\beta \sum_n \sum_j H_{bj} \|x^n - m^j\|^2). \quad (4.6)$$

Here we have introduced two constants, Z' and β, which are not needed in the ML estimate of the codebook m but which are necessary for the complete density model. The probability density function in Eq. 4.6 is given by

$$p(x|m, H) = Z \exp(-\beta \sum_j H_{bj} \|x - m^j\|^2), \quad (4.7)$$

which is a product of Gaussian densities centered at m_j, whose variances are inversely proportional to the neighborhood function values H_{bj}. Note that the discontinuity of the density is due to the discontinuity of the best-matching unit index b for the input x.

Inside a Voronoi cell, or the receptive field of unit m_b, the density function has Gaussian form:

$$p(x|x \in V_b) = Z e^{-\beta W_b} \exp\left(-\frac{1}{2s_b^2}\|x - \mu_b\|^2\right), \quad (4.8)$$

where V_b denotes the Voronoi cell around the unit m_b. The position and the variance of the kernel are denoted by μ_b and s_b^2, respectively, and W_b is a weighting coefficient. The values of the parameters are

$$\mu_b = \frac{\sum_j H_{bj} m_j}{\sum_j H_{bj}} \quad (4.9)$$

$$s_b^2 = 1/(2\beta \sum_j H_{bj}) \quad (4.10)$$

$$W_b = \sum_j H_{bj} \|m_j - \mu_b\|^2. \quad (4.11)$$

The density model consists of cut Gaussian kernels, which are centered at the neighborhood-weighted means of the reference vectors and clipped by the Voronoi cell boundaries.

The parameter W_b controls the height of the kernel; it depends on the density of the neighboring reference vectors near the centroid μ_b. The density function is not continuous at the boundaries of the Voronoi cells. See Figs. 4.3, 4.4 and 4.5 for examples of the density models. The variances of the kernels depend on the parameter β and they are equal if the neighborhood is normalized (see Sec. 4.6.1 for further discussion). In the standard SOM formulation, the border units with incomplete neighborhood have larger variances, as can be seen in Figs. 4.5 and 4.8, allowing the map to shrink into the middle of the training data distribution.

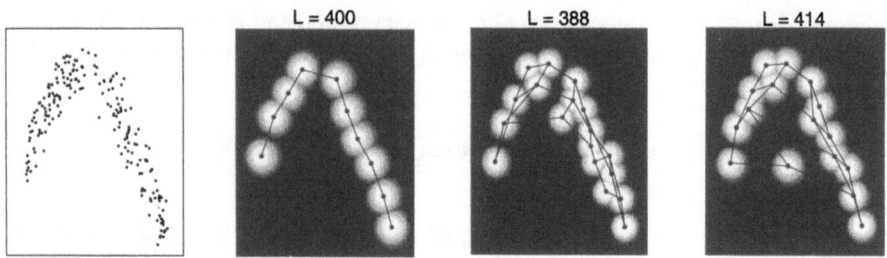

Fig. 4.4. Training data and density estimates due to different SOM topologies. L denotes the negative log-likelihood of test data. The optimal value for the parameter β is so small that the model is close to normal Gaussian mixture. Among these alternatives, the 3×6 topology (middle) produces the best model judging by the likelihood criterion.

The normalizing constant Z and the noise variance parameter β are bound together by the constraint that the integral of the density over the data space must equal one. That integral can be written as

$$\int p(x)dx = Z \sum_r e^{-\beta W_b} \int_{x \in V_r} \exp(-\frac{1}{2s_r^2}\|x - \mu_r\|^2)dx, \qquad (4.12)$$

where the integration over the data space is decomposed to the sum of integrals over each Voronoi cell. The integrals cannot be computed in closed form but they can be approximated numerically using Monte Carlo sampling. A simple way to do this is the following algorithm:

1. For each cell r, draw L samples from the normal distribution $N(\mu_r, s_r)$
2. Compute $q_r = L_r/L$, the fraction of samples that are inside the cell r.
3. The integral over V_r in Eq. 4.12 equals $q_r(2\pi s_r^2)^{d/2}$, where d is the dimension of the data space.

For a map that contains M units, this algorithm requires the computation of distances between $M \times L$ samples and the M reference vectors. Thus if M is large the computational cost of the normalization procedure exceeds that of the training algorithm itself.

In an efficient implementation the number of samples L should be chosen according to the desired accuracy. The acceptance ratio q_r varies in large range according to the neighborhood size. When the neighborhood is small, the neighborhood-weighted center μ_r is close to the reference vector m_r and s_r is likely to be small, so q_r is high. When the neighborhood is large the situation is opposite, and to achieve an equivalent accuracy L will have to be much greater. Detailed analysis of the dependence of the accuracy on L is presented in appendix A.

The maximum likelihood estimate for β can only be found by numerical optimization, by maximizing the likelihood of validation data. It is worth noting that when a numerical method such as bisection search is applied, savings can be made by allowing the accuracy to vary. Initial estimates can be very coarse, corresponding to small L, if the accuracy gradually increases towards the convergence of the search. The final accuracy should reflect the size of the validation data sample.

By equating the partial derivative of the likelihood function $\partial p(X)/\partial \beta$ with zero, an interpretation for the maximum likelihood solution β^{ML} can be found in terms of the neighborhood-weighted distortion function $D(x)$ (4.3) as follows:

$$\frac{\sum_{n=1}^{N} D(x^n)}{N} = \frac{\int D(x) \exp(-\beta^{\text{ML}} D(x)) dx}{\int \exp(-\beta^{\text{ML}} D(x)) dx}. \tag{4.13}$$

Observe that the estimated input distribution is $\hat{p}(x|\beta, H) \propto \exp(-\beta D(x))$. Heuristically, Eq. 4.13 says that, at the ML-estimate with β equal to β^{ML}, the mean value of $D(x)$ over the estimated input distribution equals the sample average of $D(x^n)$ over the input data $x^n, n = 1, \ldots, N$.

4.6   Model Selection

The SOM algorithm produces a model of the input data. The complexity of this model is determined by the number of units and the width of the neighborhood, which has a regularizing effect on the model. When the input data is a sample from a larger population, the objective is to choose the complexity such that the model generalizes as well as possible to new samples from that population. See [9] for discussion and examples of overfitting of the SOM model. The likelihood function provides a consistent way to compare the goodness of different models. In this section we discuss how this can be used for model selection in the self-organizing map.

Let us first regard the number of units as given, so neighborhood width σ is the sole control parameter. The density model allows us to select the neighborhood width σ by maximizing the likelihood of data $p(X|m, H)$. In the course of SOM training, σ is gradually decreased in some pre-specified manner, i.e. $\sigma = \sigma(t), t = 1, \ldots, K; \sigma(t + 1) < \sigma(t)$. We trust that the training algorithm will find an ML estimate for the map codebook at each value of the neighborhood width $\sigma(t)$, if it is allowed to converge every time. To construct the density model for each of these K candidate maps, we numerically optimize $\beta(t)$ as described in the previous section. This yields K different density models to compare. To choose between these we compute the likelihood values $p(X_V|m(t), \sigma(t), \beta^{\text{ML}}(t))$ for validation data X_V (which should ideally be different from that used to select $\beta^{\text{ML}}(t)$). Cross-validation can also be applied. An example of model selection is shown in Fig. 4.5. The map with $\sigma = 1.00$ maximizes the likelihood of validation data. This approach extends directly to the comparison of different size maps as well as different topologies (see Fig. 4.4). If one wishes to have a large map, it may be advisable to ease the computational requirement by finding the correct σ for a smaller map first and then simply scaling it up in proportion to the dimensions of the maps. (For example, if σ_{KL} is the optimal neighborhood width for a $K \times L$ map, then $5\sigma_{KL}$ is probably a reasonable value for a $5K \times 5L$ map.)

σ = 8.30, β = 0.10, L=2.33   σ = 4.90, β = 0.25, L=2.01

σ = 1.00, β = 2.53, L=1.86   σ = 0.20, β = 9.34, L=1.92

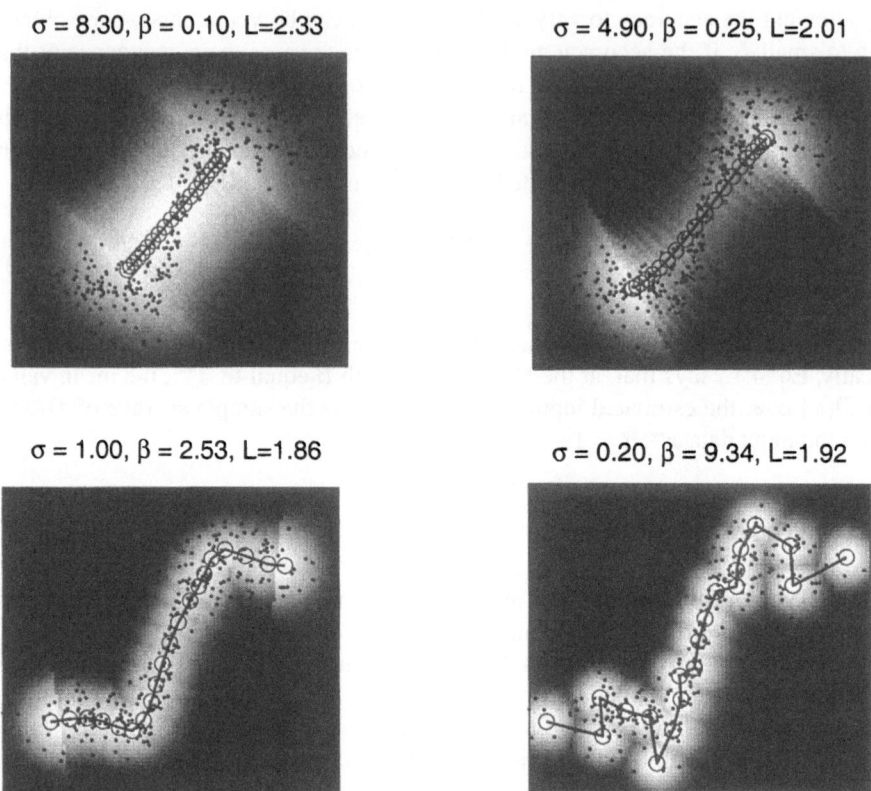

Fig. 4.5. SOM density models for different widths σ of the Gaussian neighborhood. From the total likelihood of validation data the optimal neighborhood can be chosen to avoid overfitting. L denotes the negative log-likelihood of validation data (per sample).

Because the exact value of the density function cannot be computed in closed form, it is difficult to apply methods such as Bayesian evidence to parameter selection. If the values of the function itself are approximations, then the derivatives will be even more inaccurate, and due to the numerical normalization procedure the approach would be computationally too expensive in practice.

A common application of the SOM is to look for dependencies between variables by visual inspection. In that context, the density model can be used to select the complexity of the model, but it also enables quantitative analysis. Regression or conditional expectations can be computed directly from the joint density in Eq. 4.7 by numerical integration. For example, the conditional distribution for variable x_j equals

$$p(x_j|x_{\setminus j}, m, H) = \frac{p(x|m, H)}{\int p(x|m, H) dx_j}, \qquad (4.14)$$

where $x_{\backslash j}$ denotes the vector x with element j excluded. Likewise, the regression of x_j on other variables can be computed as the conditional mean $E[x_j|x_{\backslash j}, m, H]$. It should be noted that the SOM density model may not give the best possible description of the input distribution. We have included this discussion here so as to illustrate the value of model selection.

Reducing the variance parameter to zero, $\beta \to \infty$, gives an important special case. The conditional density is then sharply peaked at the value of the "outputs" x_j in the best matching unit for the "inputs" $x_{\backslash j}$. The conditional mean $E[x_j|x_{\backslash j}]$ then gives the same value as nearest neighbor (NN) regression with the SOM reference vectors, with the neighborhood-weighted reference vectors (Eq. 4.9) as output values, producing a piecewise constant estimate. Comparison with the NN rule is interesting, because it is a close quantitative counterpart of the popular visual analysis of the SOM.

Fig. 4.6 illustrates the difference between computing the conditional mean from the density model and using the nearest neighbor rule. A random 3D data set ($\sim N(0, 1)$) is analyzed by a 6×6 SOM. We attempt to infer $E(x_2|x_1, x_3 = 0)$, the expected value of the variable x_2 given x_1, with x_3 zero. As the variables are truly independent, the answer should be $E(x_2|x_1, x_3 = 0) = E(x_2) = 0$. The optimal width of the Gaussian neighborhood function is $\sigma = 4.2$, which is a relatively large value, suggesting independent variables (a "simple" distribution). At zero neighborhood, the model is badly overfitted. Clearly, neglecting to select the correct model complexity would give unreliable results. When the complexity is right, the nearest neighbor rule can give a good approximation to the mean, though the lack of confidence intervals limits the reliability of analysis.

An example of using the conditional distributions is shown in fig. 4.7. The neighborhood width was chosen based on the maximum likelihood of test data. The data is three dimensional; there is a dependence between two of the variables, and one is independent, as follows: $x_2 = \sin(\omega x_1) + \epsilon$, $x_3 \sim N(0, 1)$. This kind of a distribution can easily be modeled by means of a two dimensional SOM; that is why no severe overfitting is observed and a small neighborhood width gives the best fit to test data. Yet it is not easy to observe the dependence from the component level display. The conditional densities, on the other hand, are easy to interpret. Nearest neighbor regression also works relatively well, since the model complexity is correct.

By visual inspection of the map it is difficult to perceive the mean or shape of the conditional distributions and thus the reliability of the conclusions is practically impossible to assess. Choosing parameter values to optimize the density estimate may not result in a mapping that is also optimal for visual display. However, the examples shown in Figs. 4.5 and 4.6 indicate that this method will outperform any prefixed heuristic rule. In any case, the results of visual inspection should be validated by other, more reliable techniques.

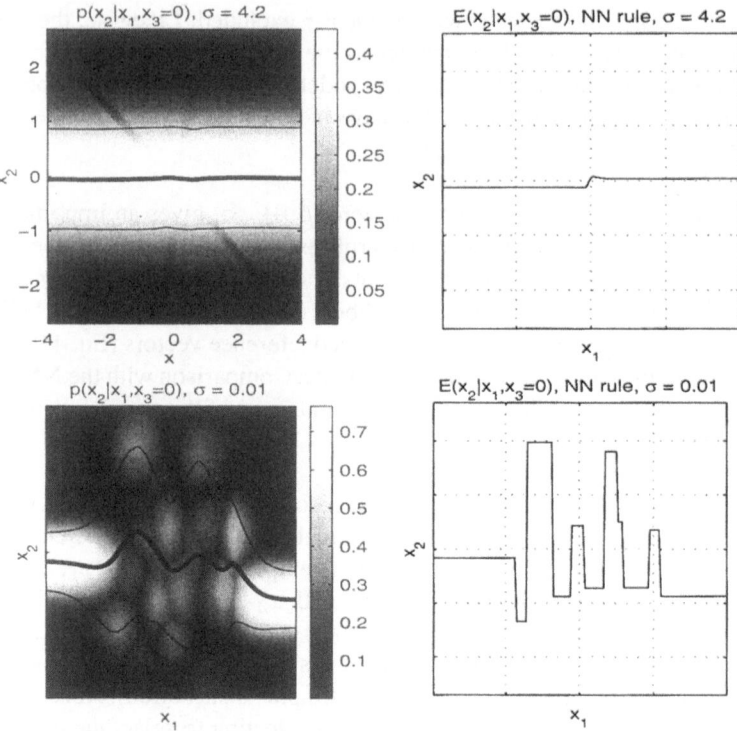

Fig. 4.6. Conditional densities from a SOM trained on random independent data. Upper row: the conditional density and the nearest neighbor prediction for optimal neighborhood $\sigma = 4.2$. Lower row: conditional density and the nearest neighbor prediction for small neighborhood $\sigma = 0.01$. The black lines show the means and standard deviations computed from the densities.

4.6.1 Border Effects

In typical implementations of the SOM, the neighborhood function is the same for each map unit. This causes problems near the borders of the map, where the neighborhood function gets clipped and thus becomes asymmetric. The effect is that, for no obvious reason, data samples which are outside the map are given less significance than those within the map. As a result, units close to the border have larger kernels and allow data points to reside farther away from the map units. Consequently, the border units are pulled towards the center of the map, and the map does not extend close to the edges of the input distribution until the neighborhood is relatively small and the regularization is loose. This leads to decrease of the likelihood for maps with large neighborhood (or increase of the quantization error), biasing the optimal width of the neighborhood towards smaller values.

This effect can be alleviated by normalizing the neighborhood function at the edges of the map. In the case of the sequential algorithm, it suffices to normalize

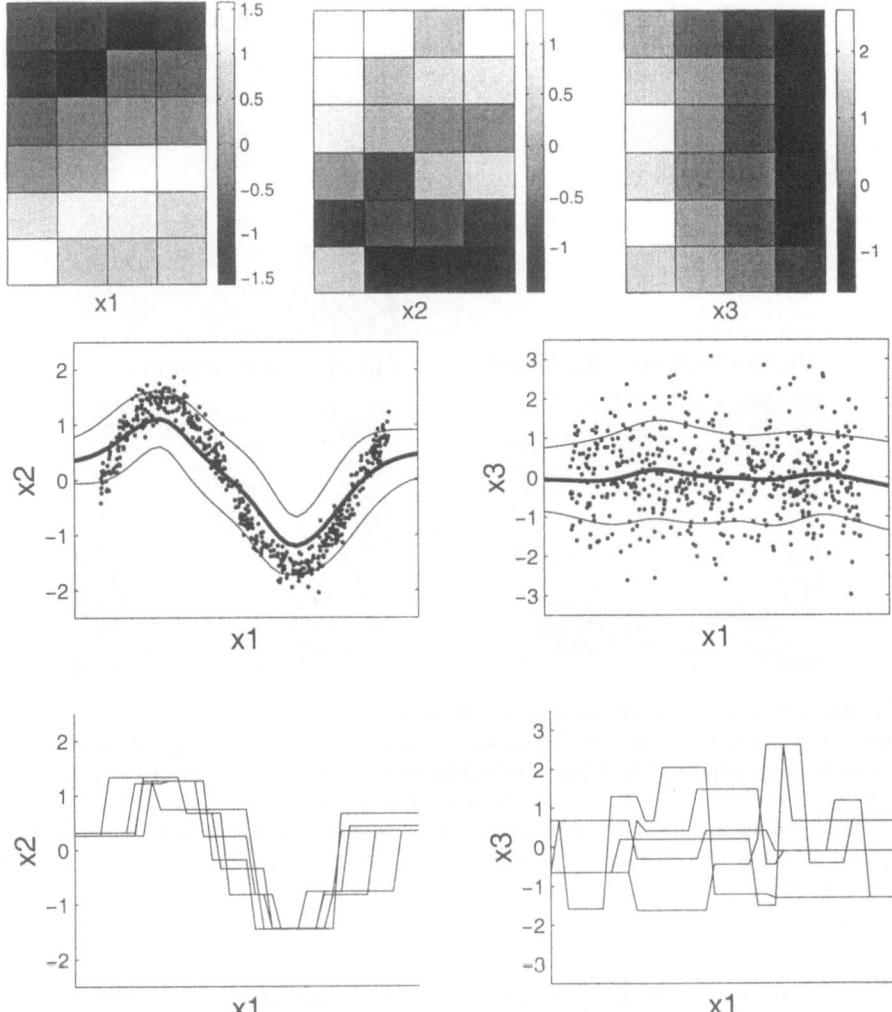

Fig. 4.7. Example of the use of SOM for data analysis. Top: all three component levels of a SOM trained down to optimal neighborhood width 0.2. Mid row: Training data and the means and standard deviations of the conditional densities $p(x_2|x_1)$ and $p(x_3|x_1)$, integrated over x_3 and x_2, respectively. Bottom: nearest neighbor estimates based on the best matching units using five different values for x_3 and x_2, respectively.

the neighborhood function such that its sum is the same in each part of the map. When using the batch algorithm, the portion of the neighborhood function that gets clipped off due to the finite size of the map lattice can be transferred to nearest edge units. Normalization of the neighborhood function is of particular importance, if the minimum distortion rule (4.4) is applied to winner selection. We see from Eq. 4.10

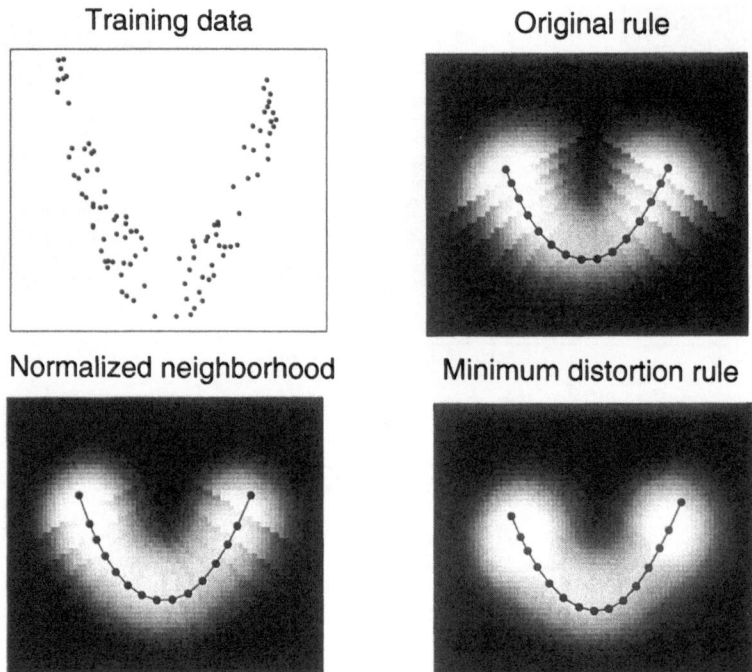

Fig. 4.8. Effect of the minimum distortion training rule on the density model. The neighborhood width is the same $\sigma = 2.8$ for all the maps. (This value is larger than would be optimal, as the purpose of the figure is to highlight the differences between these cases.) Note also, how the density kernel positions do not coincide with the reference vector positions. Especially in the middle of the maps the kernels are shifted upwards from the unit positions, according to Eq. 4.9. The jaggedness of the Voronoi cell boudaries is due to the discrete grid where the density is evaluated.

that when the sum of the neighborhood function is constant throughout the map, all cells have equal noise variance.

In practice we find that the minimal distortion rule will produce very similar results as the original rule in terms of model selection. The equalization of the neighborhood volume near map borders smoothes out the density model somewhat, as it allows better fit for the data sample with larger neighborhood, as can be seen from Fig. 4.8. Obviously, the continuous density model, with minimum distortion rule, is favorable in many respects.

4.7   Conclusions

We have presented the probability density model associated with the SOM algorithm. Also we have discussed some difficulties that arise in the application of the SOM to statistical data analysis. We have shown how the probability density model can be used to find the maximum likelihood estimate for the SOM parameters, in order to optimize the generalization of the model to new data. The parameter search involves a considerable increase in the computational cost, but in serious data analysis, the major concern is the reliability of the conclusions.

It should be stressed that although the density model is based on the error function which is not defined in all cases, in practice the only restriction in the application of the density function to data analysis is that the algorithm should be allowed to converge. Unfortunately, maximizing the likelihood of data is not directly related with ensuring the reliability of the visual representation of the SOM. Especially when the data dimension is high, the units that code the co-occurrences of all correlated variables cannot be grouped together on the map, and thus the conditional densities become distributed around the map. Such effects cannot be observed by visual inspection of the component levels, no matter how the model hyperparameters are set. Those effects can be revealed, to some extent, by calculating the conditional densities, or the conditional means and the confidence intervals.

The association of a generative probability density model with the SOM enables comparison of the SOM and other similar methods, like the Generative Topographic Mapping. If the theoretical difficulties of the SOM are avoided by adopting the minimum distortion winner selection rule, the main difference that remains is the hard vs. soft assignment of data to the units. The hard assignments of the SOM are perhaps easier to interpret and visualize. In the SOM the activation of the units (ie., the posterior probability of the kernels given one data point) is always one for the winning unit and zero for the others, or a unimodal activation bubble of the shape of the neighborhood around the winning unit, depending on the interpretation. With soft assignments the posterior probability may be multimodal (when two distant regions in the latent space are folded close to each other in the input space), and thus the activation is more difficult to visualize. Note, however, that this multimodal response gives visual indication of the folding, which may also be valuable. Apparently, the choice of methods depends on the application goals, and in real data analysis it is reasonable to apply different methods to decrease the effect of artefacts of the methods. [1]

---

[1] MATLAB® routines for evaluating the SOM probability density are available at http://www.lce.hut.fi/research/sompdf/.

Bibliography on Chapter 4

1. C. M. Bishop. *Neural Networks for Pattern Recognition*. Oxford University Press, 1995.
2. C. M. Bishop, M. Svensen, and C. K. I. Williams. GTM: a principled alternative to the self-organizing map. In C. von der Malsburg, W. von Seelen, J. C. Vorbruggen, and B. Sendhoff, editors, *Artificial Neural Networks—ICANN 96. 1996 International Conference Proceedings*, pages 165–70. Springer-Verlag, Berlin, Germany, 1996.
3. C. M. Bishop, M. Svensén, and C. K. I. Williams. GTM: The generative topographic mapping. *Neural Computation*, 10:215–234, 1998.
4. M. Cottrell, P. Gaubert, P. Letremy, and P. Rousset. Analyzing and representing multidimensional quantitative and qualitative data : Demographic study of the Rhone valley. the domestic consumption of the canadian families. In E. Oja and S. Kaski, editors, *Kohonen Maps*, pages 1–14. Elsevier, Amsterdam, 1999.
5. E. Erwin, K. Obermayer, and K. Schulten. Self-organizing maps: Ordering, convergence properties and energy functions. *Biol. Cyb.*, 67(1):47–55, 1992.
6. T. Heskes. Energy functions for self-organizing maps. In E. Oja and S. Kaski, editors, *Kohonen maps*, pages 303–315. Elsevier, 1999.
7. T. Kohonen. Self-organizing formation of topologically correct feature maps. *Biol. Cyb.*, 43(1):59–69, 1982.
8. T. Kohonen. *Self-Organizing Maps*, volume 30 of *Springer Series in Information Sciences*. Springer, Berlin, Heidelberg, 1995. (Second Extended Edition 1997).
9. J. Lampinen and T. Kostiainen. Overtraining and model selection with the self-organizing map. In *Proc. IJCNN'99*, Washington, DC, USA, July 1999.
10. J. Lampinen and A. Vehtari. Bayesian approach for neural networks – review and case studies. *Neural Networks*, 14(3):7–24, April 2001. (Invited article). In press.
11. S. P. Luttrell. Code vector density in topographic mappings: scalar case. *IEEE Trans. on Neural Networks*, 2(4):427–436, July 1991.
12. E. Oja and S. Kaski. *Kohonen Maps*. Elsevier, Amsterdam, 1999.
13. H. Ritter and K. Schulten. Kohonen self-organizing maps: exploring their computational capabilities. In *Proc. ICNN'88 International Conference on Neural Networks*, volume I, pages 109–116, Piscataway, NJ, 1988. IEEE Service Center.
14. M. E. Tipping and C. M. Bishop. Mixtures of principal component analysers. Technical report, Aston University, Birmingham B4 7ET, U.K., 1997.
15. A. Utsugi. Hyperparameter selection for self-organizing maps. *Neural Computation*, 9(3):623–635, 1997.
16. T. Villmann, R. Der, M. Herrmann, and T. M. Martinetz. Topology preservation in self-organizing feature maps: exact definition and measurement. *IEEE Transactions on Neural Networks*, 8(2):256–66, 1997.
17. I. Welling, E. Kähkönen, M. Lahtinen, K. Salmi, J. Lampinen, and T. Kostiainen. Modelling of occupants' subjective responses and indoor air quality in office buildings. In *Proceedings of the Ventilation 2000, 6th International Symposium on Ventilation for Contaminant Control*, volume 2, pages 45–49, Helsinki, Finland, June 2000.

A   APPENDIX: Accuracy of the MCMC Method

In the algorithm described in Sec. 4.5 one picks L samples and computes S, the number of samples that satisfy a certain condition. The task is to estimate q, the probability of a single sample satisfying the condition, based on S and L. S follows the binomial distribution

$$p(S|q, L) = q^S (1 - q)^{L-S} \qquad (4.15)$$

In Bayesian terms, the posterior distribution of q for given S and L is

$$p(q|S, L) = \frac{p(S|q, L)p(q)}{\int p(S|q, L)p(q)dq} = \frac{q^S(1 - q)^{L-S}}{\int q^S(1 - q)^{L-S}dq}, \qquad (4.16)$$

where the prior distribution $p(q)$ is uniform. The integral in the denominator yields

$$\int q^S (1 - q)^{L-S} dq = \sum_{k=0}^{L-S} \frac{(-1)^k}{S + k + 1} \binom{L - S}{k}. \qquad (4.17)$$

Moments of the posterior can be written as serial expressions, for example

$$E(q) = \int q p(q|S, L)dq = \frac{\sum_{k=0}^{L-S} \frac{(-1)^k}{S+k+2} \binom{L-S}{k}}{\sum_{k=0}^{L-S} \frac{(-1)^k}{S+k+1} \binom{L-S}{k}}, \qquad (4.18)$$

and from the first and second moments we can derive an expression for the variance of the estimate $\hat{q} = E(q)$. Unfortunately, the computation easily runs into numerical difficulties due to the alternating sign $(-1)^k$. Exact values of the binomial coefficients are required, and these can be difficult to obtain if, say, $L > 100$.

To consider an approximation to the variance, observe that when $S = L$ the formulas simplify considerably and the variance can be written as

$$\nu_0 = Var(\hat{q}|S = \{0, L\}) = \frac{L + 1}{(L + 3)(L + 2)^2}. \qquad (4.19)$$

This is the minimum value of the variance for given L. The variance of the *binomial distribution* at the maximum likelihood estimate $q^{\mathrm{ML}} = S/L$ equals $q^{\mathrm{ML}}(1 - q^{\mathrm{ML}})/L$. We can combine these results to get a fairly good approximation

$$Var(\hat{q}) \approx \nu^*(\hat{q}, L) = \nu_0 + \hat{q}(1 - \hat{q})/L, \qquad (4.20)$$

which slightly over-estimates the variance. A more precise approximation can be obtained directly from Eq. 4.16 by numerical integration. Picking Monte Carlo samples from the distribution $N[\hat{q}, \sqrt{\nu^*(\hat{q})}]$ truncated to the range $[0, 1]$ produces good results, except maybe at the very edges of the range, but there the exact value ν_0 can be applied. Let us write the sum in Eq. 4.12 as $F = \sum w_i q_i$, where $w_i = e^{(-\beta W_b)}(2\pi s_i^2)^{(d/2)}$. The relative standard error of an estimate of F equals

$$\epsilon = \frac{\Delta \hat{F}}{\hat{F}} = \frac{\sum w_i \Delta \hat{q}_i}{\sum w_i \hat{q}_i} = \frac{\sum w_i \sqrt{Var(\hat{q}_i)}}{\sum w_i \hat{q}_i}. \qquad (4.21)$$

Hence, to achieve a given accuracy ϵ, L should be increased until

$$\frac{\sum w_i \sqrt{Var(\hat{q}_i)}}{\sum w_i \hat{q}_i} < \epsilon. \tag{4.22}$$

5 Growing Multi-Dimensional Self-Organizing Maps for Motion Detection

Udo Seiffert

Abstract. The standard Self-Organizing Map consists of a two-dimensional rectangular grid of neurons. For many applications this represents a very good target to reduce the dimensionality of the input data. However, occasionally a multi-dimensional layer, keeping more than two dimensions of the input data, might be more advantageous. This sometimes also called hypercube topology can be considered as the universal case of the standard topology. This chapter gives an introduction and demonstrates basic properties by means of applications from motion picture analysis.

5.1 Introduction

Self-Organizing Maps (SOM) [1] are frequently used as a trainable and adaptable visualisation tool. For this purpose, a target of a two-dimensional plane seems to be very suitable. After all man is used to two-dimensional graphs and drawing planes. Consequently it is not astonishing, that neurons within the Kohonen layer are usually arranged in a two-dimensional topology. However, there is no reason against the extension of this structure to a multi-dimensional hypercube, especially when it seems to be promising and advantageous to perform a higher dimensional visualisation.

The importance of analysing moving scenes has been rapidly increased during the last few years. Due to the availability of very powerful computer hardware and image processing equipment, as well as the development of innovative approaches in *Soft-Computing* and its applications, it has become increasingly possible to process extensive video streams. One interesting aspect refers to procedures and algorithms to detect motion and to estimate the motion parameters of objects moving within an image sequence [2, 3]. The motivation for this reaches from in a wider sense automatic supervision (production, security, traffic, vehicle control, weather) [4] to fields without an immediate interest in the motion itself (image compression and transmission, MPEG, videophone) [5-8].

The basis of technical implementations of motion detection systems is, in most cases, a monocular image sequence. That means a sequence of single static images taken in constant time steps. The image sequence does not contain the motion parameters explicitly. However, it contains a brightness or colour distribution changing from image to image. If this is not only due to noise, but is also a directed flow, several motion parameters can be determined [3].

In general it seems to be very useful to implement some basic algorithms adapted from biological vision systems, to improve the performance of technical procedures. Some approaches are using neural nets in standard designs (Backpropagation, Self-Organizing Map, Adaptive Resonance Theory networks). Often some well-known and in the context of Artificial Intelligence (AI) conventional procedures, using correlation functions or tracking algorithms to determine the movement of single or compounded objects are applied. Also a combination of conventional and AI techniques has been suggested. In almost all of the above mentioned cases some kind of pre-processing has to be performed. Some examples of existing implementations using neural networks are given in [9-14].

5.2 Motion Detection System

5.2.1 General Description

The starting point is the block matching algorithm. Each single image of the sequence is divided in image blocks of fixed size. Blocks are usually quadratic and from 8x8 up to 32x32 pixels (see Fig. 5.1).

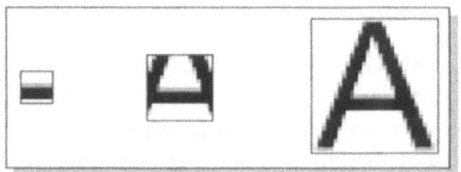

Fig. 5.1. Image blocks of 8x8, 16x16 and 32x32 pixels (from left to right). Possible complexity of block contents increases with block size.

The block matching algorithm (Fig. 5.2) assumes, that a particular object can be found in the next image of the sequence at the same or a shifted position. Assuming a block contains only one object or all objects covered by one block are moving in the same direction, a motion vector can be assigned to each block. If any a-priori information about possible motion is available, the search area in the next image can be limited. The similarity of a block with a reference block, both of size (B_x, B_y), can be computed using the *Normalised Cross Correlation Function (NCCF)*,

$$NCCF(d_x, d_y) = \frac{\displaystyle\sum_{k=1}^{B_x}\sum_{l=1}^{B_y} g_t(k, l) \cdot g_{t+1}(k + d_x, l + d_y)}{\sqrt{\displaystyle\sum_{k=1}^{B_x}\sum_{l=1}^{B_y} g_t^2(k, l)} \cdot \sqrt{\displaystyle\sum_{k=1}^{B_x}\sum_{l=1}^{B_y} g_{t+1}^2(k + d_x, l + d_y)}} \tag{5.1}$$

where the object movement is

$$d_{x_{min}} \le d_x \le d_{x_{max}}, d_{y_{min}} \le d_y \le d_{y_{max}} \tag{5.2}$$

and g_t resp. g_{t+1} are the gray or colour values of a pixel at time t resp. $t+1$.

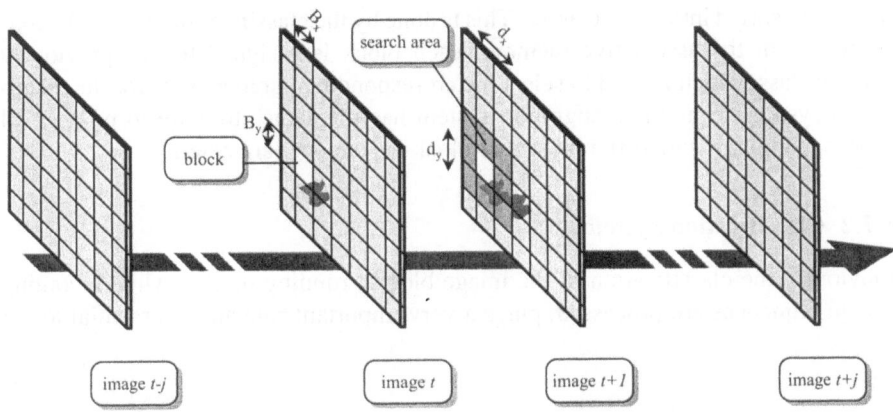

Fig. 5.2. Block matching algorithm with a considered block containing a part of the moving object in image t and a fixed search area covering the same part of the object in image $t+1$.

The smaller the block size the more limited is the possible complexity of the block content. In an 8x8 block just more or less sharp edges at different angles or colour plateaus can be found (see Fig. 5.1). Due to the relatively limited complexity of small blocks, a classification should be possible. This characterises a key assumption of the following described motion detection system (Fig. 5.3).

After taking an image, it is divided in blocks. For a typical image size of 512x512 pixels and 8x8 pixels in a block, a set of 4096 blocks is constructed. In the next step the correlation of blocks in successive images is calculated according to Eq. (5.1). If the images are distorted or deformed in one way or another, for example, moving objects are overlapping each other or illumination changes, the peak of the correlation function is also affected. Sometimes even several peaks can be observed [15, 16]. In these cases a seeming movement may occur and an error-free motion detection is not possible any more.

To overcome these negative effects, an adaptive recognition system is introduced. Its task is to estimate the motion vector for each image block suppressing the above mentioned distortions. Since many distortions depend on the image or block content, information about it has to be acquired to change the parameters of the recognition system accordingly. Suitable information may contain the texture of the currently processed block and possibly the history of the current image sequence as well as general knowledge about the expected data. However, it is not possible to provide a special parameter set for each imaginable block content. That's why they have to be sorted into a few classes. This is done by the classification system. In connection with the associative memory, each block is assigned to one previously formed class, which is used to select the corresponding parameter set for the recognition system. Thus the recognition system has the same structure to process all blocks, but is run with different sets of block dependent parameters.

5.2.2 Classification System

Obviously the classification of the image blocks, running on-line while incoming image sequences are processed, plays a very important role and is essential to get

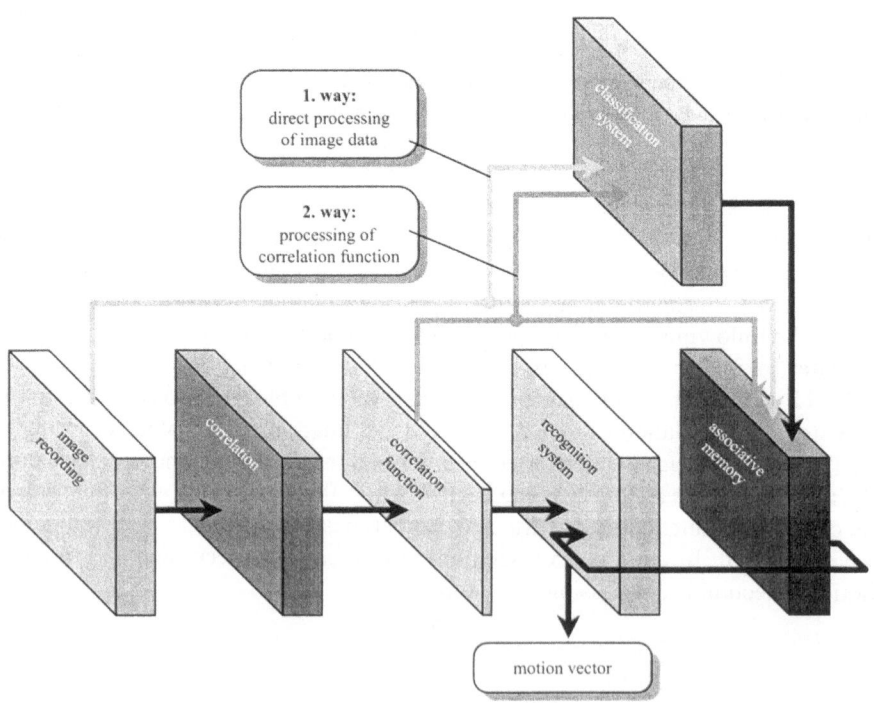

Fig. 5.3. Model of the entire motion detection system showing the signal flow including alternative ways to run the classification system.

the entire system properly working. In general there are two ways to perform this task. It depends on the input data. One potential way is to use the image blocks directly, which is fast and easy to understand. A back processing of the original blocks by means of the formed classes remains still possible. Alternatively the correlation function can serve as input for the classification system. This may be advantageous when processing very noisy data, because the data is smoothed by the correlation. On the other hand, back processing is not possible any more.

As already been stated, the existence of a *limited* number of classes is a key assumption of the whole system. This must be checked in advance. In order to answer the questions, whether there are classes at all, and if so, how many of them are needed to reproduce an image sequence in a sensible resolution, an unsupervised trained artificial neural network and in particular a Self-Organizing Map [17] seems to be very suitable. Once this *off-line* classification is successfully done, the same instruments may be used for the above described *on-line* classification. This ensures a homogeneous allocation of all classes all the time.

In this context it has to be settled, what data is supposed to belong to the same class and what not. Fig. 5.4 demonstrates the results using a standard SOM which correctly detects a significant dissimilarity between the first and the second sample, although they contain the same basic pattern (white edge) just at different positions. As to be expected, the standard two-dimensional SOM does not recognise this basic pattern[1]. Since it is impossible under the given conditions, simply to centre all incoming patterns as some kind of pre-processing, another intelligent system is required, which comes in the form of multi-dimensional SOMs.

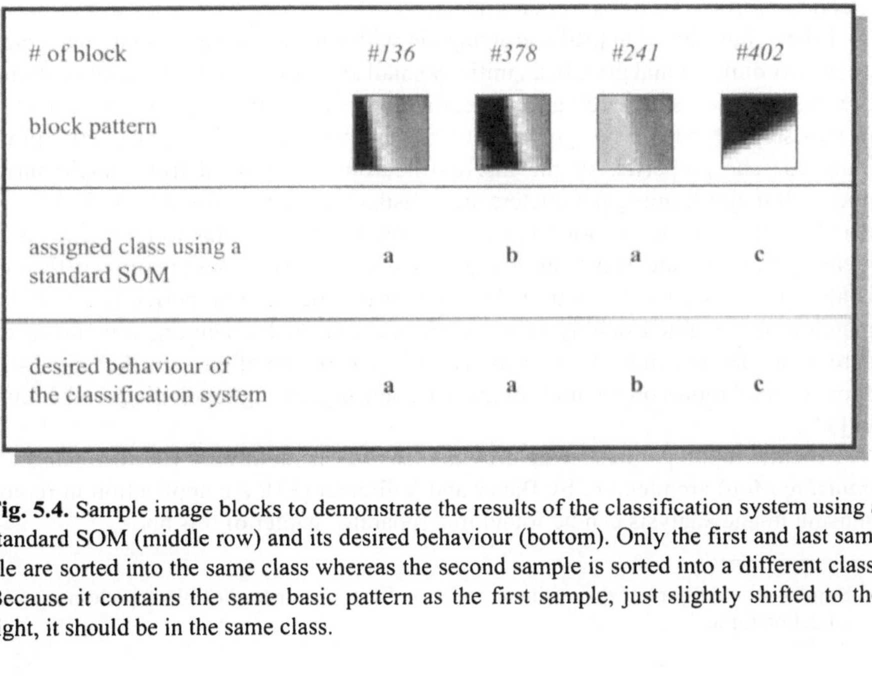

# of block	#136	#378	#241	#402
block pattern				
assigned class using a standard SOM	a	b	a	c
desired behaviour of the classification system	a	a	b	c

Fig. 5.4. Sample image blocks to demonstrate the results of the classification system using a standard SOM (middle row) and its desired behaviour (bottom). Only the first and last sample are sorted into the same class whereas the second sample is sorted into a different class. Because it contains the same basic pattern as the first sample, just slightly shifted to the right, it should be in the same class.

The definition of a desired resolution instead of a fixed number of classes is a further requirement to be met by the classification system. This can be achieved using growing network structures rather than topologies of fixed size.

So far for the definition of basic requirements of the classification system. The next section shows its realization using growing multi-dimensional Self-Organizing Maps.

5.3 Network Implementation

5.3.1 Where to Start from?

The desired task of the SOM, to classify basic patterns independently of their actual position, is neither new nor generally unsolved. There are conventional as well as neural solutions ([18], [19], [3]). However, some objectives regarding adaptability of the system, the wish for a biologically inspired technique and an intended implementation on parallel hardware, forced a neural solution with a quite simple, but massively parallel running network, as being found in biological vision systems [20-26]. As already mentioned in the previous section, an extensive pre-processing of the input data was not applicable. Thus Kohonen's Self-Organizing Maps (SOM) seemed to be a good choice as basis to solve this rather complex problem.

The original algorithm of the SOM [27], suggested by Kohonen about twenty years ago, has been frequently adapted and modified during the last two decades. Refer to the overture of this book as well as [1] and [28] for a general review and [29] for some further related work. In the context of this chapter, some special modifications seem to be of particular interest.

Fritzke introduced in [30] a growing net with a neuron arrangement on a rectangular two-dimensional grid. In an initially small array of neurons, complete columns or rows are inserted according to a specifically designed training algorithm running in two steps. First the net grows until a previously defined stopping criterion is matched. The properties of the inserted neurons are derived from neighbouring ones. All major training parameters are constant. Once the growth process has finished, a number of adaptation steps is performed to fine-tune the weights using a decreasing learning rate. If the alternative way is used and neurons are pruned, the map is also able to shrink if required. The size of the map can be perfectly adapted to match a user defined quality criterion. In most cases the neurons are spread out across the map in a much better way, compared to a map of the same but fixed size. More general topics on methods of growing self-organising topologies can be found in [31].

Another topology preserving growing network is the *GSOM* (*Growing Self-Organizing Map*) architecture by Bauer and Villmann [32]. An application in remote sensing image analysis can be found in a separate chapter of this book.

---

[1] Similar problems arise when using correlation functions instead of block patterns as classification input.

Released from the two-dimensional rectangular structure all neurons can move freely within a hyperspace in the so-called *Neural Gas* architecture suggested by Martinetz and Schulten [33]. Based on the current distance of the neurons in the input space, topology connections are inserted among them successively. There are several strategies, where to insert or to remove connections. In the original approach the number of neurons is always fixed, only the number of connections is changed during the training phase. Fritzke also added a growing feature to the Neural Gas [34].

5.3.2 Extending the Dimension of the Map

Due to the similarity criterion between a considered input vector v and any weight vector w_i,

$$D^{(i)}(v, w_i) = \|v - w_i\| \tag{5.3}$$

the original SOM is not able to solve the above mentioned problem of a local independent classification (Fig. 5.4). The winning neuron N_w is determined by

$$\|v - w_w\| = min\|v - w_i\|, \forall i. \tag{5.4}$$

The key idea to extend the dimension of a standard SOM is to have a neural structure working similar to SOMs, but with two dimensions to reflect the possible shifted positions of the basic patterns and at least one more dimension to form appropriate classes. In order to keep the input data and also the weights easy to observe by a human, the network is equipped with two-dimensional (matrix instead of vector) input layer and weight storage. These considerations lead to a three-dimensional neuron cube as shown in Fig. 5.5. Each horizontal neuron plane is used to code the possible shifted positions of basic patterns. The patterns themselves are sorted into classes along the z-direction. Each horizontal plane covers one class. Each single neuron resp. its weight matrix holds the same basic pattern at a shifted position according to the neuron's relative position within the plane. A neuron in the middle of a horizontal plane features one basic pattern in central position. The more a neuron is located on the left (right) hand side, the more left (right) shifted is the covered basic pattern. The same is true for the top and bottom of each plane, because *two* dimensions are used to code shifted basic patterns. For example, all neurons at the lower right corner contain one basic pattern at the extreme lower right position.

As in the standard SOM, the number of available neurons controls the resolution. In this case the *size* of the horizontal planes adjusts the *local resolution* and the

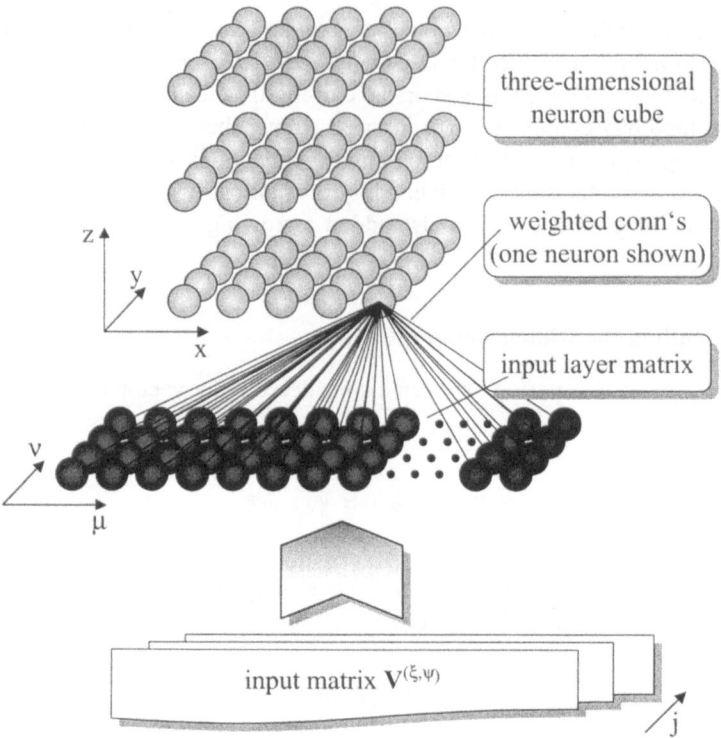

Fig. 5.5. Three-dimensional neuron cube (5x5x3) with two-dimensional input layer. The dimensions x and y are used to code shifted positions of basic patterns, which are sorted into classes in z-direction.

number of horizontal planes sets the *number of classes*. If the number of planes is step by step decreased, the classification gets more and more rough, until at last with just *one* plane, the original SOM is left, without a reasonable classification[2] at all. If, on the other hand, the size of the planes gets smaller, the local resolution becomes weaker, until a simple one-dimensional neuron column is left, which is not able to store position information.

According to the original Kohonen learning rule, the update of a particular weight w_μ of the neuron i changes from

$$w_\mu^{(i)} = w_\mu^{(i)} + \alpha \cdot \zeta(x, y) \cdot (v_\mu^{(i)} - w_\mu^{(i)}) \tag{5.5}$$

to a notation reflecting the two-dimensional (rectangular) input layer and the neu-

---

[2] In the sense of the described problem.

ron's position (x, y, z) within the three-dimensional cube with the learning coefficient α and $\zeta(x, y, z)$ being now the three-dimensional neighbourhood function

$$w_{\mu, \nu}^{(x, y, z)} = w_{\mu, \nu}^{(x, y, z)} + \alpha \cdot \zeta(x, y, z) \cdot (v_{\mu, \nu}^{(x, y, z)} - w_{\mu, \nu}^{(x, y, z)}). \quad (5.6)$$

The index (μ, ν) indicates the connection from the corresponding input element to the considered neuron (x, y, z). If a training data set V containing j samples is assumed, one objective of the learning process for each basic pattern is

$$\sum_{\chi, \psi} \zeta(x, y) \cdot (\overline{V^{(j, \chi, \psi)} - W^{(x, y)}}) = 0, \forall j. \quad (5.7)$$

In the case of convergence it follows that

$$\chi = x, \psi = y. \quad (5.8)$$

If the steady-state neighbourhood function $\zeta(x, y)$ is radially symmetric and small against the size of the input function, Eq. (5.7) can be divided into two parts, the winning neuron on the one hand, and all other neurons within the neighbourhood on the other hand. It can be written as

$$\sum_{\chi = x, \psi = y} \zeta(x, y) \cdot (\overline{V^{(j, \chi, \psi)} - W^{(x, y)}}) \;+$$

$$\sum_{\chi \neq x, \psi \neq y} \zeta(x, y) \cdot (\overline{V^{(j, \chi, \psi)} - W^{(x, y)}}) = 0, \forall j. \quad (5.9)$$

With Eq. (5.8) and at least in the case of a radial symmetry of the right hand term in Eq. (5.9) and only one basic pattern, the right hand term becomes zero. It follows

$$\overline{V^{(\chi, \psi)} - W^{(x, y)}} = 0 \quad (5.10)$$

and thus

$$V^{(\chi, \psi)} = W^{(x, y)}. \quad (5.11)$$

Practical investigations show that this architecture can be trained in this way with real-world data under weaker assumptions too (see next sections).

The input function V is distributed in incremental steps across the corresponding horizontal plane. For its components (k, l) and a shift (a, b) of the input pattern, it can be written

$$v_{k,l}^{(\chi - a,)(\psi - b)} = v_{k+a,l+b}^{(\chi, \psi)} \tag{5.12}$$

and accordingly for the weights of the trained plane

$$w_{k,l}^{(x - a, y - b)} = w_{k+a,l+b}^{(x, y)}. \tag{5.13}$$

This can also be used to pre-order the weights before the training starts to have an initial weight distribution, which respects already the final destination.

5.3.3 Why Three Dimensions are not Sufficient - the Growing QFDN

General Considerations

If one looks at the coding scheme of the proposed three-dimensional Kohonen layer, a still remaining inconsistency is to be seen. Two dimensions are used to code the local properties of the input data. With regard to the classification of input data features, apart from those local properties, it is only a one-dimensional map (z-direction). All classes are mapped to a column of planes consisting of neurons only holding all the same class (basic patterns at moved positions). However, a two-dimensional mapping in the sense of a classification is much more useful. In order to get a behaviour similar to the original Kohonen layer plus the desired feature of locally independent classes, a once more modified topology seems to be an appropriate solution.

The proposed network is a logical extension of the three-dimensional neuron cube. It consists of a two-dimensional hyperstructure of two-dimensional submaps. Continuing the discussion of the three-dimensional layer in the previous section, it could be called a four-dimensional Kohonen layer. However, with respect to the data distribution across the map it is more a *two-plus-two* structure. So it is not a four-dimensional hypercube architecture in a narrower sense. Not only for the sake of a better description the second point of view has been chosen. Thus the new network was named *Quasi-Four-Dimensional Neuroncube (QFDN)* [35, 36].

The mathematical derivation given in the previous section remains in principle valid. It can be considered as an extension from a column of horizontal planes to a matrix of submaps.

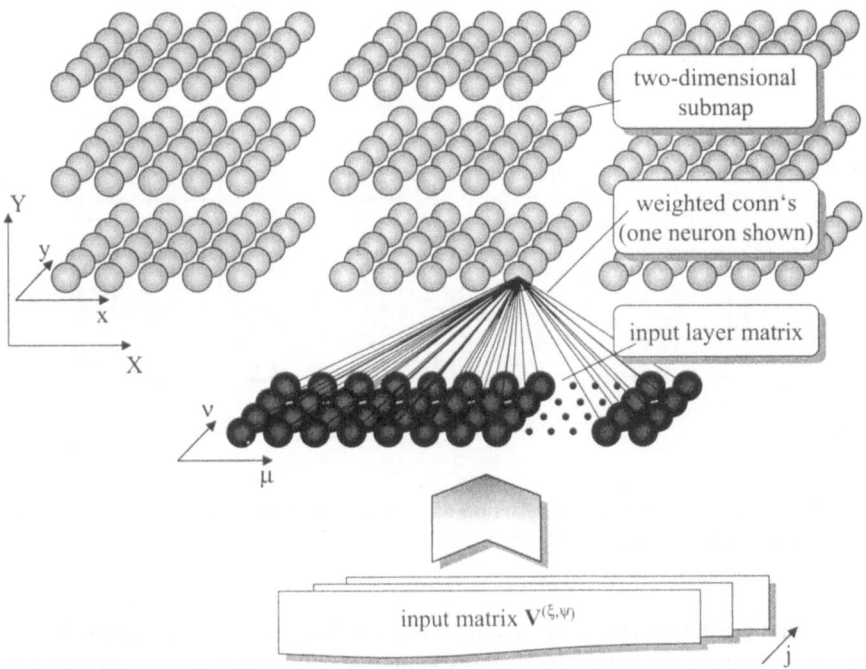

Fig. 5.6. Quasi-Four-Dimensional Neuroncube (QFDN) with an array of 3x3 submaps, each containing 5x5 neurons. As the three-dimensional layer (Fig. 5.5), this network is also equipped with a two-dimensional input layer. Each single neuron is addressed by four coordinates (X, Y, x, y), where the capital letters assign the submap and the small letters indicate the position of the neuron within its submap.

Weight Initialisation

Special attention has to be paid to the initialisation, in order to take the desired weight distribution into account. In contrast to commonly applied random weight initialisation, a special procedure, based on available a-priori information, has been developed.

Assuming a training data set V^* containing square patterns and centred within each pattern a second smaller[1] one (V), a set of initialisation weights can be constructed (see Fig. 5.7).

---

1. For example 16x16 and 8x8.

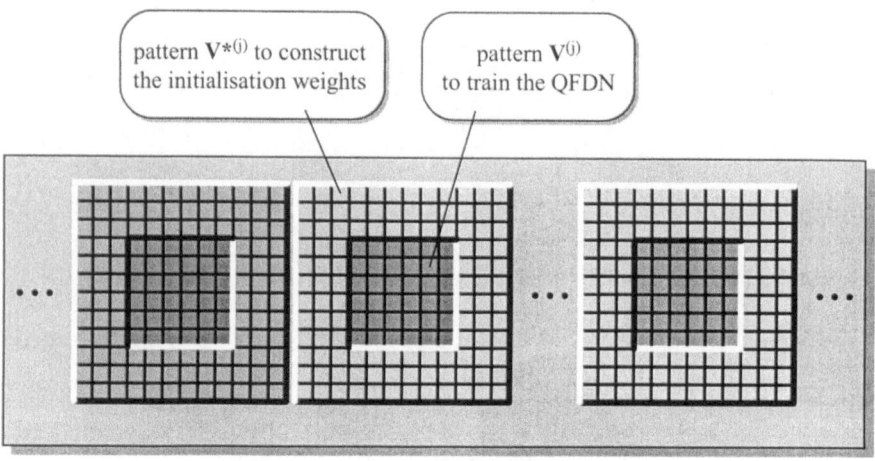

Fig. 5.7. Data set to construct initialisation weights for the QFDN (large patterns) and patterns used to train it (centred smaller patterns).

For that purpose a standard two-dimensional SOM having a small number[4] of neurons is trained using $\mathbf{V}^*$. Its weights are randomly initialised. Once this is successfully done, the trained weights of these neurons (basic weights) are used to extract the initialisation weights of the QFDN according to Fig. 5.8. A single weight element of the neuron (x, y) of submap (X, Y) is

$$w_{\chi, \psi}^{(X, Y, x, y)} = w_{basic_{p(\chi, \psi)}}^{(X, Y)} \tag{5.14}$$

with

$$p(\chi, \psi) = (\chi + (x-1) \cdot wsize(x), \psi + (y-1) \cdot wsize(y)). \tag{5.15}$$

Each basic weight matrix serves as template for all (smaller) initialisation weights belonging to the same submap. For instance, the weight matrix of the centre neuron[5] of a submap is cropped out of the middle of the basic weight matrix. Ac-

---

[4] With respect to the growing feature, a relatively small number of neurons should be chosen. Theoretically any size and distribution is imaginable. Practical investigations have shown that a square 3x3 map is very suitable. This results in 9 base weight matrices and a starting size of the QFDN of 3x3 submaps.

[5] Later on representing a basic pattern in unshifted position.

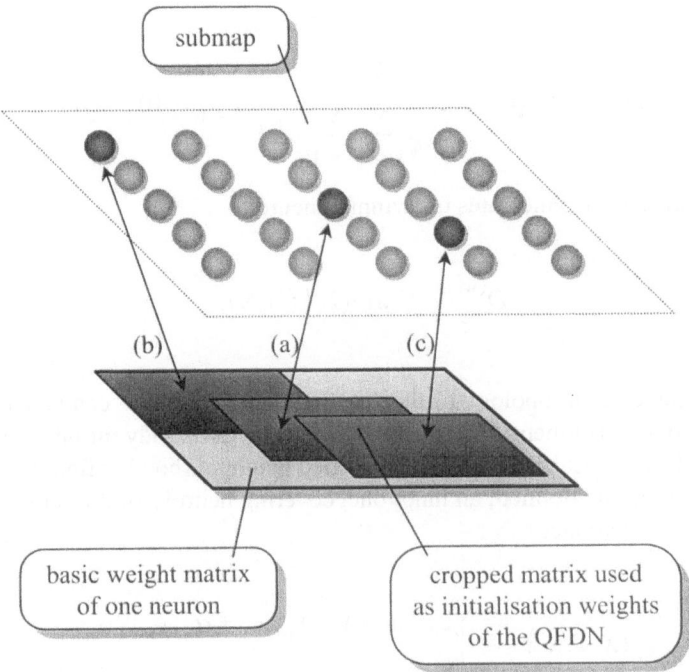

Fig. 5.8. Cropping initialisation weights of the QFDN from the previously trained basic weight matrix. The weights of the centre neuron are cropped out of the middle (a) and the weights of the other neurons are cropped according to their relative position within the sub-map (b, c).

cording to a neuron's relative position within the submap, the matrix to crop the initialisation weights is moved across the according basic weight matrix. This way the initialisation weights already contain general shift information.

Training

This phase is a combination of regular training and growing. Since the local resolution is always fixed in advance, but the number of classes is usually a-priori unknown, the growing algorithm adapts the number and distribution of entire submaps. The size of the submaps remains unchanged.

After presenting a particular training pattern $\mathbf{V}^{(j)}$, its Euclidean distance to all weight matrices[6] of all submaps $\mathbf{W}^{(i)}$ is computed

---

[6] In order to minimize computation time and use of resources, an accelerated search algorithm to determine the winning neuron has been implemented. It uses further a-priori knowledge about the network structure and data distribution and makes it not necessary to compute all neurons to find the winner.

$$D^{(i,j)}(\mathbf{V}^{(j)}, \mathbf{W}^{(i)}) = \sqrt{\sum_{k=1}^{n}\sum_{l=1}^{m}(v_{k,l}^{(j)} - w_{k,l}^{(i)})^2}, \forall i. \tag{5.16}$$

The smallest distance indicates the winning neuron

$$D^{(w,j)} = min(D^{(i,j)}), \forall i. \tag{5.17}$$

Knowing the current topology of the net, the index i resp. w can be easily transformed into a four-dimensional pointer (X, Y, x, y) as already introduced.

Special care must be taken of the modified neighbourhood definitions. Two different functions are defined; an inner one, covering neurons of the same submap to adapt the local variance

$$\zeta_n(X, Y, x, y) = \begin{cases} \zeta_n & \text{for } (X_w, Y_w, x_w \pm \alpha_n, y_w \pm \alpha_n) \\ 0 & \text{else} \end{cases} \tag{5.18}$$

and an outer area, including neurons at the same relative position within their submap to control the development of classes

$$\zeta_N(X, Y, x, y) = \begin{cases} \zeta_N & \text{for } (X_w \pm \alpha_N, Y_w \pm \alpha_N, x_w, y_w) \\ 0 & \text{else} \end{cases} \tag{5.19}$$

(see also Fig. 5.9). The coefficients α_n and α_N are positive integer values to define the sheer size of the corresponding neighbourhood areas, which are usually decreased during the training progress.

The first area is spread out around the winning neuron inside the corresponding submap and is made of single neurons. The shifted initialized weights are adapted to the really occurring positions of the current input pattern. Because the data set may not contain all shapes at all possible positions or in different occurrence frequencies, a *move* of some neurons within their submap may be necessary. This is similar to the neighbourhood functions and the weight update of the original two-dimensional SOM.

The second neighbourhood area is defined across several submaps. It contains submaps as discrete elements. The winning neuron and all neurons at the same rel-

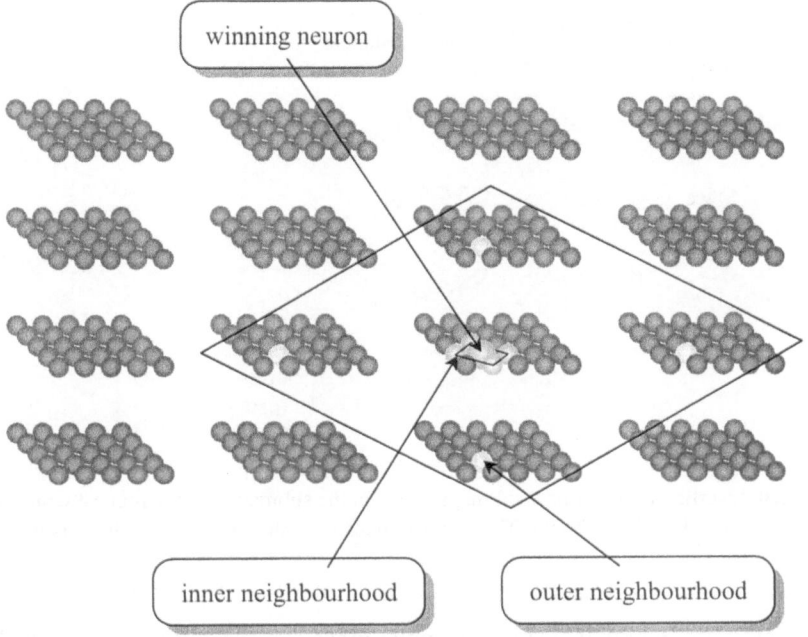

Fig. 5.9. Neighbourhood definitions within a QFDN network. Around the winning neuron an inner area across the submap and outer area including neurons at the same relative position in neighbouring submaps are defined.

ative position in its neighbouring submaps are belonging to the defined outer neighbourhood area.

The weights are updated according the following equation and the next input pattern is presented.

$$\Delta w_{\chi, \psi}^{(X, Y, x, y)} = \alpha(t) \cdot \zeta(X, Y, x, y) \cdot (v_{\chi, \psi}^{(j)} - w_{\chi, \psi}^{(X, Y, x, y)}) \quad (5.20)$$

Additionally the winning frequency of each submap $H^{(X, Y)}$ is computed. After a number of training epochs, derived from the training progress, a *size adaptation step* is performed. In general it seems to be most advantageous to insert new neurons in areas with a very high data density, where too many input patterns are represented by too few neurons[7]. A simple but yet suitable measure is the winning frequency of submaps. At this point it does not matter, whether this high density is caused by a

---

[7] Respectively submaps holding complete classes.

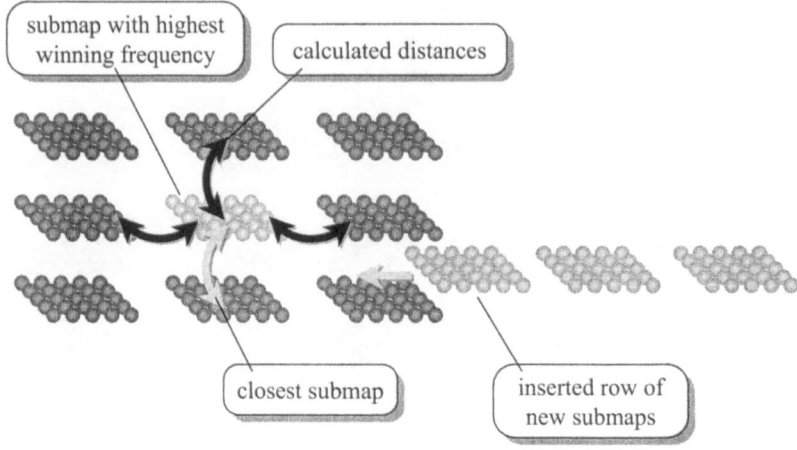

Fig. 5.10. Inserting a row of new submaps between the submap with the highest winning frequency and its closest neighbour. The new submaps have the same size as the existing ones.

heavy cluster or a high incidence of similar input patterns in the data set. The local resolution should be improved by adding new weight space.

Once the sector defined by the submap with the currently highest winning frequency is identified, is has to be checked on which edge a column or row of submaps should be inserted. Here again several strategies are possible. A very good one is to calculate the Euclidean distances between this submap and its 4 neighbours. The smallest distance (closest neighbour) indicates the insertion point.

Now the weights of the new neurons have to be set up. From the possible procedures a simple interpolation has proved to be the best choice. A new weight is calculated in the case of inserting a row according to

$$w_{\chi, \psi}^{(X, Y, x, y)} = \frac{1}{2} \cdot (w_{\chi, \psi}^{(X, Y-1, x, y)} + w_{\chi, \psi}^{(X, Y+1, x, y)}) \qquad (5.21)$$

and respectively for inserting a column to

$$w_{\chi, \psi}^{(X, Y, x, y)} = \frac{1}{2} \cdot (w_{\chi, \psi}^{(X-1, Y, x, y)} + w_{\chi, \psi}^{(X+1, Y, x, y)}). \qquad (5.22)$$

From now on the new weights are included in the regular training process. A number of training steps as described above are also performed until a further insertion step is executed or the stopping criterion is met. In contrast to the basic SOM this signals just the end of the regular combined training / growing algorithm. Now a last phase is entered. With a successive decreasing neighbourhood size (Eq. (5.18)-(5.20)) the entire network (map), now having its final size, is fine-tuned to achieve an optimal input data representation.

5.4   Applying the QFDN to the Classification Task

5.4.1   Basic Properties

This sub-section demonstrates some basic properties of the QFDN and its general suitability for the classification task of the motion detection system described in Sec. 5.2.

As to be expected, the distribution of the classes in the QFDN is better than using the three-dimensional layer with the same number of submaps in a linear topology. This is caused by the real two-dimensional classification. The centre neurons in Fig. 5.11 show the unshifted basic patterns and thus the prototype of the corresponding class. On the right hand side the weights of a rather large complete submap can be observed. The SOM typical weight distribution is also clearly to be seen. Adjacent elements are rather similar and the diagonal elements contain the most disparate information, both in the centre neurons and in the single submaps.

The general advantages of the growing feature are also noticeable. Fig. 5.12 shows an example of a comparison between the growing net and two nets with fixed size. The error temporarily increases slightly after each insertion of a row or column, because the new weights are not really trained but interpolated from their corresponding neighbours. In general the error becomes even smaller, due to the additionally available weight memory.

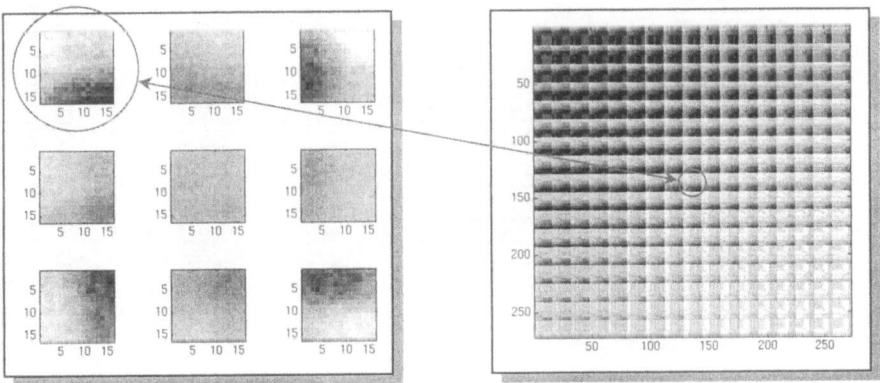

Fig. 5.11. 16x16 weight matrices of 9 centre neurons (3x3 submaps, 17x17 neurons each) and all weights of the marked submap in gray-level colour scheme.

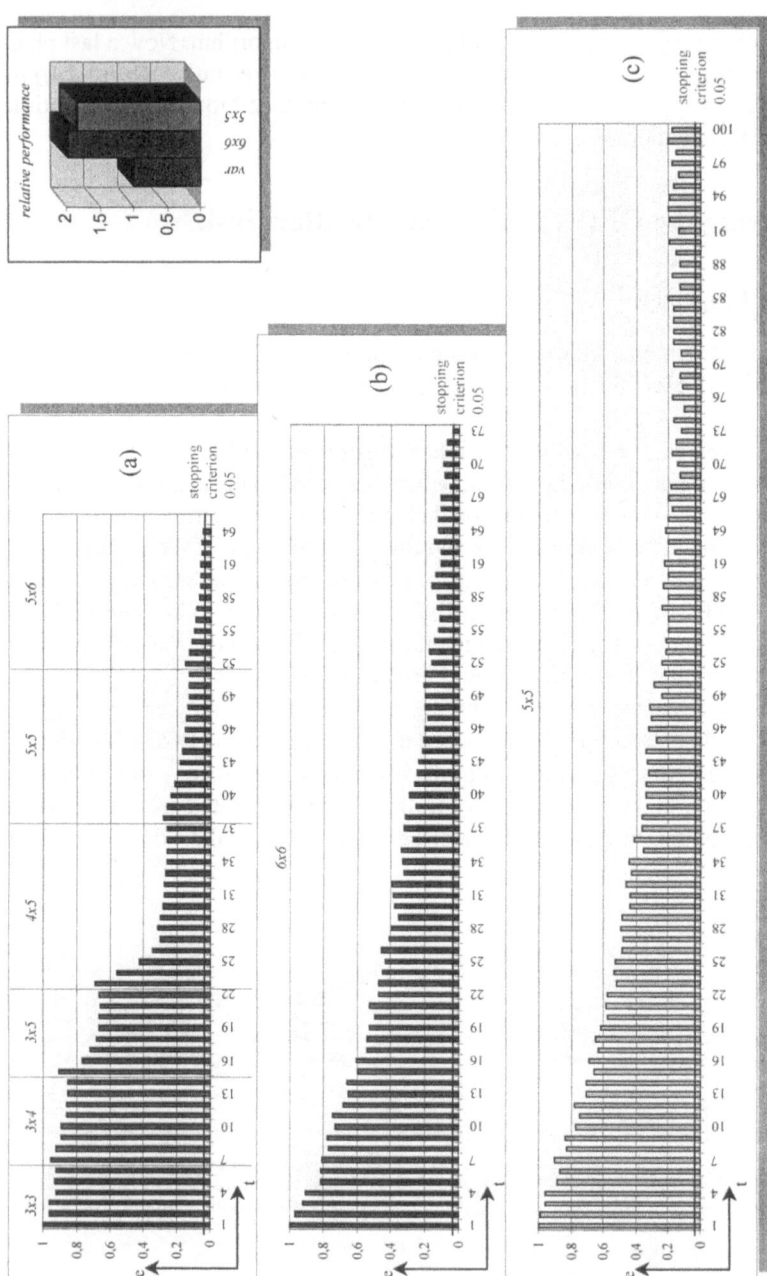

Fig. 5.12. Growing QFDN versus a standard QFDN. The growing net (top) shows the typical error progress. After an insertion step the generally decreasing error temporarily becomes higher. The two fixed size nets (centre, bottom) require significantly more computation time.

After some insertion steps the growing network reaches the desired stopping criterion. The smaller net is not able to meet this criterion at all. Its weight capacity is too small to hold all the classes in the desired resolution. The larger net can match the finishing line at a learn count less than the growing variant. However, from the beginning and during each iteration step an unnecessarily large net has to be computed. In the long run this takes more time than training the smaller and more intelligent growing net.

5.4.2 Set-Up of Classes

As already been mentioned in Sec. 5.2, the QFDN is used to solve two problems. First, an off-line set-up of classes and then within the running motion detection system an on-line classification of either image blocks or correlation functions. Two different test data sets have been constructed to meet the different requirements of both tasks.

For the first problem an image sequence containing moving objects is not really necessary, because the motion does not matter at this point. It is rather important to include as many as possible different sceneries to achieve an universally valid set of classes. Thus a set of gray-level still images of the size 512x512 pixels has been assembled (Fig. 5.13).

Fig. 5.13. Part of the still images data set to investigate the off-line set-up of classes. The images show different sceneries to obtain universally valid classes.

According to the system model (Fig. 5.3) all images are divided into blocks. In order to get all patterns at all possible positions, the cropmarks are shifted across the entire image with one pixel increment. Because all blocks have to be independent of gray-level offsets, i.e. caused by different illumination, the mean gray-level is separately subtracted from each block.

Evaluating the coordinates (X, Y, x, y) of the winning neurons after the recall of a successfully trained QFDN leads in the first instance to two major results. Vector (X, Y) points to the submap and indicates the detected class of a considered input. This is the actual result at this stage. The pair (x, y) shows the shift. Due to the topology preserving feature of SOMs, the relationship of inputs is still accessible by comparing the positions of the corresponding submaps (Fig. 5.14).

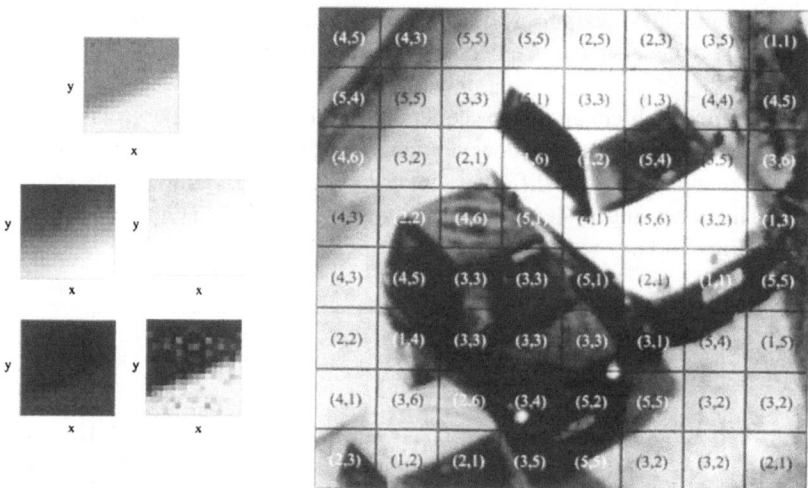

Fig. 5.14. Weight matrix of any centre neuron (top left) and 4 input examples assigned to this class (below). The big image shows some non-overlapping blocks and sample coordinates of assigned submaps. This demonstrates basic properties of the QFDN, i.e. similarity of inputs sorted into the same class and closeness of adjacent classes.

A key assumption, as specified in Sec. 5.2, is the existence of a limited number of classes. Practical investigations show, that a structure between 4x5 and 6x6 submaps is sufficient to hold all blocks resp. correlation functions in a desired and sensible resolution. Further enlargement of the map does not lead to more separate classes. That means the motion detection system has to process less than about 40 classes, which makes a successful implementation of an adaptive recognition system possible.

5.4.3 On-line Classification

This is the actual working phase processing incoming video streams. Several image sequences have been recorded. Fig. 5.15 shows a traffic sequence with several moving objects. The blocks are now cropped non-overlapping.

Once a QFDN of an appropriate size is trained, the classes are extracted and the associative memory is prepared with proper parameter sets to adaptively control the recognition system (see Sec. 5.2), the entire system can be run on-line. Based on the detected class, for each image block (which has been processed directly or as correlation function) there is a control vector

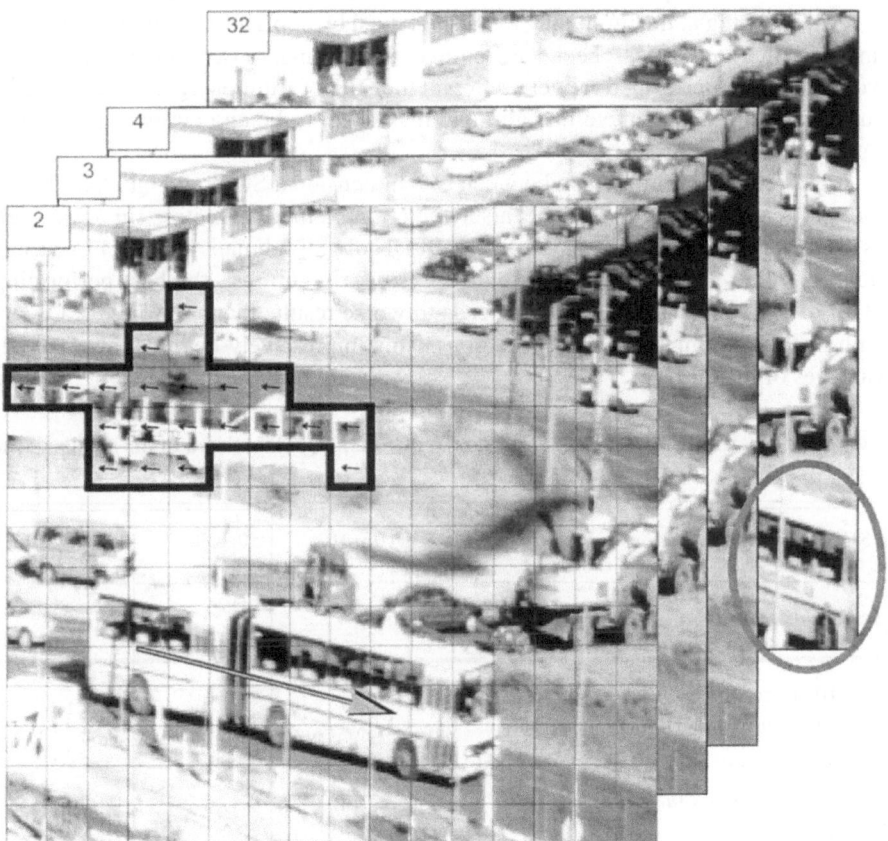

Fig. 5.15. Traffic sequence of 32 images (256x256 pixels). By means of the bus, which has partially left the observed area in image 32, a particular motion can be seen. In the second image of the sequence (in the foreground) the motion parameters are marked for blocks covering the tram. All adjoining blocks having the same motion parameters, can be compounded in order to assign more general motion parameters to parts of the images (bus).

$$\vec{z}_c^{(i)} = \left[X^{(i)} \ Y^{(i)} \ D^{(i)} \right]^T.\tag{5.23}$$

This is transmitted to the associative memory and a particular parameter set is downloaded into the recognition system, which estimates the motion parameters as shown in Fig. 5.15. There are two objects, the tram moving to the left and the bus moving to the right. Adjacent blocks having the same or very similar motion parameters can be united to a motion cluster. If the same blocks belong to the same cluster over a number of time steps, they may be considered as one compounded object. To every object a mean motion vector can be assigned, indicating for example a mean direction and velocity. This may also be used for motion based image segmentation.

After looking at the promising results obtained by applying the QFDN to the on-line classification, a crosscheck has to prove, that a standard two-dimensional SOM would not have performed so well. After all the QFDN requires a great deal of computational power to achieve this aim and may seem to be an extravagance at first sight.

Tab. 1 shows a comparison between two-dimensional SOM and QFDN. The error values indicate the number of falsely determined motion vectors (in %) as well as their root mean square error (RMS). A QFDN of 5x5 submaps and 17x17 neurons (total of 7.225 neurons) performs without error[8]. In contrast, the best performing SOM of 100x100 (10.000) neurons leads to worse results than the smallest QFDN with just 3x3 submaps and 2600 neurons. This impressively demonstrates the advantages of the QFDN.

Tab. 1. Comparison between two-dimensional SOM and QFDN related to the results of the motion vector determination.

SOM (size: *neurons x neurons*)	10 x 10	50 x 50	100 x 100
error: *false vectors in %, RMS*	6.3, 4.2	5.9, 3.7	4.7, 2.4
QFDN (size: *submaps x submaps, neurons x neurons within a submap*)	3 x 3, 17 x 17	4 x 4, 17 x 17	5 x 5, 17 x 17
error: *false vectors in %, RMS*	3.7, 1.9	0.4, 1.2	0.0, 0.0

---

[8] There are numerical differences which do not lead to misclassification or falsely detected motion parameters.

it is easier to directly recognize both classes and displacement. Furthermore the SOM needs many more training steps, although each training step is faster, which still leads to a longer training time with the SOM.

5.5 Conclusions

This chapter has shown a way of extending the original two-dimensional SOM to a multi-dimensional structure, which is tailored to handle two-dimensional input matrices and to perform a classification of basic patterns independent of their actual position within the larger input.

The presented Quasi-Four-Dimensional Neuroncube QFDN can be considered as a subset of multi-dimensional self-organising hypercube architecture. In the presented application it is a two-dimensional rectangular grid of two-dimensional submaps, which can be easily extended to three-dimensional submaps to reflect spatial patterns, i.e. moving objects in a stereoscopic observed environment. It is also possible to release the submaps from their fixed grid to construct a multi-dimensional 'Neural Gas' like network of independently moving submaps. The implementation of a growing feature enables the QFDN to optimally adapt its topology and weights to a given input data distribution and accelerates the training significantly. Due to the two-dimensional input matrix all neurons have also weight matrices rather than vectors. This keeps all neighbourhood relations within each input pattern active and makes the stored information easily accessible to a human observer.

Due to its fairly extensive topology and a usually big number of neurons connected to often many input elements, commonly huge computational power is required to run the net in acceptable speed. After all, the demanded task is very complex and a simpler network with then probably extensive pre-processing is expected to be very time consuming as well. The compact neural network solution offers the big advantage of an extremely fast implementation on parallel hardware due to its inherent parallelism. A very efficient and practicable parallel computer implementation on low-cost hardware has been developed by the author, and this enables real-time processing of the entire system.

The QFDN has been successfully applied to image processing as a powerful and adaptive classification tool. It forms a part of a motion detection system. It has solved the two problems formulated at the beginning very well.

Using the QFDN it is possible to form several classes limited to about 30-40 relatively independent classes. Due to the small block size, the variety and complexity of the possible contents in each block is low such as single edges and gray-level plateaus. This is the reason for the relatively small number of necessary classes on the one hand. On the other hand it makes it possible to train one master network to handle a wide range of image sequences with varying scenes. The classes consist of the trained blocks of the correlation functions or gray-scale image parts (image blocks) sorted by their geometrical similarity.

Based on this information the implementation of an adaptively controlled recognition system is possible. When running on-line it receives for each class a particular parameter set from an associative memory which is addressed by the QFDN. This way an image sensitive determination of motion parameters can be performed. Many pseudo-motion effects leading to false motion parameters can be suppressed.

The results compared to those of a standard two-dimensional Self-Organizing Map could be considerably improved. The QFDN has demonstrated its perfect suitability and fitness to fulfil the desired task. Other applications of the QFDN may be found in all fields, where a two-dimensional classification of two-dimensional input data is required. Beyond the motion detection system it has already been successfully applied to classification tasks of biological and medical data [37]. In general multi-dimensional Self-Organizing Maps such as the QFDN can be used as an universal method to solve classification problems of two-dimensional (matrix-shaped) input data without extensive pre-processing. Apart from being a solution to this particular problem, this may be a key feature of this work. Thus it is another universal contribution to the extensions and improvements of the basic SOM algorithm.

Bibliography on Chapter 5

1. Kohonen, T. (1995): Self-Organizing Maps, 2nd Edition. Springer, Berlin.
2. Singh, A. (1991): Optic Flow Computation. IEEE Press, Los Alamitos.
3. Fleet, D.J. (1992): Measurement of Image Velocity. Kluwer Academic Publishers, Amsterdam.
4. Koller, D., Daniilidis, K., Nagel, H.H. (1993): Model-Based Object Tracking in Monocular Image Sequences of Road Traffic Scenes. International Journal of Computer Vision, 10, 257-281.
5. Jain, J.R., Jain, A.K. (1981): Displacement Measurement and its Application in Inter-Frame Image Coding. IEEE Transactions on Communications, 29, 1799-1808.
6. Yamaguchi, H., Sugi, T., Kinuhata, K. (1987): Movement-Compensated Frame-Frequency Conversion of Television Signals. IEEE Transactions on Communications, 35, 1069-1082.
7. The Open MPEG (Moving Picture Expert Group) Consortium (1990): MPEG-1 Specification ISO 11172-1 and 11172-2.
8. The Open MPEG (Moving Picture Expert Group) Consortium (1990): MPEG-2/DVD Specification ISO 13818-1 and 13818-2.
9. Hogden, J., Saltzman, E., Rubin, P. (1993): Tracking Moving Objects with Unsupervised Neural Networks. In: Proceedings of World Congress on Neural Networks. Lawrence Erlbaum, Hillsdale, 409-412.
10. Tsukamoto, A., Lee, C.W., Tsuji, S. (1993): Detection and Tracking of Human Face with Synthesised Templates. In: Proceedings of Asian Conference on Computer Vision. IEEE Press, Osaka, 183-186.
11. Wimbauer, S., Gerstner, W., van Hemmen, J.L. (1994): Motion Detection in a Linsker Network. In: Proceedings of International Conference on Artificial Neural Networks. Springer, Berlin, 1001-1004.
12. Jiang, J. (1995): Algorithm Design of an Image Compression Neural Network. In: Proceedings of World Congress on Neural Networks. INNS Press, New York, 792-798.
13. Szu, H., Telfer, B., Garcia, J. (1996): Wavelet Transforms and Neural Networks for Compression and Recognition. Neural Networks, 4, 695-708.
14. Wang, R. (1996): A Network Model for the Optic Flow Computation of the MST Neurons. Neural Networks, 3, 411-426.
15. Michaelis, B., Schnelting, O., Seiffert, U., Mecke, R. (1995): Motion Estimation Using a Compounded SOM-MLP Network. In: Proceedings of World Congress on Neural Networks. INNS Press, New York, 103-106.
16. Michaelis, B., Schnelting, O., Seiffert, U., Mecke, R. (1996): Adaptive Filtering of Distorted Displacement Vector Fields Using Artificial Neural Networks. In: Proceedings of International Conference on Pattern Recognition, Vol. IV. IEEE Press, Los Alamitos, 335-339.
17. Lippmann, R.P. (1987): An Introduction to Computing with Neural Nets. IEEE ASSP Magazine, 4, 4-23.
18. Arbib, M.A. (1993): Brains, Machines and Mathematics. Springer, New York.
19. Kosslyn, S.M. (1994): Image and Brain. Bradford, Cambridge.
20. Rolls E.T. (1987): Information Representation, Processing and Storage in the Brain: Analysis at the Single Neuron Level. In: Changeux, J.-P., Konishi, M. (Eds.): The Neural and Molecular Bases of Learning. Wiley, New York, 503 539.
21. Ullman, S. (1979): The Interpretation of Visual Motion. MIT Press, Cambridge.
22. Van Essen, D.C., Maunsell J.H. (1983): Hierarchical Organization and Functional Streams in the Visual Cortex. Trends in Neurosciences, 6, 370-375.
23. Van Essen, D.C., Anderson C.H., Felleman, D.J. (1992): Information Processing in the Primate Visual System: An Integrated Systems Perspective. Science, 255, 419-423.

120   Udo Seiffert

24. Talairach, J., Tournoux, D.P. (1988): Coplanar Stereotaxic Atlas of the Human Brain. Thieme, New York.
25. Mumford, D. (1991): On the Computational Architecture of the Neocortex I. Biological Cybernetics, 65, 135-145.
26. Mumford, D. (1992): On the Computational Architecture of the Neocortex II. Biological Cybernetics, 66, 241-251.
27. Kohonen, T. (1982): Self-Organizing Formation of Topologically Correct Feature Maps. Biological Cybernetics, 43, 59-69.
28. Kangas, J., Kohonen, T., Laaksonen, J. (1990): Variants of Self-Organizing Maps. IEEE Transactions on Neural Networks, 1, 93-99.
29. Oja, E., Kaski, S. (Eds.) (1999): Kohonen Maps. Elesevier, Amsterdam.
30. Fritzke, B. (1995): Growing Grid - A Self-Organizing Network with Constant Neighborhood Range and Adaption Strength. Neural Processing Letters, 5, 1-5.
31. Fritzke, B. (1995): Growing Self-Organizing Networks - Why? In: Proceedings of European Symposium on Artificial Neural Networks. D-Facto Publishers, Brussels, 61-72.
32. Bauer, H.-U., Villmann, T. (1997): Growing a Hypercubical Output Space in a Self-Organizing Feature Map. IEEE Transactions on Neural Networks, 2, 218-226.
33. Martinetz, T., Schulten, K. (1991): A 'Neural Gas' Learns Topologies. In: Kohonen, T. et al. (Eds.): Artificial Neural Networks. North-Holland, Amsterdam, 397-402.
34. Fritzke, B. (1995): A Growing Neural Gas Network Learns Topologies. In: Tesauro G., Touretzky, D.S., Leen, T.K. (Eds.): Advances in Neural Information Processing Systems. MIT Press, Cambridge.
35. Seiffert U., Michaelis, B. (1998): Growing Multi-Dimensional Self-Organizing Maps. International Journal of Knowledge-Based Intelligent Engineering Systems, 2, 42-48.
36. Seiffert, U., Michaelis, B. (1998): Quasi-Four-Dimensional Neuroncube and its Application to Motion Estimation. In: Bulsari, A.B., de Canete, J.F., Kallio, S. (Eds.): Engineering Benefits from Neural Networks. SEA, Turku, 78-81.
37. Sommerkorn, G., Seiffert, U., Surmeli, D., Herzog, A., Michaelis, B., Braun, K. (1997): Classification of 3-D Dendritic Spines Using Self-Organizing Maps. In: Smith, G.D. et al. (Eds.): Artificial Neural Nets and Genetic Algorithms. Series Computer Science. Springer, Vienna, 129-132.

6 Extensions and Modifications of the Kohonen-SOM and Applications in Remote Sensing Image Analysis

Thomas Villmann and Erzsébet Merényi

Summary. Utilization of remote sensing multi- and hyperspectral imagery has shown a rapid increase in many areas of economic and scientific significance over the past ten years. Hyperspectral sensors, in particular, are capable of capturing the detailed spectral signatures that uniquely characterize a great number of diverse surface materials. Interpretation of these very high-dimensional signatures, however, has proved an insurmountable challenge for many traditional classification, clustering and visualization methods. This chapter presents spectral image analyses with Self-Organizing Maps (SOMs). Several recent extensions to the original Kohonen SOM are discussed, emphasizing the necessity of faithful topological mapping for correct interpretation. The effectiveness of the presented approaches is demonstrated through case studies on real-life multi- and hyperspectral images.

6.1 Introduction

Airborne and satellite-borne remote sensing spectral imaging has become one of the most advanced tools for collecting vital information about the surface of Earth and other planets. The utilization of these data includes areas such as mineral exploration, land use, forestry, ecosystem management; assessment of natural hazards, water resources, environmental contamination, biomass and productivity; and many other activities of economic significance, as well as prime scientific pursuits such as looking for possible sources of past or present life on other planets. The number of applications has dramatically increased in the past ten years with the advent of imaging spectrometers that greatly surpass traditional multi-spectral sensors (*e.g.,* *Landsat Thematic Mapper (TM)*). Imaging spectrometers can resolve the known, unique, discriminating spectral features of minerals, soils, rocks, and vegetation. While a multi-spectral sensor samples a given wavelength window (typically the $0.4 - 2.5\,\mu m$ range in the case of Visible and Near-Infrared imaging) with several broad bandpasses, leaving large gaps between the bands, imaging spectrometers sample a spectral window contiguously with very narrow, $10 - 20\,nm$ badpasses. Hyperspectral technology is in great demand because direct identification of surface compounds is possible without prior field work, for materials with known spectral signatures.

Spectral images consist of an array of multi-dimensional vectors assigned to particular spatial areas (pixel locations) reflecting the response of a spectral sensor at various wavelengths (see Fig. 6.1). These vectors are called spectra. A spectrum is a

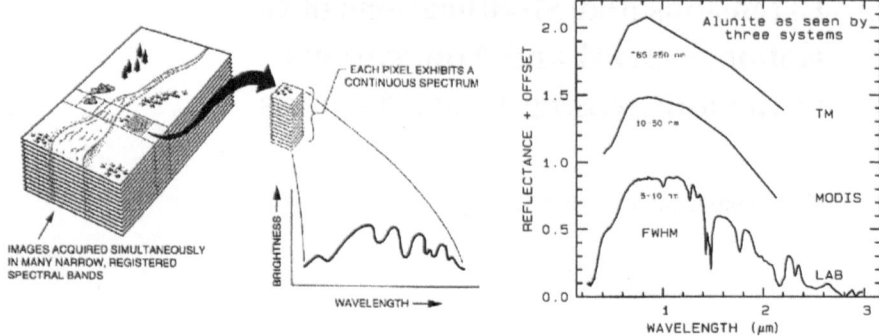

Fig. 6.1. Left: The concept of hyperspectral imaging. Figure from [9]. Right: The spectral signature of the mineral alunite as seen through the 6 broad bands of Landsat TM, as seen by the moderate spectral resolution sensor MODIS (20 bands in this region), and as measured in laboratory. Figure from [10]. Hyperspectral sensors such as AVIRIS of NASA/JPL [16] produce spectral details comparable to laboratory measurements.

characteristic pattern that provides a clue to the surface material within the respective area. Depending on the wavelength resolution and the width of the wavelength window used by a particular sensor, the dimensionality of the spectra can be as low as $5 - 6$ (such as in Landsat TM), or as high as several hundred for hyperspectral imagers.

Classification of intricate, high-dimensional spectral signatures has turned out far from trivial. Discrimination among many surface cover classes, discovery of spatially small, interesting spectral species proved to be an insurmountable challenge to many traditional clustering and classification methods. This motivates research into advanced and novel approaches. [30]. By costumary measures (such as, for example, Principle Component Analysis (PCA)) the intrinsic spectral dimensionality of hyperspectral images appears to be surprisingly low, $5 - 10$ at most. Yet dimensionality reduction to such low numbers has not been successful in terms of preservation of important class distinctions. The spectral bands, many of which are highly correlated, may lie on a low-dimensional but non-linear manifold, which is a scenario that eludes many classical approaches.

One powerful approach to these problems is the application of Self-Organizing Maps (SOMs) [24] to implement a suitable mapping procedure which should end in a topology preserving projection of the high-dimensional data onto a low-dimensional lattice. In most applications a two-dimensional SOM lattice is the common choice of lattice structure because of its easy visualization [23]. In general, this choice is not guaranteed to produce a topology preserving mapping and the interpretation of the resulting map may fail [40]. Topology preserving mapping, informally speaking, means that similar data vectors are mapped onto the same or neighbored locations in the lattice and vice versa [41]. A further aspect that should be addressed here is the problem of the detection of rarely occurring surface material classes in the images. If the SOM is used as a classifier system the distribution of the weight vectors across the SOM lattice is determined by a power function of the probability density of the data vectors [37], with the so-called magnification factor as the power. As we will see later, for magnification factors less than 1, which is the case for the standard SOM, it may be difficult or impossible to separate spectral patterns of seldom occurring surface materials.

In the last few years extensions of the standard SOM were derived to respond to the above challenges. As shown in [3] it is possible to control the magnification of the SOM by the introduction of a local learning rate that is dependent on the data density. The corresponding learning scheme can easily be implemented into the standard learning rule. A growing SOM (GSOM) approach was developed to generate a guaranteed topology preserving mapping in a simple hypercube structure of the lattice [5]. Both the GSOM and magnification control approaches were shown to be powerful instruments for visualization and classification of remote sensing spectral data.

This chapter is organized as follows: Sec. 2 is a short overview of the above extensions of the standard SOM. In section 3 the data sets are described. Finally, in section 4 application results are presented and discussed.

6.2   SOMs and Data Mapping

Self-organizing maps [24] as a special kind of neural maps project data from some (possibly high-dimensional) input space $V \subseteq \Re^{D_V}$ onto a position in some output space (neural map) A, such that a continuous change of a parameter of the input data should lead to a continuous change of the position of a localized excitation in the neural map. This property of *neighborhood preservation* depends on an important feature of the SOM, its output space topology, which has to be specified prior to learning. If the topology (i.e. dimensionality and edge length ratios) of A does not match that of the data shape, neighborhood violations are likely to occur [43]. This can be cast in a formal way by writing the output space positions as $\mathbf{r} = (i_1, i_2, i_3, ..., i_{n_j})$, $1 < i_k < n_k$ with $N = n_1 \times n_2 \times ... \times n_j$ where n_k, $k = 1, ..., j$ is the dimension of A (the length of the edge of the lattice) in the kth direction .[1] Associated with each neuron $\mathbf{r} \in A$, is a weight vector, or *pointer*, $\mathbf{w_r}$ in V. The mapping $\Psi_{V \to A}$ is realized by a winner take all rule

$$\Psi_{V \to A} : \mathbf{v} \mapsto \mathbf{s} = \operatorname*{argmin}_{r \in A} \|\mathbf{v} - \mathbf{w_r}\| \qquad (6.1)$$

whereas the reverse mapping is defined as $\Psi_{A \to V} : \mathbf{r} \mapsto \mathbf{w_r}$. The two functions together determine the map

$$\mathcal{M} = (\Psi_{V \to A}, \Psi_{A \to V}) \qquad (6.2)$$

---

[1] Other spatial arrangements are also possible, which can be described by a connectivity matrix. Here we only consider hypercubes.

realized by the SOM network. All data points $\mathbf{v} \in \Re^{D_V}$ that are mapped onto the neuron $\mathbf{r}$ make up its receptive field $\hat{\Omega}_\mathbf{r}$. The masked receptive field of neuron $\mathbf{r}$ is defined as the intersection of its receptive field with $\mathcal{V}$:

$$\Omega_\mathbf{r} = \{\mathbf{v} \in \mathcal{V} : \mathbf{r} = \Psi_{\mathcal{V} \to \mathcal{A}}(\mathbf{v})\}. \tag{6.3}$$

Therefore, the masked receptive fields $\Omega_\mathbf{r}$ are closed sets. All masked receptive fields form the Voronoi tesselation of $\mathcal{V}$. If the intersection of two masked receptive fields $\Omega_\mathbf{r}$, $\Omega_{\mathbf{r}'}$ is non-vanishing we call $\Omega_\mathbf{r}$ and $\Omega_{\mathbf{r}'}$ neighbored. The neighborhood relations form a corresponding graph structure $\mathcal{G}_\mathcal{V}$ in $\mathcal{A}$: two neurons are connected in $\mathcal{G}_\mathcal{V}$ if and only if their masked receptive fields are neighbored. The graph $\mathcal{G}_\mathcal{V}$ is called the induced Delaunay-graph (See, for example, [27] for detailed definitions). Due to the bijective relation between neurons and weight vectors, $\mathcal{G}_\mathcal{V}$ also represents the Delaunay graph of the weights.

To achieve the map $\mathcal{M}$, SOMs adapt the pointer positions during the presentation of a sequence of data points $\mathbf{v} \in \mathcal{V}$ selected from a data distribution $\mathcal{P}(\mathcal{V})$, as follows:

$$\triangle \mathbf{w}_\mathbf{r} = \epsilon h_{\mathbf{rs}}(\mathbf{v} - \mathbf{w}_\mathbf{r}). \tag{6.4}$$

$h_{\mathbf{rs}}$ is the neighborhood function, usually chosen to be of Gaussian shape:

$$h_{\mathbf{rs}} = \exp\left(-\frac{\|\mathbf{r} - \mathbf{s}\|^2}{2\sigma^2}\right) \tag{6.5}$$

Note that $h_{\mathbf{rs}}$ is dependent on the best matching neuron (6.1).

Topology preservation in SOMs is defined as the preservation of the continuity of the mapping from the input space onto the output space, more precisely it is equivalent to the *continuity* of $\mathcal{M}$ between the *topological spaces* with properly chosen metric in both $\mathcal{A}$ and $\mathcal{V}$. For lack of space we refer to [43] for detailed considerations. The topology preserving property can be used for immediate evaluations of the resulting map, for instance for interpretation as a color space, as demonstrated in sec. 6.4.2 . Topology preservation also allows the applications of interpolating schemes such as the parametrized SOM (PSOM) [36] or interpolating SOM (I-SOM) [14]. A higher degree of topology preservation, in general, improves the accuracy of the map [4]. As pointed out in the introduction violations of topographic mapping can result in false interpretations. Several approaches were developed to judge the degree of topology preservation for a given map. Here we briefly describe a variant $\tilde{P}$ of the well known topographic product P [4]. Instead of the Euclidean distances between the weight vectors, this measure uses the respective distances $d^{\mathcal{G}_\mathcal{V}}(\mathbf{w}_\mathbf{r}, \mathbf{w}_{\mathbf{r}'})$ of minimal path lengths in the induced Delaunay-graph $\mathcal{G}_\mathcal{V}$ of the $\mathbf{w}_\mathbf{r}$. During the computation of $\tilde{P}$ for each node $\mathbf{r}$ the sequences $\mathbf{n}_j^\mathcal{A}(\mathbf{r})$ of j-th neighbors of $\mathbf{r}$ in $\mathcal{A}$ and $\mathbf{n}_j^\mathcal{V}(\mathbf{r})$ describing the j-th neighbor of $\mathbf{w}_\mathbf{r}$, have to be determined. These sequences and further averaging over neighborhood orders j and nodes $\mathbf{r}$ finally leads to

$$\tilde{P} = \frac{1}{N(N-1)} \sum_\mathbf{r} \sum_{j=1}^{N-1} \frac{1}{2j} \log(\Theta) \tag{6.6}$$

with

$$\Theta = \Pi_{l=1}^{j} \frac{d^{\mathcal{G}_\mathcal{V}}\left(\mathbf{w_r}, \mathbf{w}_{\mathbf{n}_l^A(\mathbf{r})}\right)}{d^{\mathcal{G}_\mathcal{V}}\left(\mathbf{w_r}, \mathbf{w}_{\mathbf{n}_l^\mathcal{Y}(\mathbf{r})}\right)} \cdot \frac{d_A\left(\mathbf{r}, \mathbf{n}_l^A(\mathbf{r})\right)}{d_A\left(\mathbf{r}, \mathbf{n}_l^\mathcal{Y}(\mathbf{r})\right)}. \tag{6.7}$$

$\tilde{P}$ can take on positive or negative values: if $\tilde{P} < 0$ holds the output space is too low-dimensional, and for $\tilde{P} > 0$ the output space is too high-dimensional. In both cases neighborhood relations are violated. Only for $\tilde{P} \approx 0$ does the output space approximately match the topology of the input data.[2]

Application of SOMs to very high-dimensional data can produce difficulties which may result from the so-called 'curse of dimensionality': the problem of sparse data caused by the large data dimensionality. We want to refer to two methods for overcoming this problem. The first approach, introduced by KASKI [23] uses the fact that in extremly high-dimensional data spaces the inner product of vectors often tends to be zero, i.e. many of the data vectors $\mathbf{v} \in \mathcal{V}$ seem to be nearly orthogonal to one another. Assume that the data undergo a (linear) random mapping

$$\tilde{\mathbf{x}} = \mathbf{R}\mathbf{x} \tag{6.8}$$

where $\mathbf{R}$ is a random matrix the columns of which are normalized to unity and the components of each coloumn are independent, identically and normally distributed with zero mean. Let $d_\mathbf{R}$ be the *reduced* dimension of the target space. If we assume that $d_\mathbf{R}$ is large and consider then $\mathbf{R}^T\mathbf{R} = \mathbf{I} + \epsilon$ with the identity matrix $\mathbf{I}$, one can show that the elements ϵ_{ij} are approximately normally distributed and the respective variance, denoted by σ_ϵ^2, can be approximated by $\sigma_\epsilon^2 \approx \frac{1}{d}$. Considering the inner product $\tilde{\mathbf{x}}^T\tilde{\mathbf{y}}$ of two mapped vectors $\tilde{\mathbf{x}}, \tilde{\mathbf{y}}$ we have for the deviation δ from the original inner product $\mathbf{x}^T\mathbf{y}$ the expectation value zero and the variance

$$\sigma_\delta^2 = \left[1 + \left(\sum_k x_k y_k\right)^2 - 2\sum_k x_k^2 y_k^2\right]\sigma_\epsilon^2 \tag{6.9}$$

Thus, random mapping can be applied to reduce the data dimension if the original dimension is very large. We should emphasize that the remaining dimension $d_\mathbf{R}$ must be large.[3] Hence, this scheme is not practicable in many applications for final processing but it can serve as a useful preprocessing. In particular, in the remote sensing application presented below neither the assumptions of the zero-tendency of the inner product nor the requirement of high reduced data dimensionality is not fulfilled.

---

[2] The present variant $\tilde{P}$ overcomes the problem of strongly curved maps which may be judged neighborhood violating by the original P even though the shape of the map might be perfectly justified [43].

[3] In the WEBSOM-application of KASKI the dimesion was reduced from 5781 to approximately 150 [23].

The second method introduces a local neighborhood range $\sigma_{\mathbf{r}}$ for sparse data sets in (6.5) to obtain a faithful mapping [11]. $\sigma_{\mathbf{r}}$ is determined by the inner curvature strength γ of the lattice which can be obtained by a wavelet analysis of the neuron weights. For a more detailed description we refer to [11].

6.2.1   Structure Adaptation by GSOM

The growing SOM (GSOM) approach [5] is an extension of the standard SOM. Its output is a *structure adapted* hypercube $\mathcal{A}$, produced by adaptation of both the dimensions and the respective edge length ratios of $\mathcal{A}$ during the learning, in addition to the usual adaptation of the weights. In comparison to the standard SOM, the overall dimensionality and the dimensions along the individual directions in $\mathcal{A}$ are variables that evolve into the hypercube structure most suitable for the input space topology. The GSOM starts from an initial 2-neuron chain, learns like a regular SOM, adds neurons to the output space based on the criterion described below, learns again, adds again, etc., until a prespecified maximum number $N_{\max}$ of neurons is distributed over $\mathcal{A}$. The output space topology always remains of the form $n_1 \times n_2 \times ...$, with $n_j = 1$ for $j > D_{\mathcal{A}}$, where $D_{\mathcal{A}}$ is the current dimensionality of $\mathcal{A}$. Hence, the initial configuration is $2 \times 1 \times 1 \times ...$, $D_{\mathcal{A}} = 1$. From there it can grow either by adding nodes in one of the directions that are already included in the output space or by initializing a new dimension. This decision is made on the basis of the masked receptive fields $\Omega_{\mathbf{r}}$ defined in (6.3). When reconstructing $\mathbf{v} \in \mathcal{V}$ from neuron $\mathbf{r}$, an error $\theta = \mathbf{v} - \mathbf{w}_{\mathbf{r}}$ remains decomposed along the different directions, which results from projecting the output space grid back onto the input space $\mathcal{V}$:

$$\theta = \mathbf{v} - \mathbf{w}_{\mathbf{r}} = \sum_{i=1}^{D_{\mathcal{A}}} a_i(\mathbf{v}) \frac{\mathbf{w}_{\mathbf{r}+\mathbf{e}_i} - \mathbf{w}_{\mathbf{r}-\mathbf{e}_i}}{\|\mathbf{w}_{\mathbf{r}+\mathbf{e}_i} - \mathbf{w}_{\mathbf{r}-\mathbf{e}_i}\|} + \mathbf{v}' \tag{6.10}$$

Here, $\mathbf{e}_i$ denotes the unit vector in direction i of $\mathcal{A}$ and $a_i(\mathbf{v})$ are the projection amplitudes.[4] Considering a receptive field $\Omega_{\mathbf{r}}$ and determining its main (first) principal component ω_{PCA} allows a further decomposition of $\mathbf{v}'$. Projection of $\mathbf{v}'$ onto the direction of ω_{PCA} then yields $a_{D_{\mathcal{A}}+1}(\mathbf{v})$,

$$\mathbf{v}' = a_{D_{\mathcal{A}}+1}(\mathbf{v}) \frac{\omega_{PCA}}{\|\omega_{PCA}\|} + \mathbf{v}''. \tag{6.11}$$

The *criterion for the growing* now is to add nodes in that direction which has on average the largest (normalized) expected error amplitude $\tilde{a}_i$:

$$\tilde{a}_i = \sqrt{\frac{n_i}{n_i + 1}} \sum_{\mathbf{v}} \frac{|a_i(\mathbf{v})|}{\sqrt{\sum_{j=1}^{D_{\mathcal{A}}+1} a_j^2(\mathbf{v})}}, i = 1, ..., D_{\mathcal{A}} + 1 \tag{6.12}$$

After each growth step, a new learning phase has to take place in order to readjust the map. For a detailed study of the algorithm we refer to [5].

---

[4] At the border of the output space grid, where not two, but just one neighboring neuron is available, we use $\frac{\mathbf{w}_{\mathbf{r}} - \mathbf{w}_{\mathbf{r}-\mathbf{e}_i}}{\|\mathbf{w}_{\mathbf{r}} - \mathbf{w}_{\mathbf{r}-\mathbf{e}_i}\|}$ or $\frac{\mathbf{w}_{\mathbf{r}+\mathbf{e}_i} - \mathbf{w}_{\mathbf{r}}}{\|\mathbf{w}_{\mathbf{r}-\mathbf{e}_i} - \mathbf{w}_{\mathbf{r}}\|}$ to compute the backprojection of the output space direction $\mathbf{e}_i$ into the input space.

6.2.2   Magnification Control in SOMs

A further extension of the basic SOM concerns the so-called magnification. The standard SOM distributes the pointers $\mathbf{W} = \{\mathbf{w_r}\}$ according to the input distribution

$$P(\mathbf{W}) \sim P(\mathcal{V})^\alpha \qquad (6.13)$$

with the magnification factor $\alpha = \frac{2}{3}$ [37], [25].[5] The first approach to influence the magnification of a learning vector quantizer, proposed in [12] is called the *mechanism of conscience*. For this purpose a bias term is added in the winner rule (6.1):

$$\Psi_{\mathcal{V}\to A} : \mathbf{v} \mapsto s(\mathbf{v}) = \underset{i\in A}{\mathrm{argmin}} \left(\|\mathbf{v} - \mathbf{w}_i\| - \gamma \left(\frac{1}{N} - p_i \right) \right) \qquad (6.14)$$

where p_i is the actual winning probability of the neuron i and γ is a balance factor. Hence, the winner determination is influenced by this modification. The algorithm should converge such that the winning probabilities of all neurons are equalized. This is related to a maximization of the entropy and consequently the resulting magnification is equal to unity.[6] However, an arbitrary magnification cannot be achieved. Therefore, BAUER ET AL. in [3] introduced a local learning parameter ϵ_r with $\langle \epsilon_r \rangle \propto P(\mathcal{V})^m$ in (6.4), where m is an additional control parameter. Equation (6.4) now reads as

$$\triangle \mathbf{w_r} = \epsilon_s h_{rs} (\mathbf{v} - \mathbf{w_r}) . \qquad (6.15)$$

Note that the learning factor ϵ_s of the winning neuron s is applied to all updates. This local learning leads to a similar relation as in (6.13):

$$P(\mathbf{W}) \sim P(\mathcal{V})^{\alpha'} \qquad (6.16)$$

with $\alpha' = \alpha(m+1)$ and allows a magnification control through the choice of m. In particular, one can achieve a resolution of $\alpha' = 1$, which maximizes mutual information [26,45].

6.2.3   Data Mining and Knowledge Discovery Using SOMs

If a proper SOM is trained according to the above mentioned criteria several methods for representation and post-processing can be applied. In case of a two-dimensional lattice of neurons many visualization approaches are known. The most common method for the visualization of SOMs is to project the weight vectors in

---

[5] This result is valid for the one-dimensional case and higher dimenional ones which separate.

[6] VAN HULLE points out that adding a conscience algorithm to the SOM does not equate to equiprobabilistic mapping, in general [21]. However, for *very high dimensions*, a minimum distortion quantizer (such as the conscience algorithm) approaches an equiprobable quantizer ([21] - page 93).

the first dimension of the space spanned by the principle components of the data and connecting these units to the respective nodes in the lattice that are neighbored [23]. However, if the shape of the SOM lattice is hypercubical there exist several more ways to visualize the properties of the map. Here we concentrate only on those that are of interest in the applications presented later in this chapter. An extensive overview can be found in [39].

One interesting evaluation is the so-called U-matrix introduced by ULTSCH ET AL. [38]. The elements $U_{rr'}$ are the distances between the respective weight vectors $\mathbf{w_r}$ and $\mathbf{w_{r'}}$ where $\mathbf{r}$ and $\mathbf{r'}$ are neighbored in $\mathcal{A}$

$$\mathbf{U_{rr'}} = \|\mathbf{w_r} - \mathbf{w_{r'}}\| \qquad (6.17)$$

U can be used to determine clusters within the weight vector set and, hence, within the data space. Assuming that the map $\mathcal{M}$ is approximately topology preserving, large values of U indicate cluster boundaries. If the lattice is a two-dimensional array the U-matrix can easily be viewed and gives a powerful tool for cluster analysis.

Another visualization technique can be used if the lattice $\mathcal{A}$ is three-dimensional. The data points then can be mapped onto neuron $\mathbf{r}$ can be identified by the color combination *red, green* and *blue* assigned to the location $\mathbf{r}$. In this way we are able to assign a color to each data point according to equation (6.1) and similar colors will encode groups of input patterns that were mapped close to one another in the lattice $\mathcal{A}$ [42]. It should be emphasized that for a proper interpretation of this color visualization, as well as for the analysis of the U-matrix, topology preservation of the map $\mathcal{M}$ is a strict requirement. The topology preserving property of $\mathcal{M}$ must be proven prior to any evaluation of the map.

If we regard the SOM as a preprocessing method, the data can be analyzed in the lower-dimensional neuron space defined by the GSOM-generated lattice. Beside the above mentioned U-matrix approach, a cluster algorithm such as Ward-clustering may be applied, taking the neighborhood relations into account explicitly [44]. An additional counter-propagation layer can also be added to the SOM, to learn a classification task in a supervised manner. This approach is faster than a MLP learning in many applications [19].

6.3   Remote Sensing Spectral Images

Spectral images can formally be described as a matrix $S = \mathbf{v}^{(x,y)}$, where $\mathbf{v}^{(x,y)} \in \Re^{D_\mathcal{V}}$ is the vector of spectral information associated with pixel location (x, y). The elements $v_i^{(x,y)}$, $i = 1 \ldots D_\mathcal{V}$ of spectrum $\mathbf{v}^{(x,y)}$ reflect the responses of a spectral sensor at a suite of wavelengths (see Fig. 6.1). The spectrum is a characteristic pattern that provides a clue to the surface material within the area defined by pixel (x, y). The individual 2-dimensional image $S_i = v_i^{(x,y)}$ at wavelength i is called the ith image band.

The data space $\mathcal{V}$ spanned by Visible-Near Infrared reflectance spectra is $[0 - noise, U + noise]^{D_\mathcal{V}} \subseteq \Re^{D_\mathcal{V}}$ where $U > 0$ represents an upper limit of the measured scaled reflectivity and *noise* is the maximum value of noise across all spectral channels and image pixels. The data density $\mathcal{P}(\mathcal{V})$ may vary strongly within this space. Sections of the data space can be very densely populated while other parts may be extremely sparse, depending on the materials in the scene and on the spectral bandpasses of the sensor. According to this model traditional multi-spectral imagery has a low $D_{\mathcal{V}}$ value while $D_{\mathcal{V}}$ can be several hundred for hyperspectral images. The latter case is of particular interest because the great spectral detail, complexity, and very large data volume pose new challenges in clustering, cluster visualization, and classification of images with such high spectral dimensionality [29].

In addition to dimensionality and volume, other factors, specific to remote sensing, can make the analyses of hyperspectral images even harder. For example, given the richness of data, the goal is to separate many cover classes, however, surface materials that are significantly different for an application may be distinguished by very subtle differences in their spectral patterns. The pixels can be mixed, which means that several different materials may contribute to the spectral signature associated with one pixel. Training data may be scarce for some classes, and classes may be represented very unevenly.

Noise is far less problematic than the intricacy of the spectral patterns, because of the high Signal-to-Noise Ratios $(500 - 1,500)$ that present-day hyperspectral imagers provide. For this discussion, we will omit noise issues, and additional effects such as atmospheric distortions, illumination geometry and albedo variations in the scene, because these can be addressed through well-established procedures prior to clustering or classification.

6.3.1   Low-Dimensional Data: LANDSAT TM Multi-spectral Images

LANDSAT-TM satellite-based sensors produce images of the Earth in 7 different spectral bands. The ground resolution in meters is 30×30 for bands $1 - 5$ and band 7. Band 6 (thermal band) has a spatial resolution of 60×60 only and it is often dropped from analyses. The LANDSAT TM bands were strategically determined for optimal detection and discrimination of vegetation, water, rock formations and cultural features within the limits of broad band multi-spectral imaging. The spectral information, associated with each pixel of a LANDSAT scene is represented by a vector $\mathbf{v} \in \mathcal{V} \subseteq \Re^{D_{\mathcal{V}}}$ with $D_{\mathcal{V}} = 6$. The aim of any classification algorithm is to subdivide this data space into subsets of data points, with each subset corresponding to specific features such as wood, industrial region, etc. The feature categories are specified by prototype data vectors (training spectra).

In the present contribution we consider two LANDSAT TM images. The first one is the north–east region of the city Leipzig in Germany[7]. The second one is from the Colorado area, U.S.A. [8] For the Colorado image we also have a manually generated label map (ground truth image) for comparison. The labels indicate several regions of different vegetation.

---

[7] obtained from UMWELT-FORSCHUNGSZENTRUM Halle-Leipzig, Germany

[8] Thanks to M. Augusteijn (Univerity of Colorado) for providing this image.

A Grassberger-Procaccia analysis [15] of the Leipzig image yields $D^{\mathcal{GP}} \approx 1.7$ as an estimation for the intrinsic spectral dimension. The costumary Principal Component Analysis results in the following vector of eigenvalues:

$$ev = (274.06, 76.19, 39.78, 11.92, 8.27, 6.28)^T \qquad (6.18)$$

Application of the same two procedures to the Colorado image yields $D^{\mathcal{GP}} \approx$ 3.1414 and

$$ev = (4.93, 0.68, 0.29, 0.05, 0.02, 0.02)^T, \qquad (6.19)$$

respectively.

6.3.2   Hyperspectral Data: The Lunar Crater Volcanic Field AVIRIS Image

A Visible-Near Infrared $(0.4 - 2.5 \ \mu\text{m})$, 224-band, 20 m/pixel AVIRIS image of the Lunar Crater Volcanic Field (LCVF), Nevada, U.S.A., was analyzed in order to study SOM performance for high-dimensional remote sensing spectral imagery. (AVIRIS is the Airborne Visible-Near Infrared Imaging Spectrometer, developed at NASA/Jet Propulsion Laboratory. See http://makalu.jpl.nasa.gov for details on this sensor and on imaging spectroscopy.) The LCVF is one of NASA's remote sensing test sites, where images are obtained regularly. A great amount of accumulated ground truth from comprehensive field studies [2] and research results from independent earlier work such as [13] provide a detailed basis for the evaluation of the results presented here.

Fig. 6.2 shows a natural color composite of the LCVF with labels marking the locations of 23 different surface cover types of interest. This $10 \times 12 \,\text{km}^2$ area contains, among other materials, volcanic cinder cones (class A, reddest peaks) and weathered derivatives thereof such as ferric oxide rich soils (L, M, W), basalt flows of various ages (F, G, I), a dry lake divided into two halves of sandy (D) and clayey composition (E); a small rhyolitic outcrop (B); and some vegetation at the lower left corner (J), and along washes (C). Alluvial material (H), dry (N,O,P,U) and wet (Q,R,S,T) playa outwash with sediments of various clay content as well as other sediments (V) in depressions of the mountain slopes, and basalt cobble stones strewn around the playa (K) form a challenging series of spectral signatures for pattern recognition (see in [29]). A long, NW-SE trending scarp, straddled by the label G, borders the vegetated area. Since this color composite only contains information from three selected image bands (one Red, one Green, and one Blue), many of the cover type variations remain undistinguished. They will become evident in the cluster and class maps below.

After atmospheric correction and removal of excessively noisy bands (saturated water bands and overlapping detector channels), 194 image bands remained from the original 224. These 194-dimensional spectra are the input patterns in the following analyses.

The spectral dimensionality of hyperspectral images is not well understood and it is an area of active research. While many believe that hyperspectral images are

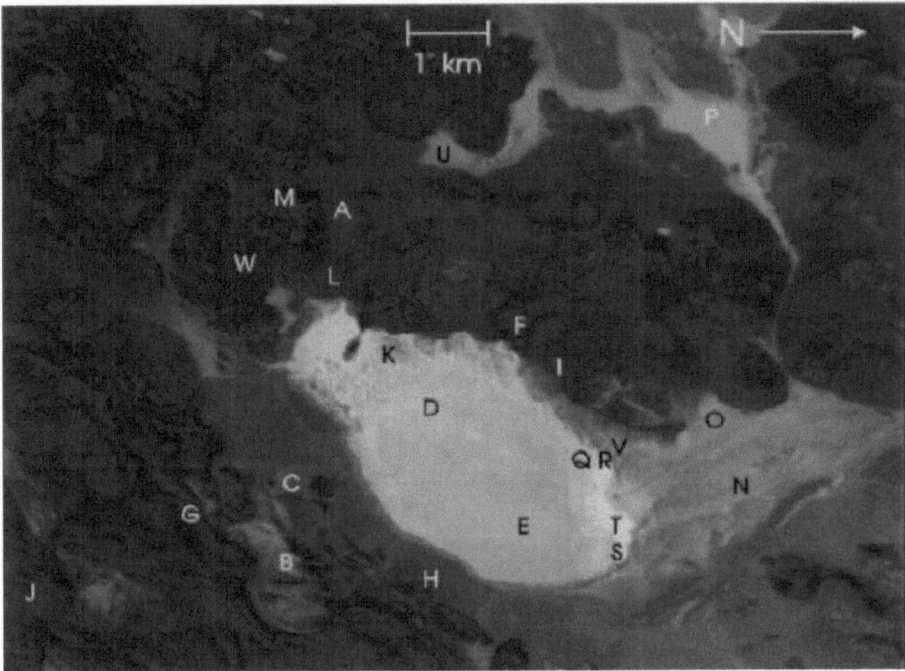

Fig. 6.2. The Lunar Crater Volcanic Field. RGB natural color composite from an AVIRIS, 1994 image. The original image comprises 224 image bands over the 0.4 - 2.5 μm range, 512 x 614 pixels, altogether 140 Mbytes of data. Labels indicate different cover types described in the text. The ground resolution is 20 m/pixel.

highly redundant because of band correlations, others maintain an opposite view, which also manifests in the vigorous development of hyperspectral sensors and commercialization of hyperspectral data services. Few investigations exist into the intrinsic dimensionality (ID) of hyperspectral images. Linear methods such as PCA or determination of mixture model endmembers [1] [22] usually yield 3 − 8 "endmembers". BRUSKE [8] finds the spectral ID of the LCVF AVIRIS image (Fig. 6.2) to be between 3 and 7, using a non-linear neural network based approach (Optimally Topology Preserving Maps), whereas the Grassberger-Procaccia analysis [15] estimates the intrinsic dimension as $D^{\mathcal{GP}} \approx 3.06$. These surprisingly low numbers, that increase with improved sensor performance [17], result from using statistical thresholds for the determination of what is "relevant", regardless of application dependent criteria.

   The number of relevant components increases dramatically when specific goals are considered such as what cover classes should be separated or what known properties of the surface can be postulated. With an associative neural network, Pendock [35] extracted 20 linear mixing endmembers from a 50-band (2.0 − 2.5 μm) segment of an AVIRIS image of Cuprite, Nevada (another well-known remote sensing test site), setting only a rather general surface texture criterium. Benediktsson et al. [6] performed feature extraction on an AVIRIS geologic scene of Iceland, which resulted in 35 bands. They used an ANN (the same network that performed the classification itself) for Decision Boundary Feature Extraction (DBFE). The DBFE is claimed to preserve all features that are necessary to achieve the same accuracy as in the original data space, by the same classifier for predetermined classes. However, no comparison of classification accuracy was made using the full spectral dimension to support the DBFE claim. In this particular study a relatively low number of classes, 9, were of interest, and the question posed was to find the number of features to describe those classes. Separation of a higher number of classes may require more features.

It is not clear how feature extraction should be done in order to preserve relevant information in hyperspectral images. Later in this chapter it is demonstrated that selection of 30 bands from the LCVF image in Fig. 6.2 by any of several methods leads to a loss of a number of the originally determined 23 cover classes. Wavelet compression studies on an earlier image of the the same AVIRIS scene [34] conclude that various schemes and compression rates affect different spectral classes differently, and none was found overall better than another, within $25\% - 50\%$ compressions (retaining $75\% - 50\%$ of the wavelet coefficients). In a study on simulated, 201-band spectral data, [7] show slight accuracy increase across classifications on 20-band, 40-band, and 60-band subsets. However, they base the study on only two vegetation classes, the feature extraction is a progressive hierarchical subsampling of the spectral bands, and there is no comparison with using the full, 201-band case. Comparative studies using full spectral resolution and many classes are lacking, in general, because few methods can cope with such high-dimensional data technically, and the ones that are capable (such as Minimum Distance, Parallel Piped) often perform too poorly to merit consideration.

Undesirable loss of relevant information can result using any of these feature extraction approaches. In any case, finding an optimal feature extraction requires great preprocessing efforts just to taylor the data to available tools. An alternative is to develop capabilities to handle the full spectral information. Analysis of unreduced data is important for the establishment of benchmarks, exploration and novelty detection (such as in the case of hard-earned data in planetary exploration); as well as to allow for the distinction of significantly greater number of cover types, according to the potential provided by modern imaging spectrometers.

6.4   SOM-Applications

6.4.1   Analysis of LANDSAT TM Images

One way to get good results for visualization of the clusters of LANDSAT TM data is to use a SOM dimension $D_A = 3$ [18] and interpret the positions of the neurons $\mathbf{r}$ in the lattice A as vectors $\mathbf{r} = \mathbf{c} = (r, g, b)$ in the color space C, where r, g, b are the intensities of the colors red, green and blue, respectively [18]. Such assignment of colors to winner neurons immediately yields a pseudo-color cluster

Fig. 6.3. LANDSAT-TM six-band spectral images. Clusters of the Leipzig image using the $7 \times 6 \times 6$ standard SOM (top) and the $14 \times 6 \times 3$ GSOM-solution (bottom). (Color version available on request from villmann@informatik.uni-leipzig.de)

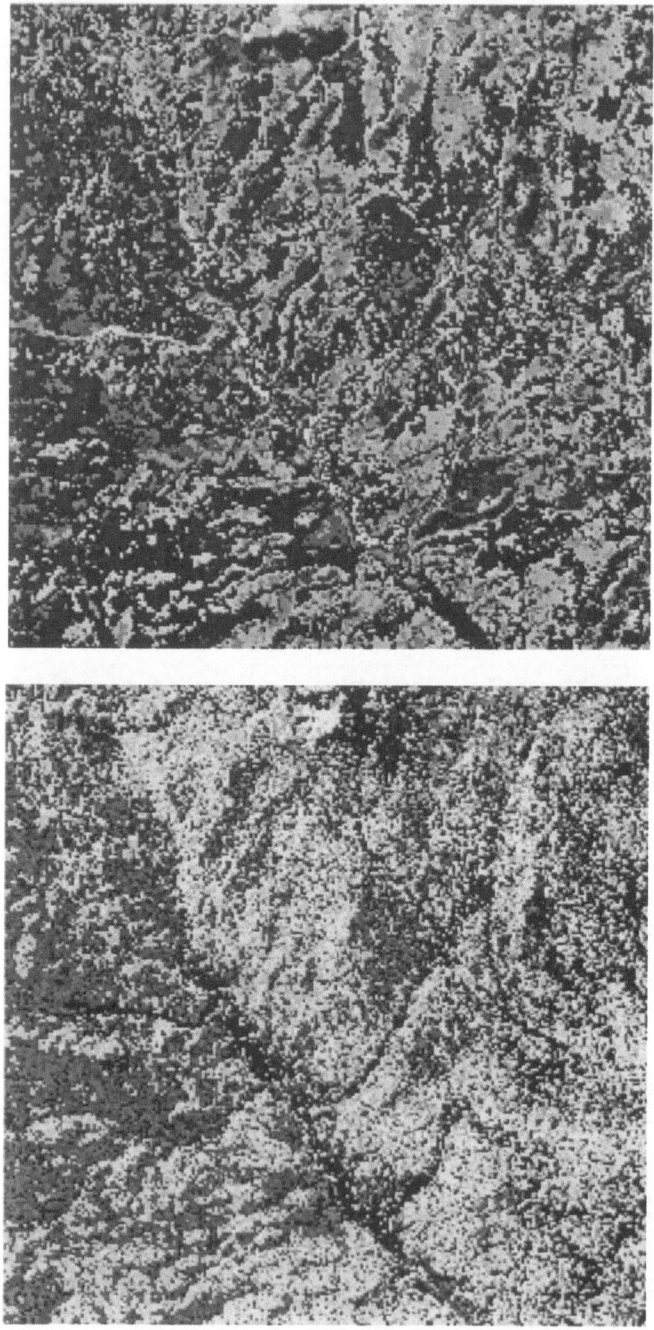

Fig. 6.4. Cluster maps of the Colorado image. The RGB color composite using bands 2, 3, and 4 (top), and the $12 \times 7 \times 3$ GSOM-solution derived from all six bands. (Color version available on request from villmann@informatik.uni-leipzig.de)

map of the original image for visual interpretation. Since we are mapping the data clouds from a 6-dimensional input space onto a three-dimensional color space dimensional conflicts may arise and the visual interpretation may fail. However, in the case of topologically faithful mapping this color representation, prepared using all six LANDSAT TM image bands, contains considerably more information than a costumary color composite combining three TM bands (frequently bands 2, 3, and 3). [18].

In the first LANDSAT example we investigate an image of the north–east region of Leipzig (described in sec. 6.3.1), using the GSOM approach. For comparison we also trained several regular SOMs with fixed output spaces the dimesion of which range from 1 to 4, and determined the respective $\tilde{P}$–values, shown in Table 6.1. The

$N_{\max}$	lattice structure	$\tilde{P}$
256	256	-0.189 ± 0.00612
256	16×16	-0.0642 ± 0.00031
252	$7 \times 6 \times 6$	$+0.0282 \pm 0.00024$
256	$4 \times 4 \times 4 \times 4$	$+0.0816 \pm 0.00387$

Table 6.1. Values of the topographic product $\tilde{P}$ using different, fixed output spaces for the LANDSAT satellite image of Leipzig. For each structure, 3 or more runs were averaged.

topographic product $\tilde{P}$ favors an output space dimension D_A between 2 and 3. However a clear decision cannot be made between 2 and 3. Yet, the edge length ratios of the lattice provide a further free choice. We should emphasize that in these runs, as well as in the further GSOM simulations, we applied the new learning rule (6.15) to achieve maximal mutual information as pointed out in sec. 6.2.2 . The GSOM algorithm was applied in several runs with different values of $N_{\max}$ (maximum number of neurons allowed). The results obtained are depicted in Table 6.2. The favored

$N_{\max}$	lattice structure	$\tilde{P}$
128	$12 \times 5 \times 2$	0.0047
256	$14 \times 6 \times 3$	0.0050
512	$15 \times 6 \times 4$	0.0051

Table 6.2. GSOM results for the Leipzig image.

quasi two-dimensional structure is supported by the Grassberger-Procaccia analysis described above. Fig. 6.3 shows the visualization of the best GSOM solution with respect to the $\tilde{P}$–value. The $\tilde{P}$–values obtained by the GSOM are better than the respective values for the fixed lattice structures. Furthermore, for all values of N_{max}, the GSOM yields approximately the same structure, and the length ratios of the edges are in a good agreement with the PCA eigenvalues (see (6.18)). However, in general, the standard linear PCA fails, as shown by the second LANDSAT image from the Colorado area. The PCA for this image suggests a one-dimensional structure (see (6.19)). The GSOM generates a $12 \times 7 \times 3$ lattice ($N_{\mathrm{max}} = 256$) in agreement with the Grassberger-Procaccia analysis ($D^{\mathcal{GP}} \approx 3.1414$), which corresponds to a $\tilde{P}$-value of 0.0095 indicating good topology preservation (see Fig. 6.4).

6.4.2   SOM Analyses of Hyperspectral Imagery

A systematic supervised classification study was conducted on the LCVF image (Fig. 6.2), to simultaneously assess loss of information due to reduction of spectral dimensionality, and to compare performances of several traditional and an SOM-based hybrid ANN classifier. The 23 geologically relevant classes indicated in Fig. 6.2 represent a great variety of surface covers in terms of spatial extent, the similarity of spectral signatures [29], and the number of available training samples. The full study, complete with evaluations of classification accuracies, is described in [31]. Average spectral shapes of these 23 classes are also shown in [29].

Fig. 6.5, top panel, shows the best classification, produced by an SOM-hybrid ANN using all 194 spectral bands that remained after preprocessing. This ANN first learns in an unsupervised mode, during which the input data are clustered in the hidden SOM layer. After the SOM converges, the output layer is allowed to learn class labels. The preformed clusters in the SOM greatly aid in accurate and sensitive classification, by helping prevent the learning of inconsistent class labels. As mentioned in sect. 6.2.3 such hybrid SOM constructions can also help faster learning. Detailed description of this classifier is given in several previous scientific studies, which produced improved interpretation of high-dimensional spectral data compared to earlier analyses [20] [32] [33]. Training samples for the supervised classifications were selected based on knowledge of and interest in geologic units. The SOM hidden layer was not evaluated and used for identification of spectral types (SOM clusters), prior to training sample determination. Hence, Fig. 6.5 reflects the geologist's view of the desirable segmentation.

In order to apply Maximum Likelihood and other covariance based classifiers, the number of spectral channels needed to be reduced to 30, since the maximum number of training spectra that could be identified for *all* classes was 31. Dimensionality reduction was performed in several ways, including PCA, equidistant subsampling, and band selection by a domain expert. Band selection by domain expert proved most favorable. Fig. 6.5, bottom panel, shows the Maximum Likelihood classification on the LCVF data, reduced to 30 bands. A number of classes (notably the ones with subtle spectral differences, such as N, Q, R, S, T, V, W) were entirely lost. Class K (basalt cobbles) disappeared from most of the edge of the playa, and only traces of B (rhyolitic outcrop) remained. Class G and F were greatly overestimated.

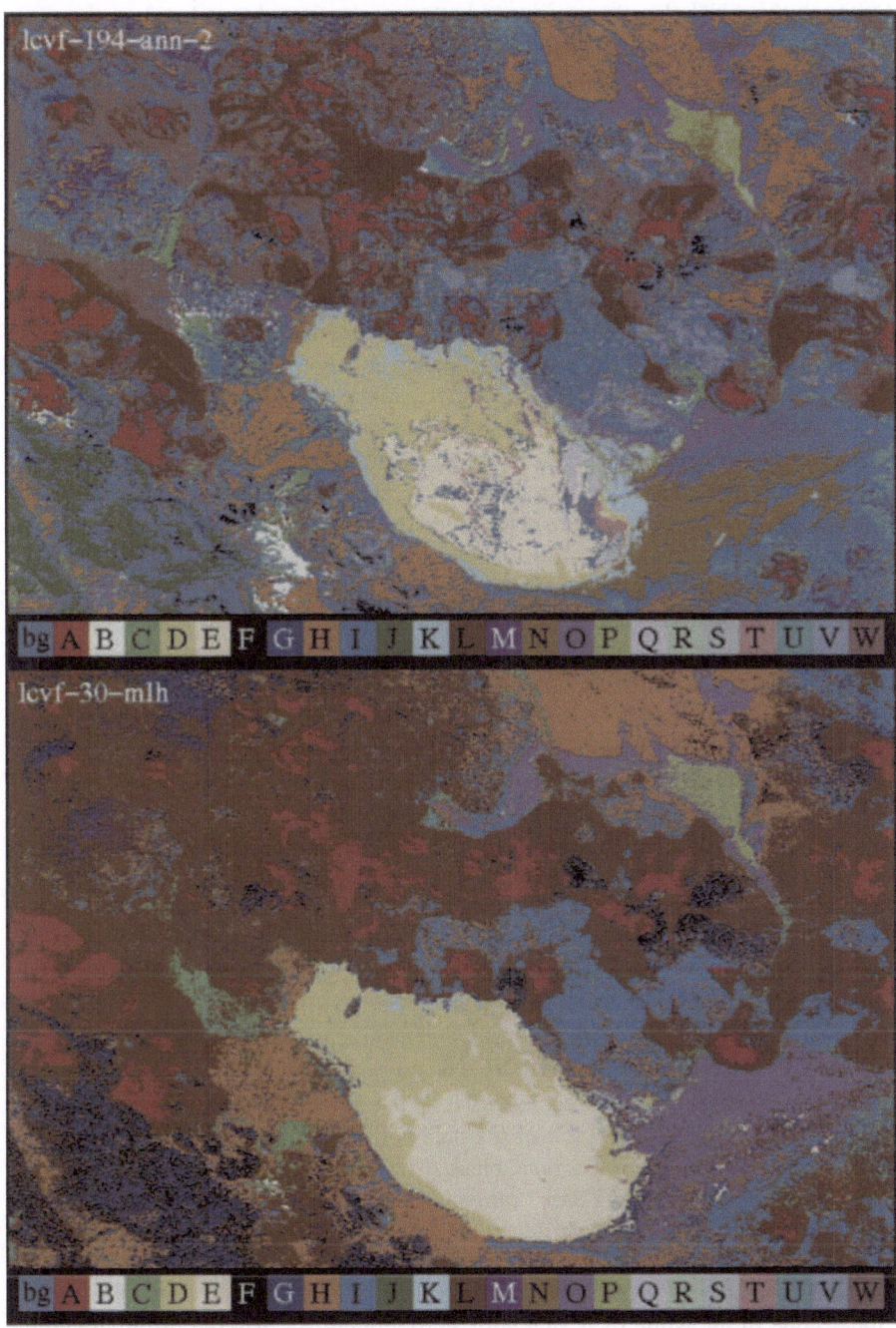

Fig. 6.5. Top: SOM-hybrid supervised ANN classification of the LCVF scene, using 194 image bands. Bottom: Maximum Likelihood classification of the LCVF scene. 30, strategically selected bands were used due to the limited number of training samples for a number of classes. Considerable loss of class distinction occurred compared to the ANN classification. 'bg' stands for background (unclassified pixels).

Although the ANN classifier produced better results (not shown here) on the same 30-band reduced data set than the Maximum Likelihood, a marked drop in accuracy occurred compared to classification on the full data set. This emphasizes that accurate mapping of "interesting", spatially small geologic units is possible from full hyperspectral information and with appropriate tools.

Discovery in Hyperspectral Images with SOMs The previous section demonstrated the power of the SOM in helping discriminate among a large number of predetermined surface cover classes with subtle differences in the spectral patterns, using the full spectral resolution. It is even more interesting to examine the SOM's preformance in terms of detection of clusters in high-dimensional data. Fig. 6.6 displays a 40×40 SOM extended by the conscience algorithm of DeSieno according to equation (6.14). The input data space was the entire 194-band LCVF image. Groups of neurons, altogether 32, that were found to be sensitized to groups of similar spectra in the 194-dimensional input data, are indicated by various colors. The boundaries of these clusters were determined by a somewhat modified version of the U-matrix described in (6.17). Areas where no data points (spectra) were mapped are the grey corners with uniformly high fences, and are relatively small. The black background in the SOM lattice shows areas that have not been evaluated for cluster detection. The spatial locations of the image pixels mapped onto the groups of neurons in Fig. 6.6, are shown in the same colors in Fig. 6.7. Color coding for clusters that correspond to classes or subclasses of those in Fig. 6.5, top, is the same as in Fig. 6.5, to show similarities. Colors for additional groups were added.

The first observation is the striking correspondence between the supervised ANN class map in Fig. 6.5, top panel, and this clustering: the SOM detected all classes that were known to us as meaningful geological units. The "discovery" of classes B (rhyolitic outcrop, white), F (young basalt flows, dark grey and black, some shown in the black ovals), G (a different basalt, exposed along the scarp, dark blue, one segment outlined in the white rectangle), K (basalt cobbles, light blue, one segment shown in the black rectangle), and other spatially small classes such as the series of playa deposits (N, O, P, Q, R, S, T) is significant. This is the capability we need for sifting through high-volume, high-information-content data to alert for interesting, novel, or hard-to-find units. The second observation is that the SOM detected more, spatially coherent, clusters than the number of classes that we trained for in Fig. 6.5. The SOM's view of the data is more refined and more precise than that of the geologist's. For example, class A (red in Fig. 6.5) is split here into a red (peak of cinder cones) and a dark orange (flanks of cinder cones) cluster, that make geologic sense. The maroon cluster to the right of the red and dark orange clusters at the bottom of the SOM fills in some areas that remained unclassified (bg) in the ANN class map, in Fig. 6.5. An example is the arcuate feature at the base of the cinder cone in the white oval, that apparently contains a material different enough to merit a separate spectral cluster. This material fills other areas too, consistently at the foot of cinder cones (another example is seen in the large black oval). Evaluation of further refinements are left to the reader. Evidence that the SOM mapping in Fig.

6. approximates an equiprobabilistic mapping (that the magnification factor for the SOM in Fig. 6. is close to 1), using DeSieno's algorithm, is presented in [28].

As mentioned above, earlier investigations showed the intrinsic spectral dimensionality of the LCVF data set in the range of $3 - 7$ [8]. The Grassberger-Procaccia analysis [15] yields $D_{\mathcal{A}}^{\mathcal{GP}} \approx 3.06$ corroborating the above results, i.e. the data are highly correlated, therefore a drastic dimensionality reduction may be possible. However, a faithful mapping is necessary to preserve the information contained in the hyperspectral image. For this purpose, the magnification control and the growing SOM (GSOM) procedure, as extensions of the standard SOM, are suitable tools. The GSOM produced a lattice of dimensions $8 \times 6 \times 6$, a radical dimension reduction. This is in agreement with the Grassberger-Procaccia analysis above. The resulting false color visualization of the spectral clusters is depicted in Fig. 6.8. It shows approximately the same quality as the $2d$-SOM vector quantized but manually labeled image (supervised classification) in Fig. 6.5, top panel.

6.5   Conclusion

Self-Organizing Maps have been showing great promise for the analyses of remote sensing spectral images. With recent advances in remote sensor technology, very high-dimensional spectral data emerged and demand new and advanced approaches to cluster detection, visualization, and supervised classification. While standard SOMs produce good results, the high dimensionality and large amount of hyperspectral data call for very careful evaluation and control of the faithfulness of topological mapping performed by SOMs. Faithful topological mapping is required in order to avoid false interpretations of cluster maps created by an SOM. This chapter summarized several advances that were made in the past few years, and that ensure strict topology preservation through mathematical considerations. Two of these extensions to the standard Kohonen SOM, the Growing Self-Organizing Map and magnification control, were discussed in detail, along with their relationship to other SOM extensions such as the DeSieno conscience mechanism, or to independent analyses such as the Grassberg-Procaccia analysis for the determination of intrinsic dimensionality. Case studies on real multi- and hyperspectral images were presented that support our theoretical discussions. While it is outside the scope of this contribution, as a final note we want to point out that full scale investigations such the LCVF study in this chapter also have to make heavy use of advanced image processing tools and user interfaces, to handle great volumes of data efficiently, and for effective graphics/visualization. References to such tools are made in the cited literature on data analyses.

Acknowledgements

E.M. has been supported by the Applied Information Systems Research Program of NASA, Office of Space Science, NAG54001 and NAG59045. Contributions by Dr. William H. Farrand, providing field knowledge and data for the evaluation of the LCVF results are gratefully acknowledged.

Bibliography on Chapter 6

1. J. B. Adams, M. O. Smith, and A. R. Gillespie. Imaging spectroscopy: Interpretation based on spectral mixture analysis. In C. Peters and P. Englert, editors, *Remote Geochemical Analysis: Elemental and Mineralogical Composition*, pages 145–166. Cambridge University Press, New York, 1993.
2. R. E. Arvidson and M. D.-B. and. et al. Archiving and distribution of geologic remote sensing field experiment data. *EOS, Transactions of the American Geophysical Union*, 72(17):176, 1991.
3. H.-U. Bauer, R. Der, and M. Herrmann. Controlling the magnification factor of self-organizing feature maps. *Neural Computation*, 8(4):757–771, 1996.
4. H.-U. Bauer and K. R. Pawelzik. Quantifying the neighborhood preservation of Self-Organizing Feature Maps. *IEEE Trans. on Neural Networks*, 3(4):570–579, 1992.
5. H.-U. Bauer and T. Villmann. Growing a Hypercubical Output Space in a Self-Organizing Feature Map. *IEEE Transactions on Neural Networks*, 8(2):218–226, 1997.
6. J. A. Benediktsson, J. R. Sveinsson, and et al. Classification of very-high-dimensional data with geological applications. In *Proc. MAC Europe 91*, pages 13–18, Lenggries, Germany, 1994.
7. J. A. Benediktsson, P. H. Swain, and et al. Classification of very high dimensional data using neural networks. In *IGARSS'90 10th Annual International Geoscience and Remote Sensing Symp.*, volume 2, page 1269, 1990.
8. J. Bruske and E. Merényi. Estimating the intrinsic dimensionality of hyperspectral images. In *Proc. Of European Symposium on Artificial Neural Networks (ESANN'99)*, pages 105–110, Brussels, Belgium, 1999. D facto publications.
9. J. Campbell. *Introduction to Remote Sensing*. The Guilford Press, U.S.A., 1996.
10. R. N. Clark. Spectroscopy of rocks and minerals, and principles of spectroscopy. In A. Rencz, editor, *Manual of Remote Sensing*. John Wiley and Sons, Inc, New York, 1999.
11. R. Der, G. Balzuweit, and M. Herrmann. Constructing principal manifolds in sparse data sets by self-organizing maps with self-regulating neighborhood width. In *ICNN 96. The 1996 IEEE International Conference on Neural Networks (Cat. No. 96CH35907)*, volume 1, pages 480–483. IEEE, New York, NY, USA, 1996.
12. D. DeSieno. Adding a conscience to competitive learning. In *Proc. ICNN'88, Int. Conf. on Neural Networks*, pages 117–124, Piscataway, NJ, 1988. IEEE Service Center.
13. W. H. Farrand. *VIS/NIR Reflectance Spectroscopy of Tuff Rings and Tuff Cones*. PhD thesis, University of Arizona, 1991.
14. J. Goppert and W. Rosenstiel. The continuous interpolating self-organizing map. *Neural Processing Letters*, 5(3):185–92, 1997.
15. P. Grassberger and I. Procaccia. Maesuring the strangeness of strange attractors. *Physica*, 9D:189–208, 1983.

16. R. O. Green. Summaries of the 6th Annual JPL Airborne Geoscience Workshop. 1. AVIRIS Workshop, Pasadena, CA, March 4–6 1996.

17. R. O. Green and J. Boardman. Exploration of the relationship between information content and signal-to-noise ratio and spatial resolution. In *Proc. 9th AVIRIS Earth Science and Applications Workshop*, Pasadena, CA, February, 23–25 2000.

18. M. H. Gross and F. Seibert. Visualization of multidimensional image data sets using a neural network. *Visual Computer*, 10:145–159, 1993.

19. J. A. Hertz, A. Krogh, and R. G. Palmer. *Introduction to the Theory of Neural Computation*, volume 1 of *Santa Fe Institute Studies in the Sciences of Complexity: Lecture Notes*. Addison-Wesley, Redwood City, CA, 1991.

20. E. S. Howell, E. Merényi, and L. A. Lebofsky. Classification of asteroid spectra using a neural network. *Jour. Geophys. Res.*, 99(10):847–865, 1994.

21. M. M. V. Hulle. *Faithful Representations and Topographic Maps*. Wiley Series and Adaptive Learning Systems for Signal Processing, Communications, and Control. Wiley Sons, New York, 2000.

22. R. S. Inc. *ENVI v.3 User's Guide*, 1997.

23. S. Kaski, J. Nikkilä, and T. Kohonen. Methods for interpreting a self-organized map in data analysis. In *Proc. Of European Symposium on Artificial Neural Networks (ESANN'98)*, pages 185–190, Brussels, Belgium, 1998. D facto publications.

24. T. Kohonen. *Self-Organizing Maps*. Springer, Berlin, Heidelberg, 1995. (Second Extended Edition 1997).

25. T. Kohonen. Comparison of SOM point densities based on different criteria. *Neural Computation*, 11(8):212–234, 1999.

26. R. Linsker. How to generate maps by maximizing the mutual information between input and output signals. *Neural Computation*, 1:402–411, 1989.

27. T. Martinetz and K. Schulten. Topology representing networks. *Neural Networks*, 7(3):507–522, 1994.

28. E. Merényi. "Precision mining" of high-dimensional patterns with self-organizing maps: Interpretation of hyperspectral images. In P. Sinčak and J. Vasčak, editors, *Quo Vadis Computational Intelligence? New Trends and Approaches in Computational Intelligence (Studies in Fuzziness and Soft Computing, Vol. 54.* Physica-Verlag, 2000.

29. E. Merényi. Self-organizing ANNs for planetary surface composition research. In *Proc. Of European Symposium on Artificial Neural Networks (ESANN'98)*, pages 197–202, Brussels, Belgium, 1998. D facto publications.

30. E. Merényi. The challenges in spectral image analysis: An introduction and review of ANN approaches. In *Proc. Of European Symposium on Artificial Neural Networks (ESANN'99)*, pages 93–98, Brussels, Belgium, 1999. D facto publications.

31. E. Merényi, W. H. Farrand, and et al. Efficient geologic mapping from hyperspectral images with artificial neural networks classification: a comparison to conventional tools. *IEEE TGARS*, in preparation, 2001.

32. E. Merényi, E. S. Howell, and et al. Prediction of water in asteroids from spectral data shortward of 3 microns. *ICARUS*, 129(10):421–439, 1997.

33. E. Merényi, R. B. Singer, and J. S. Miller. Mapping of spectral variations on the surface of mars from high spectral resolution telescopic images. *ICARUS*, 124(10):280–295, 1996.

34. T. Moon and E. Merényi. Classification of hyperspectral images using wavelet transforms and neural networks. In *Proc. of the Annual SPIE Conference*, page 2569, San diego, CA, 1995.

142   Thomas Villmann and Erzsébet Merényi

35. N. Pendock. A simple associative neural network for producing spatially homogeneous spectral abundance interpretations of hyperspectral imagery. In *Proc. Of European Symposium on Artificial Neural Networks (ESANN'99)*, pages 99–104, Brussels, Belgium, 1999. D facto publications.
36. H. Ritter. Parametrized self-organizing maps. In S. Gielen and B. Kappen, editors, *Proc. ICANN'93 Int. Conf. on Artificial Neural Networks*, pages 568–575, London, UK, 1993. Springer.
37. H. Ritter and K. Schulten. On the stationary state of Kohonen's self-organizing sensory mapping. *Biol. Cyb.*, 54:99–106, 1986.
38. A. Ultsch. Self organized feature maps for monitoring and knowledge aquisition of a chemical process. In S. Gielen and B. Kappen, editors, *Proc. ICANN'93, Int. Conf. on Artificial Neural Networks*, pages 864–867, London, UK, 1993. Springer.
39. J. Vesanto. SOM-based data visualization methods. *Intelligent Data Analysis*, 3(7):123–456, 1999.
40. T. Villmann. Benefits and limits of the self-organizing map and its variants in the area of satellite remote sensoring processing. In *Proc. Of European Symposium on Artificial Neural Networks (ESANN'99)*, pages 111–116, Brussels, Belgium, 1999. D facto publications.
41. T. Villmann. Topology preservation in self-organizing maps. In E. Oja and S. Kaski, editors, *Kohonen Maps*, number ISBN 951-22-3589-7, pages 279–292, Amsterdam (Holland), June 1999. Helsinki, Elsevier.
42. T. Villmann. Neural networks approaches in medicine – a review of actual developments. In *Proc. Of European Symposium on Artificial Neural Networks (ESANN'2000)*, pages 165–176, Brussels, Belgium, 2000. D facto publications.
43. T. Villmann, R. Der, M. Herrmann, and T. Martinetz. Topology Preservation in Self-Organizing Feature Maps: Exact Definition and Measurement. *IEEE Transactions on Neural Networks*, 8(2):256–266, 1997.
44. T. Villmann, W. Hermann, and M. Geyer. Data mining and knowledge discovery in medical applications using self-organizing maps. In R. Brause and E. Hanisch, editors, *Medical Data Analysis*, pages 138–151, Berlin, New York, Heidelberg, 2000. Lecture Notes in Computer Science 1933, Springer–Verlag.
45. T. Villmann and M. Herrmann. Magnification control in neural maps. In *Proc. Of European Symposium on Artificial Neural Networks (ESANN'98)*, pages 191–196, Brussels, Belgium, 1998. D facto publications.

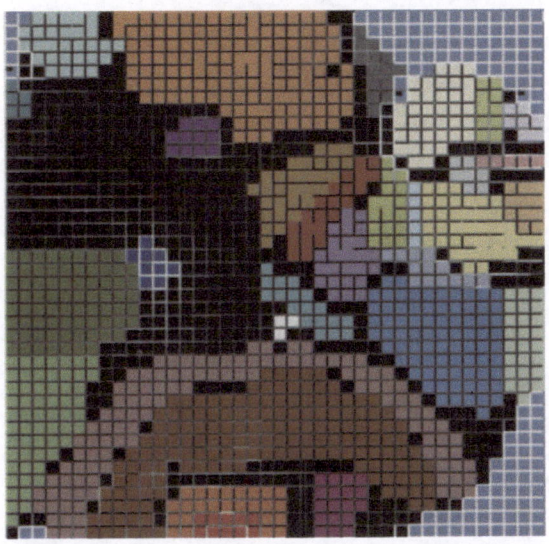

Fig. 6.6. Clusters identified in a 40 × 40 SOM. The SOM was trained on the entire 194-band LCVF image, using the DeSieno [12] algorithm.

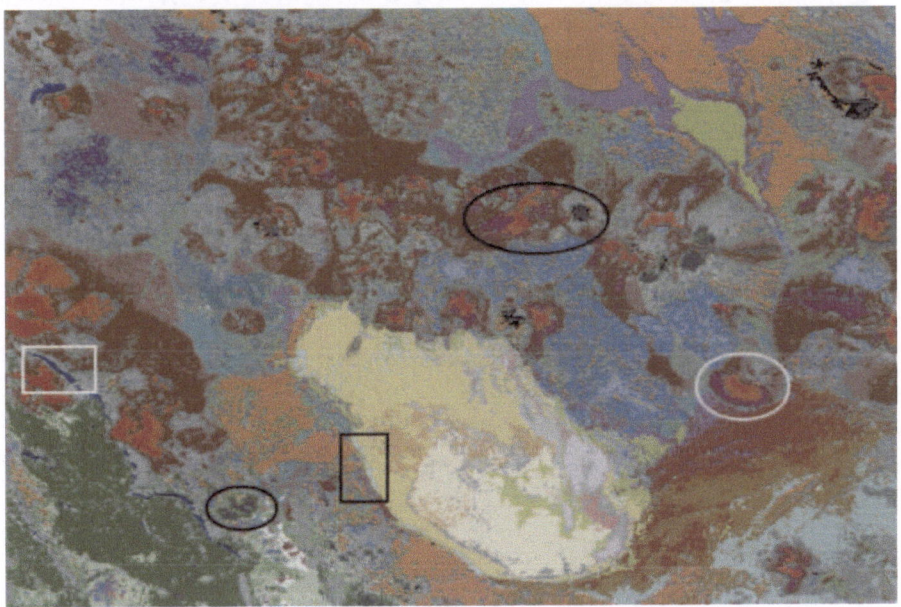

Fig. 6.7. The clusters from Fig.6.6 remapped to the original spatial image, to show where the different spectral types originated from. The relatively large, light grey areas correspond to the black, unevaluated parts of the SOM in Fig.6.6. Ovals and rectangles highlight examples discussed in the text.

Fig. 6.8. GSOM generated false color cluster map of the same 194 band hyperspectral image of the Lunar Crater Volcanic Field, Nevada, USA, as in Fig.6.5. It shows similar groups as seen in the supervised classification map in Fig. 6.5, top panel.

7 Modeling Speech Processing and Recognition in the Auditory System Using the Multilevel Hypermap Architecture

Bernd Brückner and Thomas Wesarg

Abstract. The Multilevel Hypermap Architecture (MHA) is an extension of the Hypermap introduced by Kohonen. By means of the MHA it is possible to analyze structured or hierarchical data (data with priorities, data with context, time series, data with varying exactness), which is difficult or impossible to do with known self-organizing maps so far.

In the first section of this chapter the theoretical work of the previous years about the MHA and its learning algorithm are summarized. After discussion of a simple example, which demonstrates the behavior of the MHA, results from MHA applications for classification of moving objects and analysis of images from functional Magnetic Resonance Imaging (fMRI) are given.

In the second section one application using the MHA within a system for speech processing and recognition will be explained in detail. Our approach to the implementation of this system is the simulation of the human auditory system operations in hearing and speech recognition using a multistage auditory system model. The goal of this system is to combine two different abstraction levels, a more biological level for peripheral auditory processing and the abstract behavior of an artificial neural network. The multistage model consists of the coupled models of neural signal processing at three different levels of the auditory system.

A model of peripheral auditory signal processing by the cochlea forms the input stage of the overall model. This model is capable of generating spatio-temporal firing rate patterns of the auditory nerve for simple acoustic as well as speech stimuli.

An uniform lateral inhibitory neural network (LIN) system performs an estimation of the spectrum of the speech stimuli by spatial processing of the cochlear model's neural response patterns.

Finally, the Multilevel Hypermap Architecture is used for learning and recognition of the spectral representations of the speech stimuli provided by the LIN system.

7.1 The Multilevel Hypermap Architecture

7.1.1 Overview

One type of Learning Vector Quantization (LVQ) is the Hypermap principle introduced by Kohonen [1]. This principle can be applied to both LVQ and SOM algorithms. In the Hypermap the input pattern is recognized in several separate phases: the recognition of the context around the pattern to select a subset of nodes is followed by a restricted recognition in this subset. This architecture speeds up searching in very large maps and may carry out stabilizing effects, especially if different inputs have very different dynamic ranges and time constants [2].

The modification and extension of the Hypermap, the Multilevel Hypermap Architecture (MHA), are described in [3–5]. The learning mechanism of the context-dependent data is also illustrated. The system model is shown in Fig. 7.1.

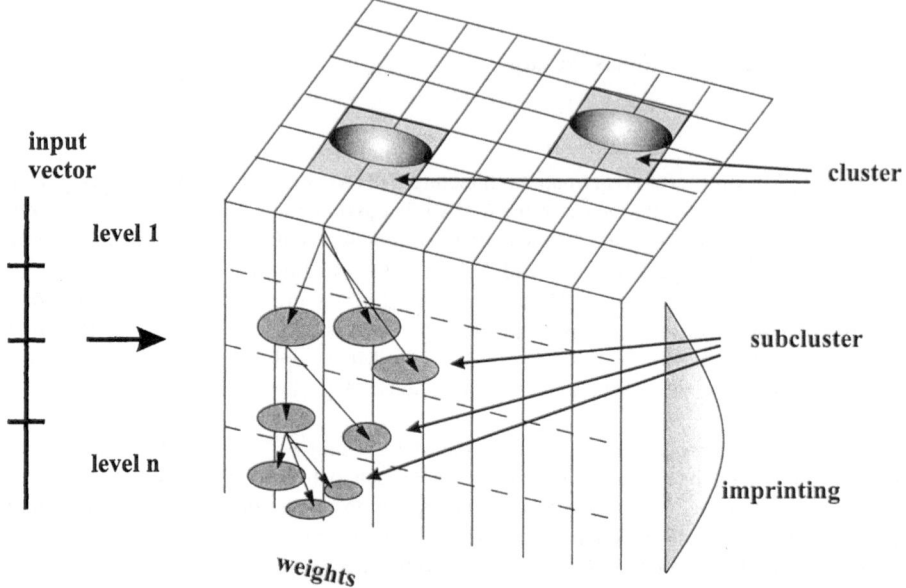

Fig. 7.1. Multilevel Hypermap Architecture

Instead of two levels proposed in the Hypermap [1], the data and the context level, the MHA supports several levels of data relationship and therefore the input vector consists also of an arbitrary number of levels. In the MHA there is the same number of levels in the weight vector of each unit and these levels are related to the corresponding levels of the input vector. A varying number of levels for the units of the map is supported.

The MHA is trained with the different levels of the input vector whose representation is a hierarchy of encapsulated subsets of units, the so called clusters and subclusters, which define different generalized stages of classification.

The learning algorithm as described in the next section has the following steps:

- Find a first level with a subset S_j of nodes with an error under a given threshold of that level. The sizes of the thresholds should be decreased according to the order of the levels to obtain encapsulated subsets S_j ("imprinting").
- Find the best match $\mathbf{m}_c$ between all nodes in the subset and adapt the weights accordingly.
- Test for a priori knowledge, i.e. if there is a significant higher error in one level in comparison with the other levels of this input vector, then an adaptive learning algorithm saves these trained states [4].

Classification is achieved by finding the best matching node for each level of the hierarchy and by determining the square mean error of matching.

7.1.2   The Learning Algorithm

Let the input vector of one level l_j be $\mathbf{x}_{l_j}$ and one processing unit $\mathbf{m}_{i,l_j}$ then, in a first phase, one has to find a first level with a subset S_j of nodes for which

$$\|\mathbf{x}_{l_j} - \mathbf{m}_{i,l_j}\| \leq \delta_j, \tag{7.1}$$

with δ_j being the threshold of that level. Then it is necessary to find the best match $\mathbf{m}_c$ for all nodes in the subset and to adapt the weights accordingly.

The learning algorithm of the Multilevel Hypermap Architecture depends also on the a priori knowledge. The a priori knowledge is defined by the mean error

$$\sigma = \frac{1}{n}\sum_{j=1}^{n}\|\mathbf{x}_{l_j} - \mathbf{m}_{c,l_j}\| \tag{7.2}$$

If the difference $\|\mathbf{x}_{l_j} - \mathbf{m}_{c,l_j}\| - \sigma$ is greater than a given ϵ for one or more levels l_j, then there exists a node with a priori knowledge and the affiliated levels of the input vector represent no regular data for training (i.e. artifacts). In the learning phase of the MHA this will be very helpful to stabilize a training set of data like in the case of the speech processing application described later in this chapter.

In the normal case the adaptation of the weights is done by

$$\mathbf{m}_{c,l_j}(t+1) = \mathbf{m}_{c,l_j}(t) + \alpha'(l_j)\alpha(t)[\mathbf{x}_{l_j}(t) - \mathbf{m}_{c,l_j}(t)], \tag{7.3}$$

where

$$c = arg\min_{i}\{\|\mathbf{x}_{l_j} - \mathbf{m}_{i,l_j}\|\}, \tag{7.4}$$

$$\alpha'(l_j) = e^{-\|l_i - l_j\|}, \text{ the "imprinting" coefficient}, \tag{7.5}$$

and

$$\alpha(t) = c_0 e^{-D(t)}, \ D(t) \ distance \ function \tag{7.6}$$

The sizes of the thresholds δ_j should be decreased according to the order of the levels to obtain encapsulated subsets S_j. This behavior is mainly supported by the "imprinting" coefficient $\alpha'(l_j)$. Therefore the "imprinting" coefficient is responsible for the topological order of the subclusters in the MHA.

If a priori knowledge has been detected then (7.3) is modified by (7.7)

$$\mathbf{m}_{c,l_j}(t+1) = \mathbf{m}_{c,l_j}(t) + \alpha''(l_j)\alpha'(l_j)\alpha(t)[\mathbf{x}_{l_j}(t) - \mathbf{m}_{c,l_j}(t)], \tag{7.7}$$

where

$$\alpha''(l_j) = e^{-\|\mathbf{x}_{l_j} - \mathbf{m}_{c,l_j}\| - \sigma} \tag{7.8}$$

for all these levels l_j (for $t > t_0$).

Classification is achieved by finding the best matching node for each level of the hierarchy and by determining the square mean error of matching. To protect the algorithm from initial disordering the a priori knowledge learning step should start after a given time t_0. In principle the algorithm handles different numbers of levels in the input vector.

7.1.3   Example of Classification by MHA

In order to demonstrate the learning and classification of the Multilevel Hypermap Architecture a simple example of processing sample data is used. Fig. 7.2 shows the input data and the structure of the MHA.

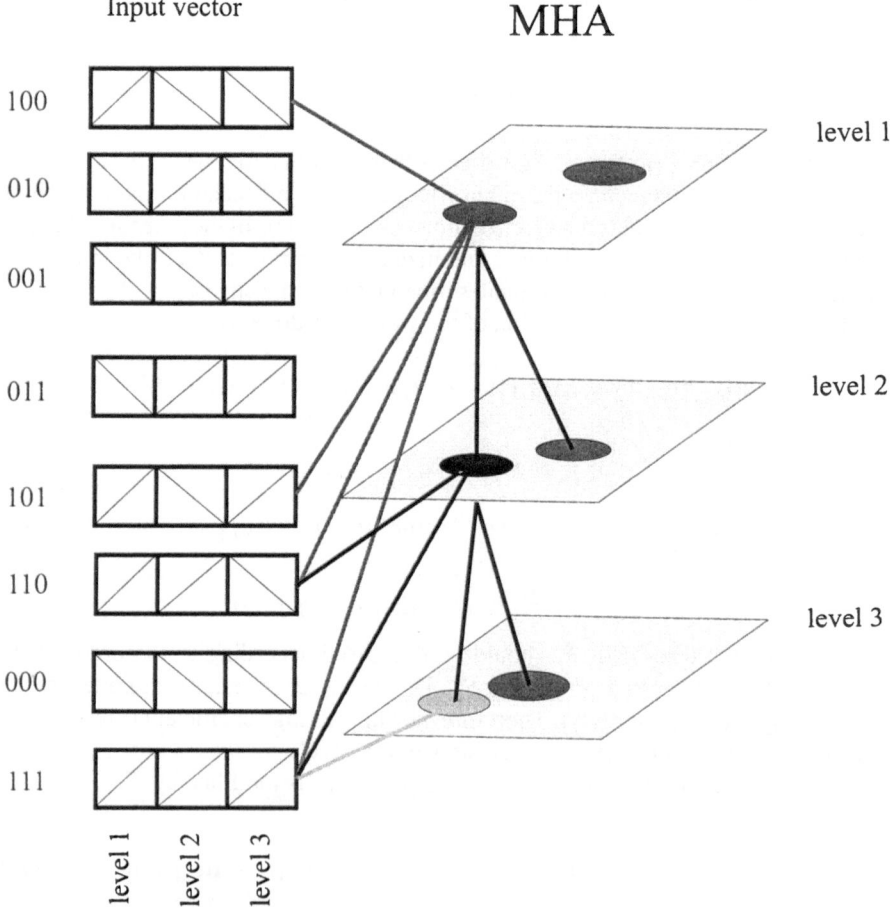

Fig. 7.2. Example showing the classification output of the MHA.
Transformed binary data is used for training. The network learning algorithm forms a hierarchical order of clusters and subclusters related to the binary value of the data.

In our example the MHA has a dimension of 25 by 25 units and 3 levels. These 3 levels should be used for a classification of triples of the binary digits 0 and 1 to illustrate the ability of hierarchical clustering. Therefore the input vector has 3 parts, in which the binary digits are realized as ramp signals to be useful for the LVQ based learning algorithm.

Suppose the values in one level of the input vector $y_i = f(i)$, which represent one binary digit, are $y_i = c_0 i$ for the digit 1 and $y_i = c_0(N - i)$ for the digit 0, whereby $i = 0, .., N$ and $c_0 > 0$ constant. Therefore in each level are $N + 1$ values which form ramp signals as illustrated in (Fig. 7.2), increasingly for the binary digit 1 and decreasingly for 0.

After initiation of training, on the first level all input vectors with the digit 1 and all with 0 are separated in one cluster each, that means, on the first level of the MHA only two clusters are formed. When the training process reaches the next level, this separation procedure will continue for each cluster. Therefore, there are two subclusters in each cluster related to the digits 1 and 0 respectively.

The formation of only two subclusters for each 2. level subcluster also takes place on the 3. level of the MHA in the training process. The result will be a hierarchical clustering of the input data as demonstrated in Fig. 7.2. A computational result of the classification by the MHA learning algorithm for our sample data is shown in Fig. 7.3.

Important for the occurrence of the hierarchical structure of clusters and subclusters in that way is the so called imprinting coefficient (7.5). Without imprinting on the lower level(s) the hierarchical relationship would disappear and there will be independent data relationships on all levels of the MHA. In the case of our example this would mean that only two clusters on each level of the MHA are found and the lower subclusters in the hierarchy will have no topological relationship to the upper ones.

On the other hand this feature can be useful to train a multiple set of unrelated data in one training process; each data set related to one level of the MHA.

In contrast to the MHA it is impossible to detect the hierarchical order using a conventional SOM or LVQ network. In the case of our example one cluster for each of the 8 input vectors is formed, that means, for the neural network of such type the input vectors are totally different from each other. The classification of data with hierarchies is of course the main advantage of the MHA learning algorithm.

7.1.4 Applications Using the MHA

In the previous years some real world applications using the MHA were reported in the literature. Beside the system for speech processing and recognition, which will be described in Sec. 2 of this chapter, mainly two other applications are remarkable.

The first application deals with an implementation of the Modified Hypermap Architecture for classification of image objects within moving scenes.

In our experiments the MHA network was trained with data which represent features of typical objects in images of a moving scene. These features were extracted from a block matching algorithm, the so called object-specific MAD features. The

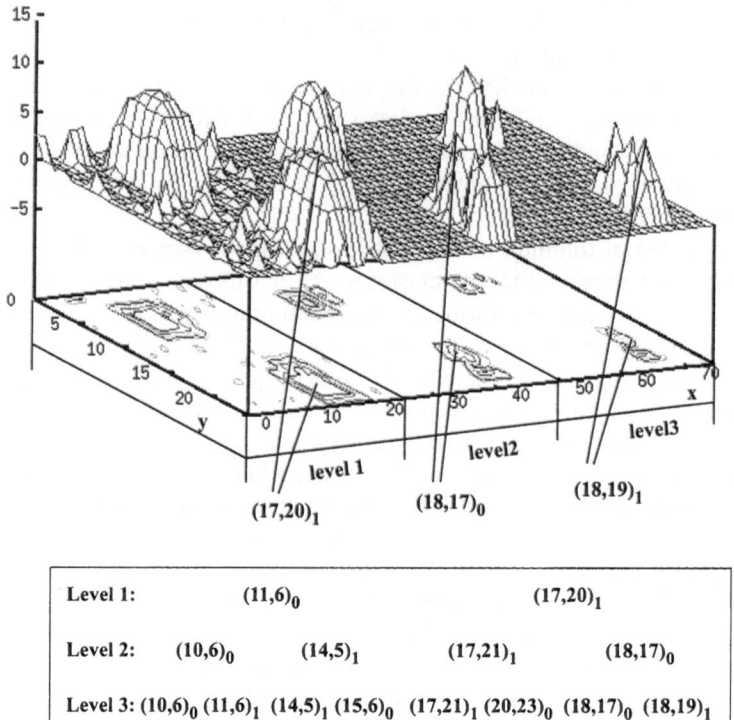

Level 1:		$(11,6)_0$			$(17,20)_1$	
Level 2:	$(10,6)_0$		$(14,5)_1$	$(17,21)_1$		$(18,17)_0$
Level 3:	$(10,6)_0$ $(11,6)_1$	$(14,5)_1$ $(15,6)_0$	$(17,21)_1$ $(20,23)_0$ $(18,17)_0$ $(18,19)_1$			

Fig. 7.3. Classification for a simple binary data set.
The binary input triples are classified in a hierarchical order. On the first level there are only two best matching nodes for the digits 0 and 1 respectively. In the error surface the triple 101 is indicated for illustration. The indices of the network coordinates of the best matching nodes describe the allocation to the learned binary digit. The result box shows the best matching nodes in the level related hierarchical order.

Mean Absolute Difference (MAD) is used as a similarity criterion. The trained features, represented by the object-specific MAD, are used as an associative memory. The performance capability and suitability of the Modified Hypermap Architecture for the problems of classification and its application as an associative memory are shown for selected two-dimensional measuring situations. The recall from this associative memory is usable for recognition of the two-dimensional object-specific MAD features for an improvement of motion estimation. More details about this work can be found in [6].

The second application and at the same time a newer investigation with the MHA is related to the analysis of fMRI images of auditory cortex activity which were obtained from acoustic stimulation [7].

In the auditory cortex of awake animals and humans, responses to the same repetitive auditory stimulus will strongly habituate, dishabituate and change with general alertness, context, cognitive performances and learning. The analysis of these non-stationarities is further complicated by the fact that the representation of a given stimulus in an auditory cortex field is not topographically stable over time. One approach to explore such stimulus related functional maps in the brain is the use of functional Magnetic Resonance Imaging (fMRI). Besides statistical methods, artificial neural networks are also used for analysis of fMRI data sets.

Several different acoustic stimuli (potpourri of various sounds, series of tones with shifting frequency, tone pairs with different frequencies) were used for the experiments with normal-hearing subjects. Subjects were scanned in a Bruker Biospec 3T/60 cm system. For the details of these experiments please refer [8].

In a first step of the analysis of acoustically evoked fMRI data, the MHA had to build hierarchical clusters of periodically similar data. In a second step these clusters were compared with the stimulus signal to detect non-stationarities. Finally the results are compared with the statistical tests (Pearson's cross-correlation).

After classification of the fMRI data with the MHA, territories in human auditory cortex were found, that are activated in parallel by different auditory stimulation. These findings are similar to the results of the statistical tests. The detection of topographies of clusters with stationary or non-stationary activation is important. Because the stimuli were presented periodically and the periods of the stimuli were built up in the multi-level structure of the MHA, the non-stationarities in these periods were represented by the hierarchy of clusters found in the MHA after training.

7.1.5   Advances in the MHA

By means of MHA it is possible to analyze structured or hierarchical data, i.e.

- data with priorities, e.g. projection of hierarchical data structures in data bases
- data with context (data bases, associative memories)
- time series, e.g. speech, moving objects
- data with varying degrees of exactness, e.g. sequences of measured data

One advantage of the MHA is the support for both, the classification of data and the projection of the structure in one unified map. The resulting hierarchy has some redundancy like in biological systems.

Because of its feature to store data relationships beside classification of objects future extensions of the MHA will deal with more "cognitive" learning elements.

We are very interested in a wide-ranging use of the Multilevel Hypermap Architecture. Therefore the MHA software containing the main modules, the documentation and some examples are available on our FTP server (ftp://ftp.ifn-magdeburg.de/mha).

7.2   MHA Application for Speech Recognition

7.2.1   The Multistage Model

The aim of the system architecture for speech processing and recognition is divided into two abstraction levels. First the system model should have a close relationship to the biological archetype. But furthermore this system should be able to process and recognize speech data in a technical way. Therefore the speech recognition system was realized as a multistage model. A schematic of the multistage auditory model is shown in Fig. 7.4.

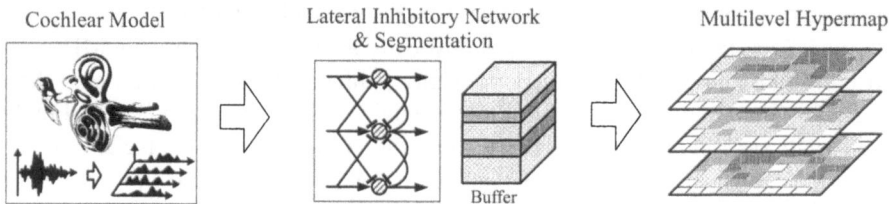

Fig. 7.4. Overall structure of the three-stage speech processing and recognition model.

The first stage of the multistage model is the digital time-domain model of the human cochlea developed by Kates [9]. It was designed to represent normal peripheral auditory function and its degradation due to auditory impairment but is also applicable to the field of automatic speech recognition.

Digitalized speech signals obtained by recording the analogue waveform of speech sound with a condensor microphone and 48-kHz ADC conversion, form the input (tympanic membrane pressure) to the middle-ear front-end of the cochlear model. The model output consists of the instantaneous firing rates of the auditory nerve fibers.

The composition of the frequency spectrum of an acoustic stimulus represents a primary cue for its identification by the auditory system (Shamma [10]). Therefore the aim of the second stage of the overall model is to find and highlight characteristic properties like peaks and edges in the complex structure of the firing rate patterns in the auditory nerve and to generate an estimate of this spectrum as a cue for learning and recognition by a structured neural network like the Multilevel Hypermap Architecture. There are different models of how the central auditory system makes an estimate of this spectrum from the responses of the auditory nerve fibers which can be classified into spatial, temporal and spatio-temporal processing models.

For the purpose of spectral estimation one of the fundamental neural network topologies, the lateral inhibition topology is used. A two-layer uniform Lateral Inhibition Network (LIN) system model (Shamma [10]) which belongs to the class of spatio-temporal processing models and seems to play an important role for spectral estimation within the central auditory system, is used for the spectral estimation task within the multistage model.

A nonrecurrent network is responsible for the detection of spectral peaks in the spatio-temporal firing rate patterns, and a network with recurrent inhibition is utilized for further sharpening of the averaged firing rate profiles. The extracted features depend on the spectral content and sound intensity of the speech stimuli and enable the segmentation of the LIN output. The segmentation process is an artificial step in the model which is realized by finding segment boundaries by a segment detector neuron.

The third stage is a structured multi-layer formal neural network realized as a modification of the Hypermap (Brückner et al. [4]). The Multilevel Hypermap Architecture is trained with segmented LIN output of speech signals. Each segment of the LIN output time sequence which represents the spectro-temporal properties of a whole syllable or a part of a syllable is related to one layer in the input vector. The whole time sequence represents the input vector for one complete speech component, i.e. a word or a part of a sentence. After training the network is able to classify untrained speech signals, i.e. can recognize several words spoken by different subjects.

7.2.2 The Cochlear Model

The digital time-domain model of the human cochlea developed by the cochlear model of Kates [9] is utilized for modeling the processing of sound in the auditory periphery.

Natural sound processing in the cochlea is realized by the conversion of the sound pressure waveform into neural spatio-temporal excitation patterns on the auditory nerve at many different places along the cochlear length dimension. The information of an acoustic signal is represented on the auditory nerve by the distribution of neural activity with distance along the cochlea and by the temporal structure of the firing patterns of the auditory nerve fibers.

The place-domain continuous cochlea is modeled as 112 discrete sections covering the frequency range of 100 Hz to 16 kHz and reflecting the tonotopic organization of the cochlear components. The sample rate of the mechanical and neural signals in the cochlear model is 48 kHz.

Each section consists of the following main structural and functional components:

- the mechanical signal processing,
- the mechanical-to-neural signal transduction and
- the level-dependent feedback system, i.e. the dynamic range compression concerning cochlear input/output relationships and the adaptive adjustment of the mechanical filter behavior.

A block diagram for one section of the cochlear model is shown in Fig. 7.5.

There are three aspects of mechanical signal processing in the cochlear model (Kates [9]):

154   Bernd Brückner and Thomas Wesarg

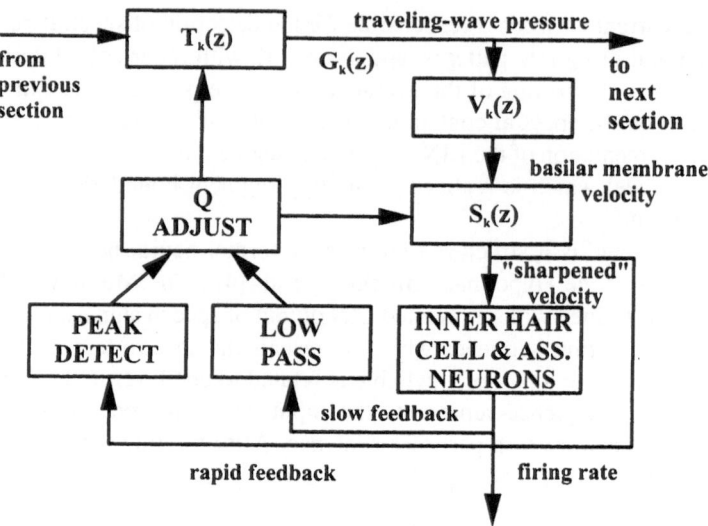

Fig. 7.5. Block diagram of the k. section of the cochlear model.

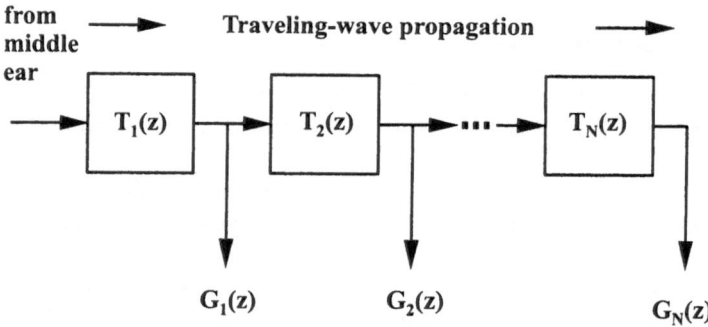

Fig. 7.6. Traveling-wave propagation in the overall cochlear model realized as a cascade of digital resonant filter sections $T_k(z)$.

- the propagation of sound pressure traveling-waves on the cochlear partition (see Fig. 7.6) realized as a cascade of discrete resonant filter sections $T_k(z)$,
- the transformation of the traveling-wave pressure to basilar membrane velocity at a particular cochlear location by the high pass velocity transformation filter $V_k(z)$ and
- the additional mechanical filtering of the velocity output at a particular location in the cochlea by the second filter $S_k(z)$ representing cochlear micromechanics. The second filters probably result from the resonance of the motion between the basilar and tectorial membrane.

The mechanical-to-neural signal transduction system at a particular cochlear location whose input is the velocity output from the second filter, is realized by a complex of an inner hair cell and four attached neurons and their fibers. This complex converts the cochlear partition's motion into an instantaneous neural firing rate pattern as the sum of the four fiber's firing rates normalized by the number of fibers and thus representing a mean spatio-temporal neural activity at a specific cochlear place. The four fibers have different spontaneous rates, excitation thresholds and saturation behavior, with two fibers being low-spontaneous and two fibers being high-spontaneous rate fibers and allow together with the level-dependent feedback system for an adequate coverage of the auditory dynamic range of about 120 dB.

A variable-resistance hair-cell model, based on the corresponding models of Allen [11] and Davis [12], is used for the implementation of the transduction behavior. Additionally a nonhomogeneous Poisson process based discharge generator can be used for the determination of spike times from the neural firing rate.

The sound pressure level-dependent feedback system represents the function of the outer hair cells and provides an active compression mechanism in the mechanical system. It consists of two feedback paths allowing for simultaneous rapid and slow adaptation of the Q-parameters of the traveling-wave propagation filters $T_k(z)$ and the second filters $S_k(z)$. A dynamic range compression ratio of 2.5:1 under steady-state conditions is used in the model.

7.2.3 The Lateral Inhibition Network

An important function of the auditory system is to discriminate and recognize complex sounds based on their spectral composition (Shamma [10]). Therefore an uniform lateral inhibitory neural network (LIN) system model (Shamma [10]) is implemented performing an estimation of the spectrum of the speech stimuli by spatial processing of the cochlear model's neural response patterns.

Lateral inhibition is one of the fundamental neural network topologies and is assumed to be involved in auditory information processing within central auditory networks, e.g. the auditory brainstem nuclei. The LIN system model is used as an intermediate neural information processing stage between the cochlear model and the artificial MHA network within the multistage model and thus allows for a spectrum based speech classification and identification by the Multilevel Hypermap Architecture.

The spectral estimates are the main spectral components of the acoustic stimuli and depend on the spectral content and the sound intensity of speech stimuli. They are derived by detecting and highlighting the peaks and edges in the spatio-temporal auditory nerve firing rate patterns which represent perceptually significant features of acoustic stimuli such as the fundamental frequency and their harmonics and the formants of voiced speech.

The spatio-temporal firing rate output patterns of the LIN system constitute the neural representation of the spectral estimates of the acoustic stimuli. Furthermore, the speech evoked LIN output patterns are less complex compared with the complex structure of the speech evoked response patterns of the auditory nerve. The response pattern of the cochlear model is little suitable as a direct input for learning and classification algorithms. The preprocessing with neural filter networks is required to detect and highlight characteristic properties like peaks and edges in the spatio-temporal signal.

The LIN system model consists of two lateral inhibition network models, a linear nonrecurrent network with a simple feed-forward architecture (LIN.I), and a nonlinear recurrent network (LIN.II).

Lateral inhibition among the network elements results in transfer functions of both networks having a spatial high-pass transfer characteristics. The nonrecurrent network is the input layer of the LIN system processing the auditory nerve response patterns directly to detect and sharpen the spectral peaks and edges in the cochlear spatio-temporal firing rate patterns.

The recurrent network is the LIN system's output layer and processes the LIN.I output firing rates to select and thus highlight the local peaks of the LIN.I output using a winner-take-all strategy.

Fig. 7.7 shows the architecture and the typical inhibitory and excitatory interconnections of the neurons within the two different lateral inhibition networks.

Both lateral inhibition networks consist of 112 tonotopically arranged neurons receiving their inputs from the corresponding 112 discrete sections of the cochlear model. They use a simple neuron model. The intracellular potentials in the neurons of the networks are given by

$$y_i(t) = e_i(t) - \sum_{j=1,j!=i}^{N} v_{ij} e_j(t) \qquad (7.9)$$

for the nonrecurrent (LIN.I), and

$$y_i(t) = e_j(t) - \sum_{j=1,j!=i}^{N} w_{ij} g(y_j(t)) \qquad (7.10)$$

for the recurrent network (LIN.II), where N is the number of neurons, e represents the input vector and v and w are the vectors containing the inhibitory weights. The relation between the intracellular potential y_i and the firing rate output of a neuron z_i is described by a sigmoidal function $g(y_i)$) which is given by

$$z_i(t) = g(y_i(t)) = \frac{z_{max}}{1 + e^{-b(y_i(t)-y_0)}} \qquad (7.11)$$

where y_0 and b are constants reflecting the spontaneous firing rate and the maximum slope of the firing rate versus intracellular potential function of the neuron, respectively. To illustrate the kind of neural auditory information processing by the LIN system model the spatio-temporal output patterns of both lateral inhibition networks are shown in Fig. 7.8.

Both networks are simulated as separate filter stages whose inhibitory weights cause a typical high pass behavior. The LIN.I filter already detects spectral peaks which represent perceptually significant features of the stimuli such as formants of voiced speech. The recurrent network is used for further sharpening of these features.

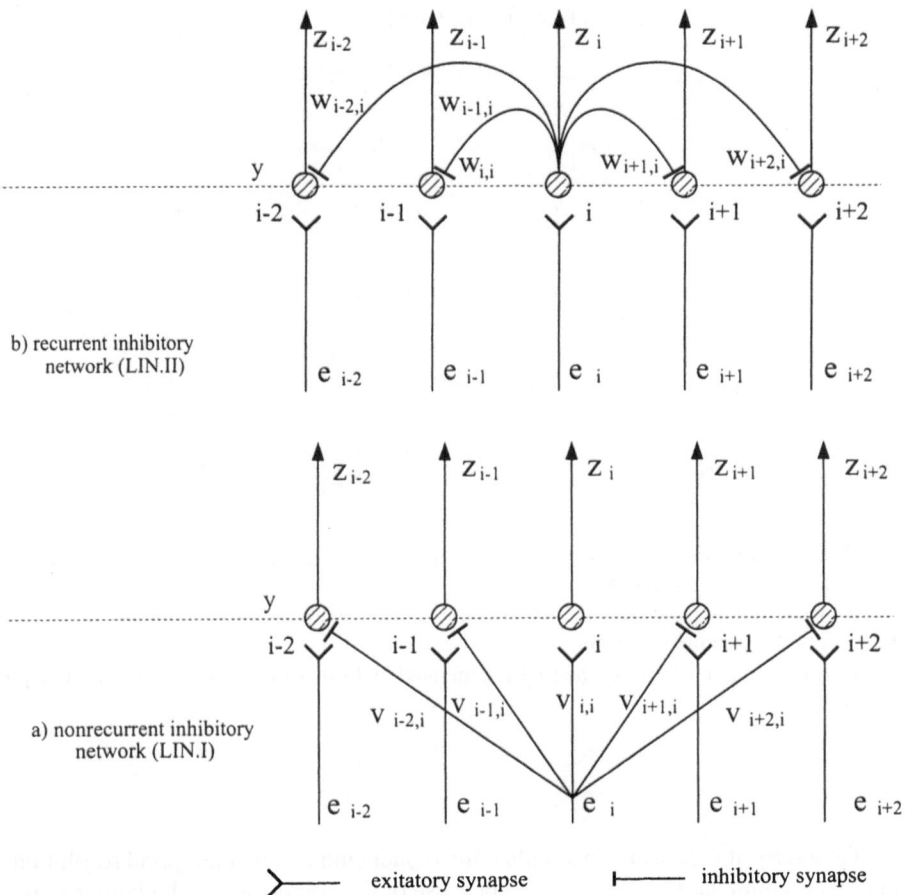

b) recurrent inhibitory
network (LIN.II)

a) nonrecurrent inhibitory
network (LIN.I)

Fig. 7.7. Schematics of the LIN system model.
a) Linear nonrecurrent lateral inhibitory network.
b) Nonlinear recurrent lateral inhibitory network.
Two types of lateral inhibitory interconnections, implemented in the filter stages of the LIN
system model with the the profiles of inhibitory weights v in the nonrecurrent and w in the
recurrent network.

Due to the computationally expensive simulations of the nonlinear LIN.II a sample rate reduction is performed from the 48000 samples/sec at the LIN.I input (corresponding to the sample rate of the acoustic stimuli) to $\approx$ 200 samples/sec at the
LIN.I output. The sample rate reduction is also motivated by the physiological observation that phase-locking and temporal resolution is less in the central auditory
system than in the auditory periphery and deteriorates significantly in the response
of central auditory neurons, which encode only averaged outputs (Shamma [10]),
such as shown in Fig. 7.8.

158   Bernd Brückner and Thomas Wesarg

a) output pattern of LIN.I

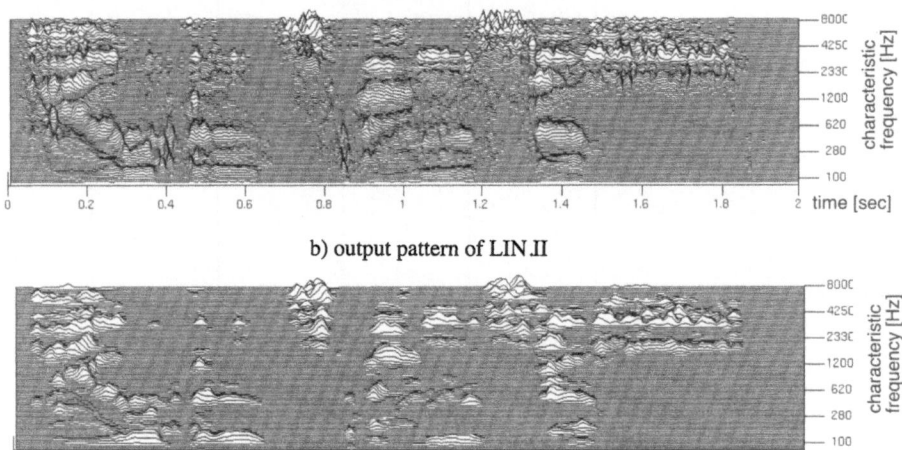

b) output pattern of LIN.II

Fig. 7.8. Auditory processing of the cochlear output by the LIN system model for the spoken german word /Einundzwanzig/.
a) Firing rate output patterns of the LIN.I with the peaks representing perceptually significant features of the stimulus spectrum.
b) Firing rate output patterns of the LIN.II generated by further sharpening of the LIN.I output patterns.

Generally, it is possible to simulate the model with a higher temporal resolution, but this is computationally expensive because of the complicated calculations in the LIN.II filter stage. The results of the spatial processing and inhibitory filter processes of speech by the LIN system depend on the spectral content and sound intensity of the speech stimuli and allow the segmentation of the speech evoked spatio-temporal firing rate output patterns.

The segmentation of speech is implemented as an artificial interface within our multistage model. Because of the structural behavior of the MHA and its buffered input it is necessary to detect and segment speech for a structural learning process. The speech segmentation is performed by a simple detector neuron shown in Fig. 7.9. This neuron is inhibited by a set of LIN.I and LIN.II neurons and shows a typical off activity.

Fig. 7.10 shows the off activity of the detector neuron used for finding the segment boundaries in the LIN speech data. In the inactivity phase of the detector neuron the LIN output is stored into a buffer as a speech segment of speech (e.g. syllable). All segments of this buffer which form the complete speech component (e.g. word) are applicable by the MHA for further classification.

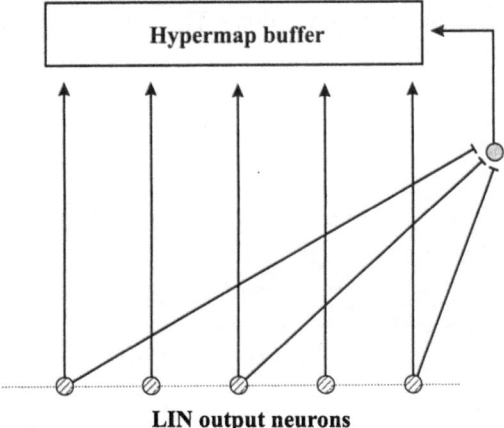

LIN output neurons

Fig. 7.9. The speech evoked LIN output is transformed into segments by a segmentation process. A detector neuron is activated by the absence of crucial output of the LIN system and is used for detecting the boundaries of the speech segments.

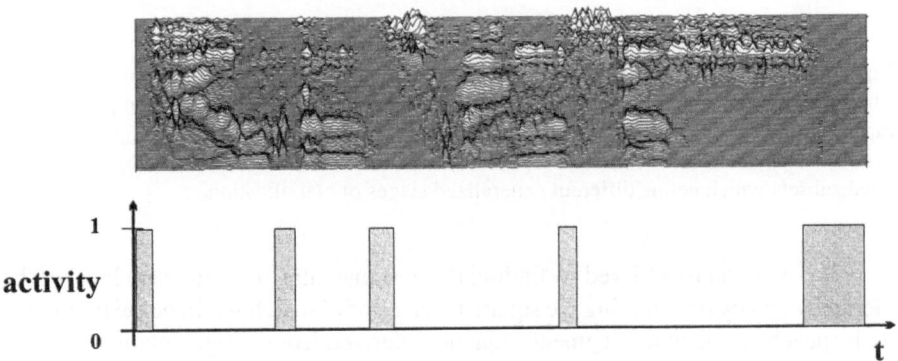

Fig. 7.10. Illustration of the segment detector neuron's operation.
The speech evoked LIN output is analyzed by the segment detector neuron. The neuron is active in case of the absence of LIN output. The inactivity phase of this neuron defines the segments of speech buffered for classification by the MHA.

7.2.4   Classification by Means of the MHA

The third stage of the model for speech processing and recognition is a structured multilevel formal neural network realized as a modification of the Hypermap described in the first section of this chapter (Brückner et al. [4,5]). With the MHA it is possible to learn time-dependent data in the form of time sequences like speech. Each part of a time sequence is related to the corresponding level in the input vector which forms a time hierarchy. The recognition of continuous speech signals is performed by a segmentation process. Each detected segment is related to a level of the network hierarchy where classification is done. The initial segment of a speech component is related to level one.

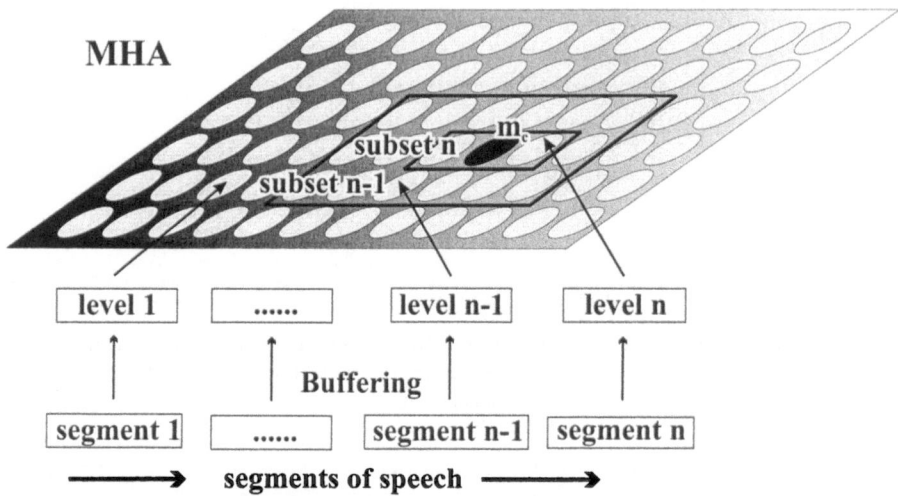

Fig. 7.11. Structure of the Multilevel Hypermap Architecture for speech recognition. The speech evoked LIN output is transformed into segments by a segmentation process. The input vector consists of a concatenation of these segments for one speech component, e.g. a syllable. The different levels of the input vector are trained and form a hierarchy of encapsulated subsets which define different generalized stages of classification.

Classification is achieved by finding the best matching node for each level of the hierarchy and by determining the square mean error of matching. In our experiments with speech we obtained segment sequences derived from a segmentation process, which stores the speech segments in a buffer. These segments represent whole syllables or parts of syllables like phonemes. In this sense the whole time sequence stored in the buffer represents one complete speech component, i.e. a word or a part of a sentence. The algorithm handles different numbers of segments in the input vector.

7.2.5 Results of Speech Processing

First of all the digitized speech signals are preprocessed and converted into neural excitation patterns by an implementation of a cochlear model (Kates [9]). Subsequently the model of a lateral inhibitory network with a highpass filter behavior highlights spectral edges of these patterns and suppresses all other activities. The edges, considered as significant features of the stimuli, are particularly stable with respect to sound level variations, despite of the limited dynamic range of the auditory nerve fibers. Consequently, the discontinuities in the complex overall response texture are largely preserved after the LIN filter process (Shamma [10]). The seg-

mentation process following is a filter algorithm indicating the segments' boundaries and is used for buffering the speech segments. This algorithm is independent
of the Hypermap Architecture.

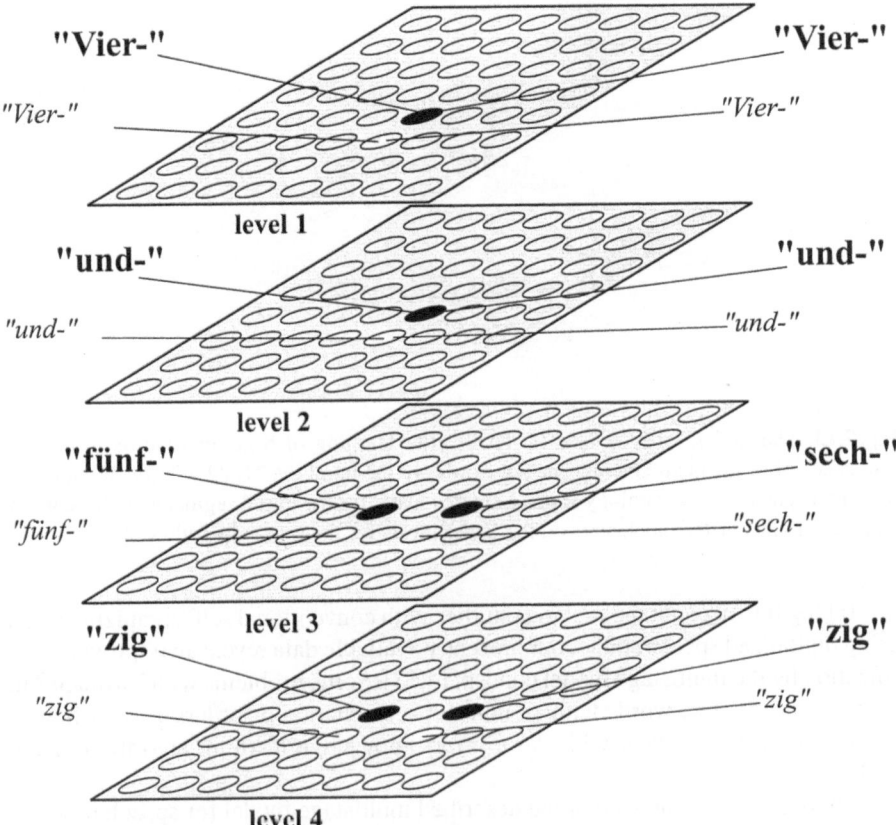

Fig. 7.12. Classification scheme of the trained speech segments by the MHA.
The german spoken numbers /Vierundfünfzig/ (54) and /Vierundsechzig/ (64) respectively
are classified. The black units show the best matching nodes in each level for both numbers (spoken numbers indicated in bold font). Beside the generalization of these words for
different speakers there is in fact also the possibility of an establishment of speaker related
classification (white units, spoken numbers indicated in cursive font).

The trained MHA is able to classify speech components from continuous speech
signals spoken by different subjects (Fig. 7.13) and can be successfully used for
speech recognition. Regardless of using the MHA for its generalization capabilities
and speaker independent speech recognition there is also the possibility of speaker
identification. This depends on the amount of speech data to classify and mainly on
the dimension used for the MHA (Fig. 7.12).

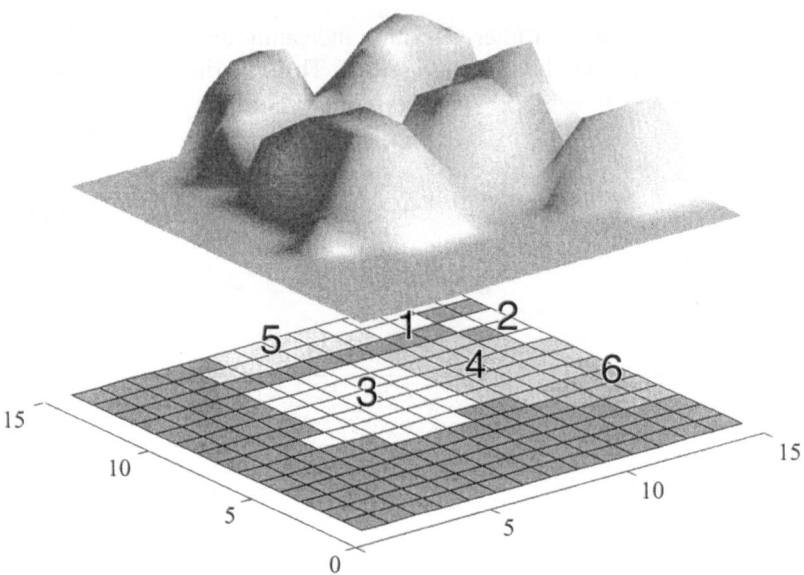

Fig. 7.13. The global error surface (4. level) after learning of 6 german words in form of sequences of segments is shown. The six words are the numbers 21, 32, 43, 54, 56 and 64 in this order. The network topology forms a single cluster for the initial segments of these words and within this cluster subclusters on the lower levels for the segments followed.

Taking into account the previous studies with conventional self-organizing maps using digitalized speech sound, our presently available data reveal an improved classification by the multistage model concept [4,13]. Some problems with "artifacts" in the training set, e.g., words with an unusual accentuation disordering the achieved classification were solved with the a priori knowledge learning algorithm of the MHA.

Although it was shown that the described multistage model for speech processing as a simulation of the auditory system is also useful as a technical application for speech recognition, there are still no tests with bigger amounts of speech data. The main reason for that is the very expansive computational power needed for the processing of the speech sound on the lower stages of the overall model. Therefore a comparison with commercial systems for speech processing and recognition is not available.

Acknowledgment

The development of the MHA was supported by LSA grant 1441B/0083 and LSA grant 851A/0023. Special thanks to **Carsten Schauer** for his work on the LIN system.

Bibliography on Chapter 7

1. T. Kohonen. The hypermap architecture. In Kohonen et al. [23], pages 1357–1360.
2. T. Kohonen. What generalizations of the self-organizing map make sense? In M. Marinaro and P.G. Morasso, editors, *ICANN'94*, pages 292–297, London, 1994. Springer Verlag.
3. B. Brückner, M. Franz, and A. Richter. A modified hypermap architecture for classification of biological signals. In I. Aleksander and J. Taylor, editors, *Artificial Neural Networks,2*, pages 1167–1170, Amsterdam, 1992. Elsevier Science Publishers.
4. B. Brückner, T. Wesarg, and C. Blumenstein. Improvements of the modified hypermap architecture for speech recognition. In *Proc. Inter. Conf. Neural Networks*, volume 5, pages 2891–2895, Perth, Australia, 1995.
5. B. Brückner. Improvements in the analysis of structured data with the multilevel hypermap architecture. In Kasabov et.al., editor, *Progress in Connectionist-Based Information Systems, Proceedings of the ICONIP97*, volume 1, pages 342–345, Singapore, 1997. Springer-Verlag.
6. C. Blumenstein, B. Brückner, R. Mecke, T. Wesarg, and C. Schauer. Using a modified hypermap for analyzing moving scenes. In Shun ichi Amari et.al., editor, *Progress in Neural Information Processing, Proceedings of the ICONIP96*, volume 1, pages 428–431, Singapore, 1996. Springer-Verlag.
7. B. Brückner, B. Gaschler-Markefski, H. Hofmeister, and H. Scheich. Detection of non-stationarities in functional mri data sets using the multilevel hypermap architecture. In *Proceedings of the IJCNN'99*, Washington D.C., 1999.
8. B. Gaschler-Markefski, F. Baumgart, C. Tempelmann, F. Schindler, H. J. Heinze, and H. Scheich. Statistical methods in functional magnetic resonance imaging with respect to non-stationary time series: auditory cortex activity. *Magn. Reson. Med.*, 38:811–820, 1997.
9. J.M. Kates. A time-domain digital cochlear model. *IEEE Transactions on Signal Processing*, 39(12):2573–2592, 1991.
10. S. Shamma. Spatial and temporal processing in central auditory networks. In C. Koch and I. Segev, editors, *Methods in Neuronal Modeling*, pages 247–289, Cambridge, 1989. The MIT Press.
11. J.B. Allen. A hair-cell model of neural response. In E. deBoer and M.A. Viergever, editors, *Mechanics of Hearing*, pages 193–202. Martinus Nijhoff, Hague, The Netherlands, 1983.
12. H. Davis. A mechanoelectrical theory of cochlear action. *Ann. Oto-Rhino-Laryngol.*, 67:789–801, 1956.
13. T. Wesarg, B. Brückner, and C. Schauer. Modelling speech processing and recognition in the auditory system with a three-stage architecture. In C. von der Malsburg et.al., editor, *Artificial Neural Networks - ICANN 96, Lecture Notes in Computer Science*, volume 1112, pages 679–684, Berlin, 1996. Springer-Verlag.
14. B. Brückner and W. Zander. Classification of speech using a modified hypermap architecture. In I. Aleksander and J. Taylor, editors, *Proceedings of the WCNN'93*, volume III, pages 75–78, Hillsdale, 1993. Lawrence Earlbaum Associates.
15. T. Kohonen. *Self-Organization and Associative Memory*. Springer-Verlag, New York, 1988.
16. T. Kohonen. *Self-Organizing Maps*. Springer-Verlag, Berlin, 1997.
17. J. Kangas. Time-dependent self-organizing maps for speech recognition. In Kohonen et al. [23], pages 1591–1594.

18. S.A. Shamma. The acoustic features of speech phonemes in a model of the auditory system: Vowels and unvoiced fricatives. *J. of Phonetics*, 16:77–91, 1988.
19. J. Kangas. the analysis of pattern sequences by self-organizing maps, 1994.
20. T. Voegtlin and P.F. Dominey. Contextual self-organizing maps: An adaptive representation of context for sequence learning, 1998-01.
21. T. Graepel, M. Burger, and K. Obermayer. Self-organizing maps: generalizations and new optimization techniques. *Neurocomputing*, 21:173–190, 1998.
22. F. Mehler and P. Wilcox. Self-organizing maps in speech recognition systems. In F. G. Bobel and T. Wagner, editors, *Proc. of the First Int. Conf. on Appl. Synergetics and Synergetic Engineering (ICASSE'94)*, pages 20–26, Erlangen, Germany, 1994.
23. T. Kohonen, K. Mäkisara, O. Simula, and J. Kangas, editors. *Artificial Neural Networks*, Helsinki, 1991. Elsevier Science Publishers.

8 Algorithms for the Visualization of Large and Multivariate Data Sets

Friedhelm Schwenker, Hans A. Kestler and Günther Palm

Abstract. In this chapter we discuss algorithms for clustering and visualization of large and multivariate data. We describe an algorithm for exploratory data analysis which combines adaptive c-means clustering and multi-dimensional scaling (ACMDS). ACMDS is an algorithm for the online visualization of clustering processes and may be considered as an alternative approach to Kohonen's self organizing feature map (SOM). Whereas SOM is a heuristic neural network algorithm, ACMDS is derived from multivariate statistical algorithms. The implications of ACMDS are illustrated through five different data sets.

8.1 Introduction

Intelligent systems learn new patterns or associations by sampling data and changing the system's parameters with respect to this incoming data. Often the sampled data are compressed into a small but representative set of prototypes. In the context of artificial neural networks learning or adaptation means parameter change, in particular, change of the synaptic weights or change of the network architecture. Typically, synaptic weight changes are describes by learning or adaptation rules formally defined in terms of differential equations (in continous time models) or difference equations (in discrete time models), whereas, the rules for changing the network architecture, e.g. the creation of an additional artifical neuron or the creation of a whole set (layer) of neurons, are given by logical if-then-else conditions.

In this chapter we study the problem of finding structures in large and high-dimensional data sets utilizing adaptive clustering and visualization procedures. Intelligent systems in many applications, e.g. mobile robots, have to deal with this problem. In particular, we discuss methods of *cluster analysis*, in which the aim is to reduce a large set of data points into a few cluster centers and *multidimensional scaling*, where a distance preserving low dimensional representation of the dataset is determined.

In many practical applications one has to explore the underlying structure of a large set objects. Typically, each object is represented by a feature vector $x \in \mathcal{X}$, where $\mathcal{X}$ is the feature space endowed with a distance measure $d_{\mathcal{X}}$. This data analysis problem can be tackled utilizing *clustering methods* (see [14] for an overview). A widely used clustering algorithm is *c-means clustering* [21,20,19] where the aim is to reduce a set of M data points $X = \{x_1, \ldots, x_M\} \subset \mathcal{X}$ into a few, but representative, cluster centers $\{c_1, \ldots, c_k\} \subset \mathcal{X}$.

A neural network algorithm for clustering is Kohonen's *selforganizing feature map (SOM)* [17]. SOM is similar to the classical sequential c-means algorithm (see

Sec. 2) with the difference that in SOM the cluster centers are mapped into a *display space* $\mathcal{Z}$ with a distance measure $d_{\mathcal{Z}}$. Each cluster center is mapped to a fixed location of the display space. The idea of this display space $\mathcal{Z}$ is that cluster centers corresponding to nearby points in $\mathcal{Z}$ have nearby locations in the feature space $\mathcal{X}$. Typically, $\mathcal{Z}$ is a 2-dimensional (or 3-dimensional) grid. Therefore it is often emphasized that Kohonen's SOM is able to combine clustering and visualization aspects. In this context it has to be mentioned that SOM is a heuristic algorithm — not derived from an objective function which incorporates both a clustering criterion and some kind of topological or neighborhood preserving measure [17,24].

Another approach for getting an overview over a high-dimensional data set $X \subset \mathcal{X}$ are *visualization methods*. In multivariate statistics several linear and nonlinear techniques have been developed. A widely used nonlinear visualization method is *multidimensional scaling (MDS)* [26,25]. MDS is a class of distance preserving mappings from the data set X into a low-dimensional *projection space* $\mathcal{Y}$ which is endowed with some distance measure $d_{\mathcal{Y}}$. Each feature vector $x_\mu \in X$ is mapped to a point $y_\mu := p(x_\mu) \in \mathcal{Y}$ in such a way that the distance matrix $D_{\mathcal{X}} := (d_{\mathcal{X}}(x_i, x_j))_{1 \leq i,j \leq M}$ in feature space $\mathcal{X}$ is approximated by the distance matrix $D_{\mathcal{Y}} := (d_{\mathcal{Y}}(y_i, y_j))_{1 \leq i,j \leq M}$ in projection space $\mathcal{Y}$.

In this chapter an algorithm, which we call ACMDS, is described. It combines c-means clustering and MDS. This procedure is able to be utilized for the online visualization of clustering processes and should be considered as an alternative approach to Kohonen's self organizing feature (SOM). Both algorithms, ACMDS and SOM, reduce the data complexity through data clustering in the feature space *data reduction* and through a distance or neighborhood preserving mapping of the cluster centers into a display or visualization space.

Throughout this chapter we restrict our considerations to the feature space $\mathcal{X} = \mathbb{R}^n$ and the projection space $\mathcal{Y} = \mathbb{R}^r$. The Euclidean distance d is used as distance measure in both spaces.

The chapter is organized as follows. The three basic algorithms *general competitive learning*, *c-means clustering* and *learning in SOM* are discussed in Sec. 2. In Sec. 3 multidimensional scaling is briefly reviewed and in Sec. 4 we describe the *ACMDS* algorithm. This is then applied to five different data sets two artifical and three real world data sets. We discuss these numerical experiments in Sec. 5.

8.2   Data Clustering and Vector Quantization

The *c-means clustering* algorithm moves a fixed set of k cluster centers into the centers of gravity of the accumulations of data points [21,20]. For the interpretation of the clustering result it is important to choose the right number of cluster centers k. If the choosen number of clusters is different from the actual number of clusters hidden within the data, the result of the clustering process has to be reconsidered.

During the c-means clustering process the current data point $x \in X$ is classified to the closest center c_{j^*}, i.e. the data point x belongs to cluster C_{j^*} if $d(x, c_{j^*}) =$

Data Matrix

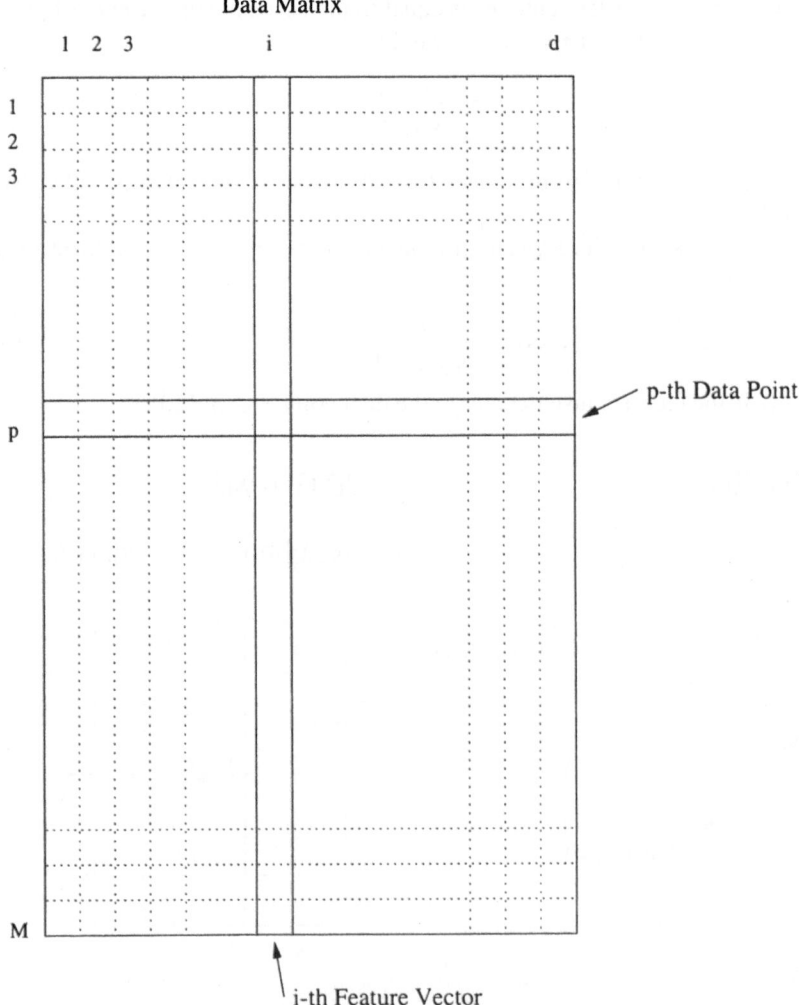

Fig. 8.1. The Data Analysis Problem: A set of M data points from a multivariate feature set may be given and has to be analysed. Typically, d the number of features/measurements, and M the number of data points is large—in many real world applications the data set grows over time and M is not known in advance. Reducing the complexity of the data set may be done through *data reduction* or/and *dimensionality reduction*. The proposed ACMDS algorithm and Kohonen's selforganizing feature map perform both types of complexity reduction simultaneously.

$\min_i d(x, c_i)$. The quantization error $H(c_1, \dots, c_k)$ defined by

$$H(c_1, \dots, c_k) = \sum_j \sum_{x \in C_j} d^2(x, c_j)$$

is minimial, if each cluster center c_j is equal to the corresponding center of gravity of cluster C_j, e.g. if for all $j = 1, \ldots, k$ hold

$$c_j = \frac{1}{|C_j|} \sum_{x \in C_j} x$$

where $|C_j|$ denotes the size of cluster C_j. This type of algorithm is called *batch c-means* [21].

We concentrate on the *sequential c-means clustering* method realized by the updating rule

$$\Delta c_{j*} = \frac{1}{|C_{j*}| + 1}(x - c_{j*}) \tag{8.1}$$

where c_{j*} is the closest cluster center to the data point $x \in X$ [20].

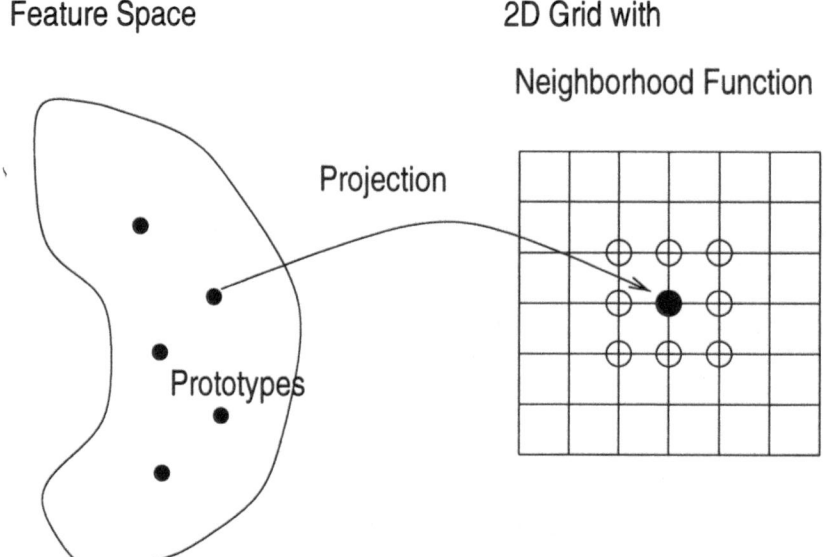

Feature Space

2D Grid with

Neighborhood Function

Projection

Prototypes

Fig. 8.2. Kohonen's selforganizing feature map: Here, the set of prototypes is adapted according to the input data of the d-dimensional feature space. These protoypes are mapped to locations of a fixed display space $\mathcal{Z}$. Typically, this display space is a regular 1D, 2D or 3D grid.

Sequential c-means clustering is closely related to learning in artificial neural networks in particular, to the paradigm of *competitive neural networks* [23]. *Kohonen's selforganizing feature map (SOM)* is a competitive learning scheme [17]. During SOM learning cluster centers c_j that are close in the display space $\mathcal{Z}$ will be adapted to the same input $x \in X$:

$$\Delta c_j = \eta_t h(r_j, r_{j*})(x - c_j) \tag{8.2}$$

where $h : \mathcal{Z} \times \mathcal{Z} \to \mathbb{R}_+$ is a neighborhood function with $h(r_j, r_{j^*}) \to 0$ for increasing distance of r_j and r_{j^*}. For the special case of $h(r_j, r_{j^*}) = 1$ for $j = j^*$ and $h(r_j, r_{j^*}) = 0$ otherwise, Kohonen's SOM updating rule is identical to sequential c-means clustering.

The simplest competitive learning rule has the structure

$$\Delta c_{j^*} = \eta_t(x - c_{j^*}) \tag{8.3}$$

where η_t is a positive sequence of learning rates and again c_{j^*} is the closest center to data point x.

This is an optimization procedure related to stochastic approximation, which asymptotically minimizes the expected value of H with respect to a given distribution of data points [11]. The decreasing learning rate η_t in (8.3) is very common in stochastic approximation, but larger (more slowly decreasing) or even constant values of η_t are also of interest, in particular in nonstationary enviroments, see [4].

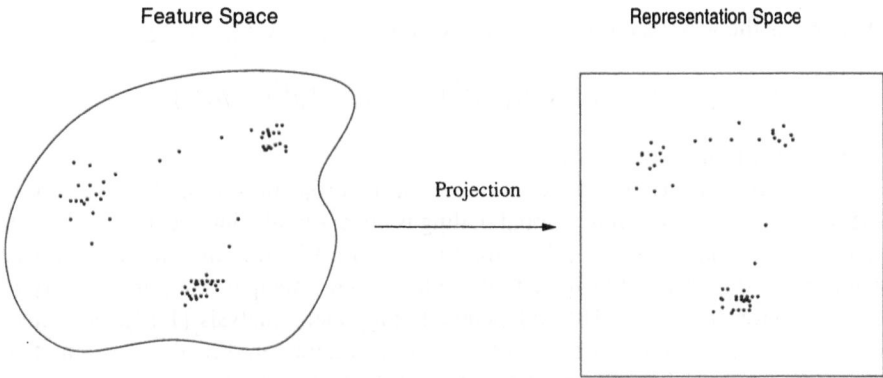

Fig. 8.3. Mapping of feature vectors from an high-dimensional feature space into a low-dimensional display or representation space. For visualization tasks this display space is typically a bounded subset of $\mathbb{R}^r$, with $r = 1, 2$ or 3. This low-dimensional projection is highly depending on the concrete application, i.e. in multidimensional scaling this projection performs a distance preserving transformation. In some cases this projection is explicitly given, e.g. in principal component analysis, independent component analysis, or projection pursuit. This is not the case for multidimensional scaling. Here the projection of a feature vector has to be calculated through an iterativ procedure.

8.3   Multidimensional Scaling

Given a set of data points $X \subset \mathcal{X}$ and a transformation $p : \mathcal{X} \to \mathcal{Y}$ a natural evaluation criterion for the distance preservation of p is some kind of difference between the matrices $D_{\mathcal{X}} := (d_{\mathcal{X}}(x_i, x_j))_{1 \leq i,j \leq M}$ in $\mathcal{X}$ and $D_{\mathcal{Y}} := (d_{\mathcal{Y}}(y_i, y_j))_{1 \leq i,j \leq M}$ in $\mathcal{Y}$.

MDS is a multivariate statistics procedure that start with a distance matrix $D_{\mathbb{R}^n}$ of M data points $X = \{x_1, \ldots, x_M\} \subset \mathbb{R}^n$ and generates a set of corresponding representation points $Y = \{p(x_1), \ldots, p(x_M)\} \subset \mathbb{R}^r$ [14,25]. This projection $p : X \to \mathbb{R}^r$ is calculated in such a way that the distance matrices $D_{\mathbb{R}^n}$ and $D_{\mathbb{R}^r}$ are similar. The difference is measured through *stress functions* defined by

$$S(x_1, \ldots, x_M) = \alpha \sum_{j,i=1}^{M} \Big(\Phi[d^2(x_i, x_j)] - \Phi[d^2(y_i, y_j)] \Big)^2 \tag{8.4}$$

where $\alpha > 0$ and $\Phi : \mathbb{R} \to \mathbb{R}$ is a strictly increasing differentiable function, i.e. $\Phi(x) = x$, $\Phi(x) = \sqrt{x}$ or $\Phi(x) = \log(x + 1)$. The consecutive points y_j may be calculated by a gradient descent algorithm

$$\Delta y_j = \eta_t \cdot \alpha \sum_{i \neq j}^{M} \delta_{ji}(y_i - y_j), \tag{8.5}$$

where δ_{ji} is the weighted difference between $d(x_i, x_j)$ and $d(y_i, y_j)$ given by

$$\delta_{ji} = \Phi'[d^2(y_i, y_j)]\Big(\Phi[d^2(x_i, x_j)] - \Phi[d^2(y_i, y_j)]\Big)$$

and $\eta_t > 0$ with $\eta_t \to 0$ as $t \to \infty$.

In this context it should be mentioned that the projection of the feature vectors performed through multidimensional scaling is not explicitly defined, but can be calculated through an iterative optimization procedure. Whereas in many other linear or nonlinear dimensionality reduction methods like principal component analysis [15,16], factor analysis [8,16], independent component analysis [1,13], or projection pursuit [6,5,12] the projection of the feature vector can directly be calculated, for instance through a matrix multiplication of the feature vector.

8.4   Clustering and Distance Preserving Low Dimensional Projection

In this approch we combine the sequential c-means clustering procedure, to detect the cluster structure in feature space, with a MDS algorithm, to get a low-dimensional representation of the cluster centers. To achieve this, the cluster centers $c_j \in \mathbb{R}^n$ move according to the sequential c-means iteration rule (8.1) and simultanously a set of low-dimensional representation centers $p_j := p(c_j)$ move in $\mathbb{R}^r$. These representation centers move in such a way that the distances $d(p_i, p_j)$ are close to the distances $d(c_i, c_j)$ of the cluster centers. This is realized by a gradient descent algorithm,

$$\Delta p_j = \eta_t \alpha \sum_{i \neq j}^{k} \delta_{ji}(p_i - p_j) \tag{8.6}$$

where δ_{ji} is

$$\delta_{ji} = \Phi'[d^2(p_i, p_j)]\Big(\Phi[d^2(c_i, c_j)] - \Phi[d^2(p_i, p_j)]\Big)$$

which minimizes the stress function

$$S(p_1, \dots, p_k) = \alpha \sum_{i,j=1}^{k} \Big(\Phi[d^2(c_i, c_j)] - \Phi[d^2(p_i, p_j)]\Big)^2 \qquad (8.7)$$

ACMDS Algorithm

```
estimate thresholds θ_new and θ_merge
set k = 0 (no prototypes)
    choose a data point x ∈ X
    calculate d_j = d(x, c_j), j = 0, ... , k
    detect the winner j* = argmin_j d_j
    if (d_j* > θ_new) or k = 0
        c_k := x and p_k according to (8.6)
        k := k + 1
    else
        adapt c_j* by (8.1) and p_j* by (8.6)
        calculate D_l = d(c_l, c_j*),  l = 0, ... , k
        detect l* := argmin_{l≠j*} D_l
        if (D_l* ≤ θ_merge)
            merge(c_l*, c_j*),  k := k - 1
    goto: choose data point
```

In addition we incorporate into the clustering algorithm a scheme for adjusting the number k of cluster centers in order to address the following two problems:

a) What is a good choice for the number of cluster centers?
b) What are the initial locations of the cluster centers?

The simplest approach to the second problem is to choose k data points from the data set at random, and initialize the centers by these data points. Another approach starts with a hierarchical clustering procedure. Running a hierachical clustering algorithm produces a sequence of partitions of the data set. From this sequence one is able to choose the partition which includes k subsets of the data set and use the center of gravity of each subset to run the c-means procedure. However, in each case k has to be chosen.

With respect to the first problem we use a modified c-means clustering procedure which allows the merging of clusters and the creation of a new cluster center if the presented data point is far away from the already existing centers. For this clustering algorithm parameters θ_{merge} and θ_{new} have to be derived from the data subset X in advance. This may done by calculating the volume of the box that encloses the data set X, or by calculating or estimating the diameter of X. Calculating the diameter is indeed an expensive procedure (in terms of compuational time as well as memory). The diameter can roughly be estimated through the maximal (or the mean) distance of the data points to the mean of data set X. For this algorithm the time complexity is linear, whereas the calculation of the diameter requires quadratic cost. Based on these data set inspections the parameters θ_{merge} and θ_{new} are estimated.

When a new cluster center is inserted, the location of the corresponding representation center p_j has to be determined, by setting the inital position to be a linear combination of its two nearest neighbors and then adapting p_j by the iteration rule (8.6).

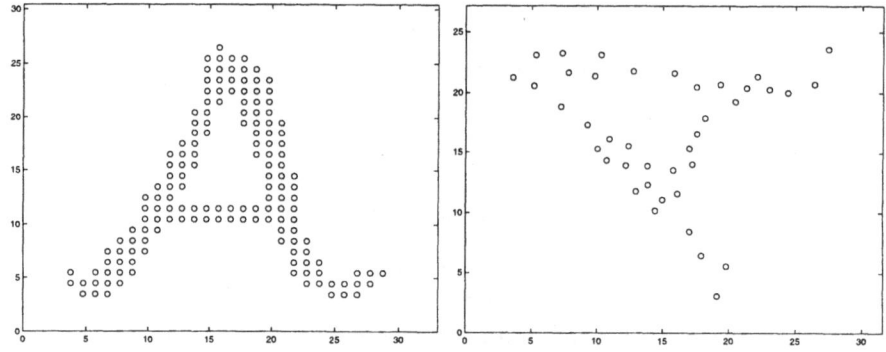

Fig. 8.4. The A-shaped data set X with 153 data points (left) and a set of 40 projection centers (right) are shown (see text).

8.5   Numerical Results and Discussion

To illustrate the behavior of the ACMDS algorithm we present results for five different data sets. They vary in size and dimension of the feature space.

8.5.1   Artifical Data Set

In Fig. 8.4 a twodimensional data set X, containing 153 samples ist shown. This A-shaped data set was made by hand using a standard icon editor. This set X was

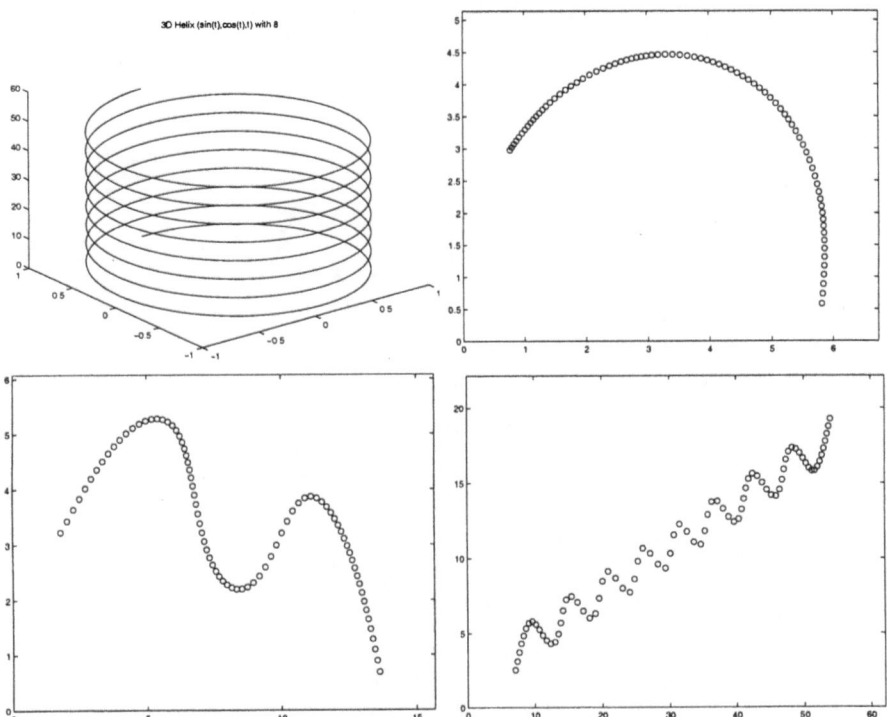

Fig. 8.5. The 3D-helix $H_8 := \{(sin(t), cos(t), t) \mid 0 \leq t \leq 16\pi\}$ and sets of 2D-projection centers for three finite subsets of H_1, H_2 and H_8 (see text).

transformed by a rotation $\phi_\alpha : \mathbb{R}^2 \to \mathbb{R}^2$ with a randomly choosen angle $\alpha \in (0, 2\pi)$ onto X_α. Then, X_α was embbed into $\mathbb{R}^{128}$ by the mapping $\phi_{ij} : \mathbb{R}^2 \to \mathbb{R}^{128}$. This is defined by $x = (x_1, x_2) \mapsto z = (z_1, \dots, z_{128})$ where $z_i := x_1$, $z_j := x_2$ and $z_k := 0$ for all k different form i and j. Obviously, the Euclidean distances between the data points $\{x_\mu\}$ and $\{z_\mu\}$ were not changed by these two transformations. After that, each component i of the data points $\{z_\mu\}$ was corrupted by a small amount of Gaussian random noise: $z_i^\mu := z_i^\mu + 0.1\mathcal{N}(0, 1)$. This set $\{z^\mu\}$ was analyzed by the ACMDS algorithm. In Fig. 8.4 (right) the set of 40 projection centers $\{p_j\}$, also A-shaped as the set X, is given.

8.5.2   3D-Helix

Next we give some results of subsets of the 3-dimensional helix

$$H := \{(sin(t), cos(t), t) \mid t \geq 0\} \subset \mathbb{R}^3.$$

In Fig. 8.5 the data set H_8, containing the first 8 loops of H is shown. From the data set H_k finite subsets X_k were equidistantly sampled with approximately 500 sample points per loop. Four data sets X_k, with $k = 1, 2, 4, 8$, were analyzed by the

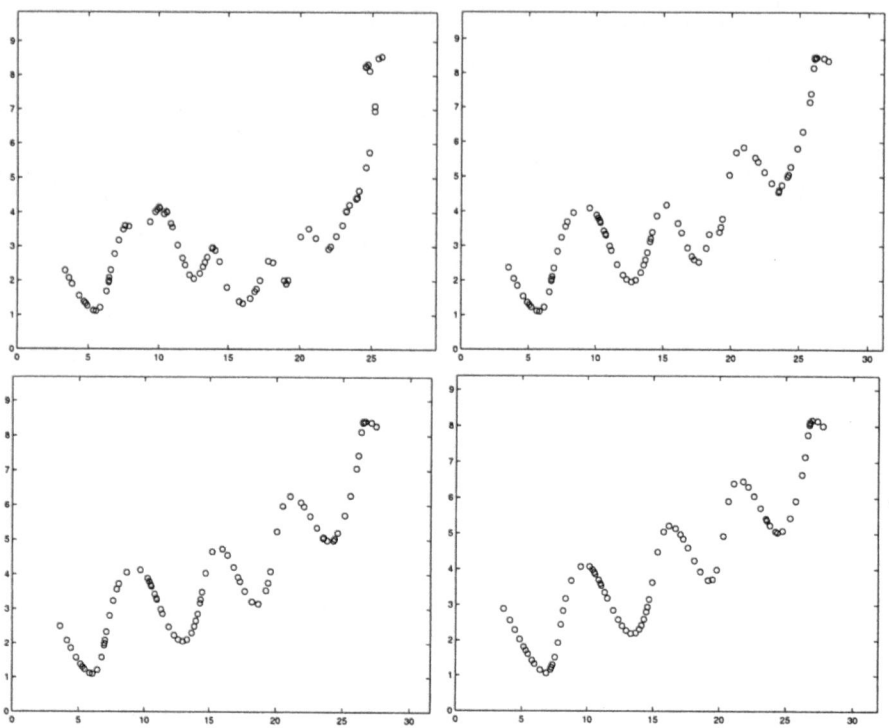

Fig. 8.6. The projection centers p_j of the data set X_4 (see text) calculated by ACMDS after 500, 1000, 1500 and 3000 online adaptation steps.

ACMDS algorithm. For the data set X_1, X_2 and X_8 sets of representation centers $p_j \in \mathbb{R}^2$ are given, see Fig. 8.5. It can be observed that the data set X_k is projected onto a sine-like curve with approximately k periods.

Fig. 8.6 shows the clustered and projected data set X_4. From the 2000 samples of X_4 80 cluster centers are projected. Snapshots after 500, 1000, 1500 and 3000 adaptation steps are given.

8.5.3   Data Set of Hand-Written Digits

The data set used in this application consisted of 10,000 hand-written digits with 1,000 samples per class (there is an independent test set with 10,000 examples). All digits were normalized in height and width. Each digit is represented by a 16 × 16 matrix (g_{ij}) where $g_{ij} \in \{0, \dots, 255\}$ is a value from a 8 bit gray scale (for details concerning the data set see [18,22]). Previously, this data set has been used for the evaluation of machine learning techniques in the STATLOG project. Details concerning this data set and the STATLOG project can be found in the final report of STATLOG [22]. Fig. 8.7 shows 60 exemplars sampled from this data base.

A set of 15 cluster centers calculated by the adaptive clustering algorithm as gray scale images is shown in Fig. 8.8. The whole set of projection centers of this

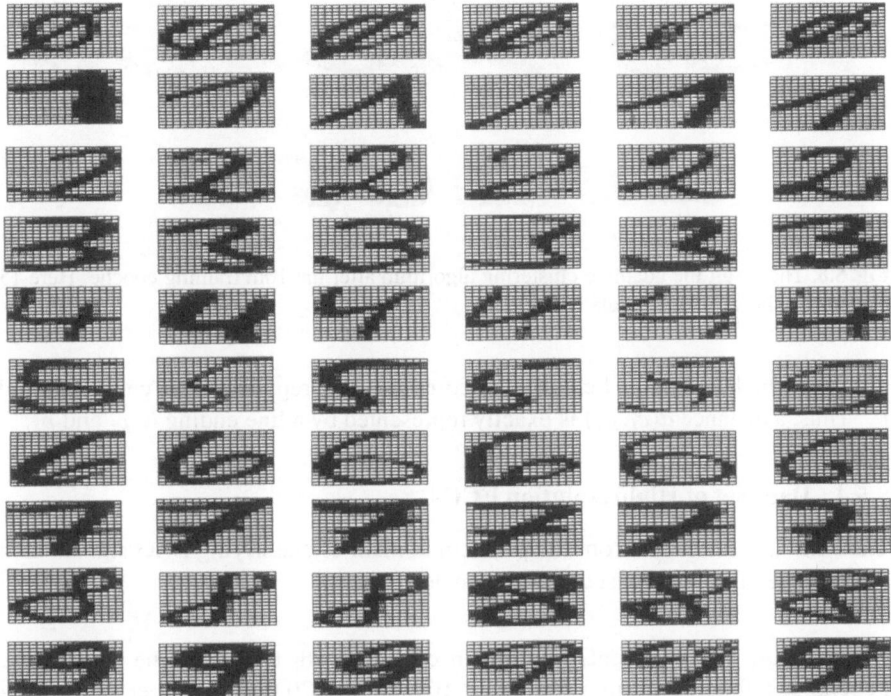

Fig. 8.7. A sample of 60 hand-written digits from the data set of 10,000 patterns. Each digit is represented by a 16×16 matrix (g_{ij}) where $g_{ij} \in \{0, \dots, 255\}$ is a value from a 8 bit gray scale.

data set is shown as a 2D-map in Fig. 8.9. Here, each projection center p_j is labeled by a class label l and frequency $freq \in [0, 1]$. Denoting the Voroni region of cluster center c_j with $V_j := \{x \in \mathbb{R}^n : d(x, c_j) = \min d(x, c_i)\}$, and $X^i := \{x \in X : x \text{ is from class } i\}$ for $i = 0, \dots, 9$ then the class label l and the frequency $freq$ are then defined by:

$$|V_j \cap X^l| = \max\{|V_j \cap X^i| : i = 0, \dots, 9\} \quad \text{and} \quad freq = \frac{|X^l \cap V_j|}{|X \cap V_j|}.$$

As described in Sec. 4, a set of representation centers $p_0, \dots, p_{14} \in \mathbb{R}^2$ is determined. This result is depicted in Fig. 8.9 and Fig. 8.10. Fig. 8.10 shows the location of the representation centers $p_0, \dots, p_{14}$, determined after the 10th training epoche. In the 2D-map each representation center p_j is labelled with its center number, the digit representing the majority in this cluster, and the corresponding fraction, i.e. (0: 7(0.90)) means that from all data points of cluster 0 about 90% are labelled with class membership 7, which can also be observed in Fig. 8.8.

In Fig. 8.10 a visualization of the stress function $S(p^1, \dots, p^k)$ is presented. Again, the 2D-map of the 15 representation centers is shown. Additionally, a distance $d(c_i, c_j)$ between the cluster centers c_i and c_j is shown by a centered line of length

Fig. 8.8. The result the adaptive clustering algorithm after the 10th training epoche. Here 15 centers are shown as gray scale images.

$d(c_i, c_j)$ which is located between the corresponding representation centers p_i and p_j. Thus, a distance $d(c_i, c_j)$ is exactly represented by a line ending in p_i and p_j.

8.5.4   Data Set of Highresolution ECGs

The next data set stems from the problem domain of classifying ECG signals with the purpose of predicting sudden cardiac death.

Background The incidence of sudden cardiac death (SCD) in the area of the Federal Republic of Germany is about 100.000 to 120.000 cases per year. Studies showed that the basis for a fast heartbeat which evolved into a heart attack is a localized damaged heart muscle with abnormal electrical conduction characteristics. These conduction defects, resulting in an abnormal contraction of the heart muscle may be monitored by voltage differences of electrodes fixed to the chest. This is the electrocardiogram (ECG) which reflects the conduction characteristics of the heart.

Damaged regions within the heart have been explored with microelectrodes and have been found to contain irregular conducting cells. This causes slow or irregular propagation of activation. In the electrocardiogram this is represented through waves that can extend beyond one heartbeat. The ECG is a save method to obtain information about the heart, i.e. there is no risk for the patient of recording an ECG. Consequently it is very desireable to use the ECG as a screening method to categorize subjects according to their risk of SCD.

High-resolution electrocardiography is used for the detection of fractionated micropotentials, which serve as a noninvasive marker for an arrhythmogenic substrate and for an increased risk for malignant ventricular tachyarrhythmias. Ventricular late potential analysis (VLP) is herein the generally accepted noninvasive method to identify patients with an increased risk for reentrant ventricular tachycardias and for risk stratification after myocardial infarction [27,7,9,10]. Techniques commonly applied in this purely time–domain based analysis are signal-averaging, high-pass filtering and late potential analysis of the terminal part of the QRS complex. The assessment of VLP's depends on three empirically defined limits of the total duration of the QRS and the duration and amplitude of the terminal low–amplitude portion of the QRS [2,3].

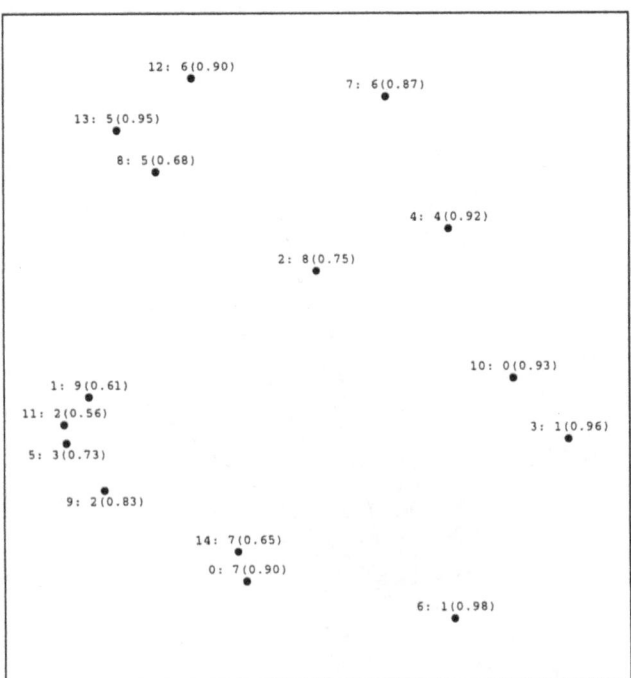

Fig. 8.9. A distance preserving 2D-mapping of the $k = 15$ cluster centers. The whole set of projection centers p_j of the data set of hand-written digits is shown as a 2D-map, where each p_j is labeled by $i : c(freq)$, with a prototype label $i \in \{0, 1, 2, \ldots, k-1\}$, a class label $c \subset \{1, \ldots, l\}$, and a frequency $freq \in [0, 1]$. Class label c is defined as the class which labels the majority of data points $x \in X \cap V_j$. Frequency $freq$ stands for the fraction of data points from this majority class c.

Subject Data and Recordings High resolution beat-to-beat recordings were obtained from 95 subjects separated into two groups: Group A consisted of 51 healthy volunteers (age 24±4.2 years, range 16-34; 31 men and 20 women) without any medication. In order to qualify as healthy, the several risk factors and illnesses had to be excluded, e.g. angina pectoris, dyspnoe, dizzyness, syncope arrhythmias, rhymatic fever, diphteria, myocarditis, influenza within the last four weeks and the cardiac risk factors hyperlipidaemia, high blood pressure, diabetes, smoking (more than 5 cigarettes a day) and the contraceptive pill. Group B consisted of 44 patients. Inclusion criteria were an inducible clinical ventricular tachycardia (>30 sec) at elec-

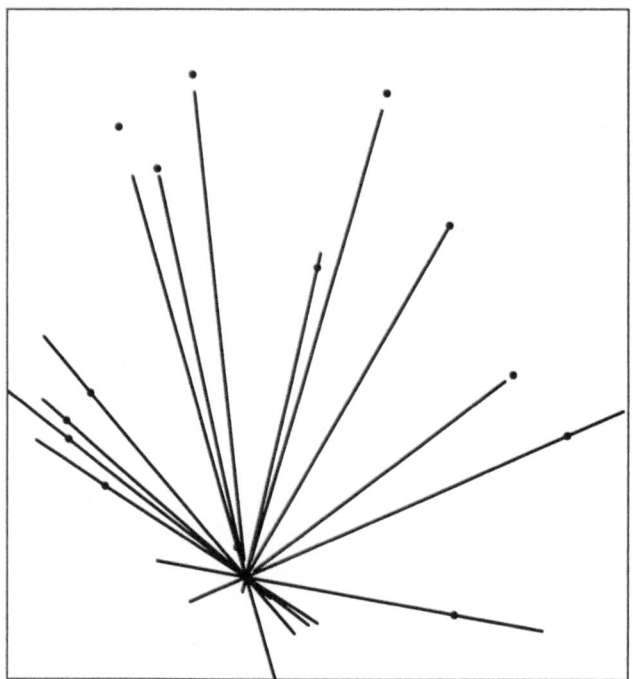

Fig. 8.10. The distance preserving 2D-mapping of the $k = 15$ cluster centers with the local stress of a single center (here for the prototype annotated with prototype label 0). The stress is visualize through centered lines between pairs (p_0, p_j) projection centers, where the length of the line is equal to the distance $d(c_0, c_j)$ between the corresponding prototypes c_0 and c_j.

trophysiologic study (EPS) with a history of myocardial infarction and coronary artery disease in angiogram.

Signal-averaged high resolution ECGs (see Fig. 8.11) were recorded from three orthogonal bipolar leads (sampling rate 2000 Hz and A/D-resolution was 16 bits filtered with 40–250 Hz bidirectional 4-pole Butterworth filter).

Features from the vector magnitude $V = \sqrt{X^2 + Y^2 + Z^2}$ are:

- Total duration of the filtered QRS:

$$QRSD := QRS_{offset} - QRS_{onset}$$

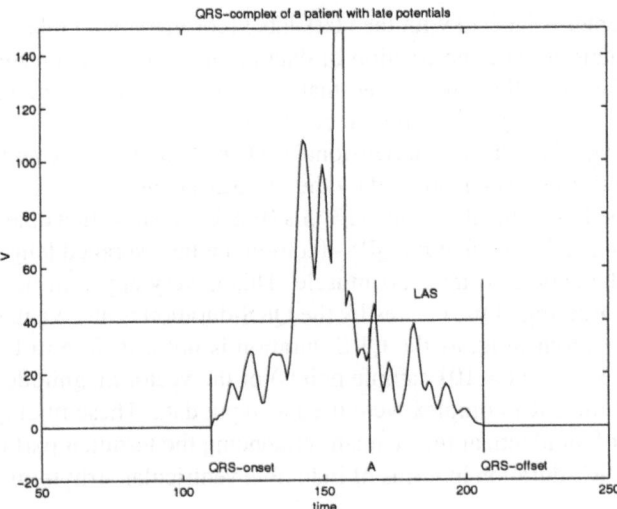

Fig. 8.11. QRS-complex of a patient with "Late Potentials". The time A is defined as $A :=$ $QRS_{offset} - 40ms$.

- RMS of the terminal 40 ms of the QRS:

$$tRMS := \sqrt{\frac{1}{QRS_{offset} - A} \sum_{i=A}^{QRS_{offset}} V_i^2}$$

- Terminal low-amplitude signal of the QRS below 40 μV:

$$LAS := QRS_{offset} - argmax\{i \mid V_i \geq 40\mu V\}$$

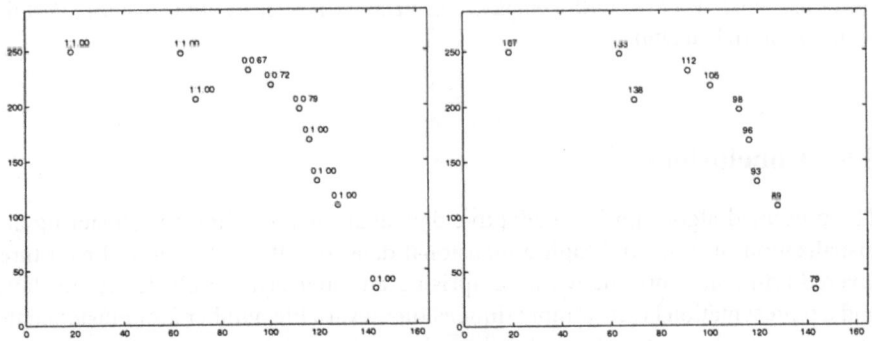

Fig. 8.12. Projection centers calculated from three features ($QRSD, tRMS, LAS$) of the QRS-complex. Annotation: Class labels of the majority (left figure); QRS_d (right figure).

First, results of the clustering and visualization procedure are shown for the three features ($QRSD, tRMS, LAS$) in Fig. 8.12. In the left part of Fig. 8.12, each projection center is labeled with the class label (coronary heart disease $=1$; healthy $=0$) of the majority and the fraction of data points within this cluster. In the right 2D-map of Fig. 8.12 the same representation centers p_j are shown, but now labeled with the mean of the QRS-duration. It can be observed that the set of representation centers is grouped around a 1-dimensional L-shaped feature space and that the chain of representation centers is ordered by the $QRSD$-labels.

These results on signal averaged ECG's (Fig. 8.12) show that conventional analysis is dominated by the feature QRS-duration, i.e the averaged length of the duration of repolarization of the heart muscle. This is very apparent in Fig. 8.12, surprisingly an ordering of the classes by the QRS-duration is also visible in Fig. 8.13. This is quite astonishing, as the QRS-duration is not a feature in this data. Here, the terminal 50 msec ($= 101$ sample points) of the vector magnitude signal V (see Fig. 8.11) of the QRS-complex were used as input data. These findings support the assumption of an identical mechanism influencing the terminal part of the QRS as well as the QRS-duration in terms of inducible ventricular arrhythmias.

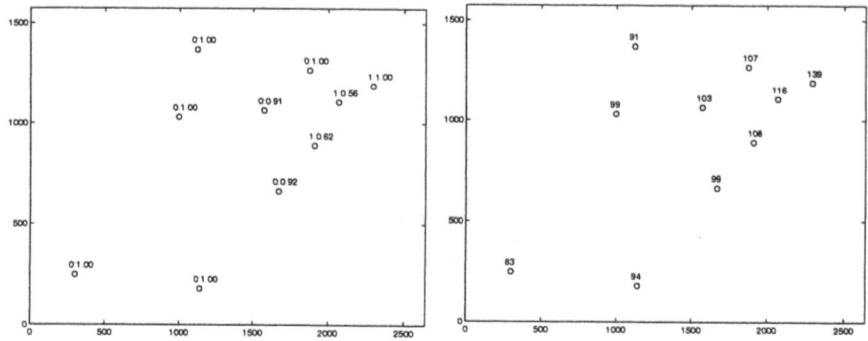

Fig. 8.13. Projection centers calculated from the terminal 50 msec ($= 101$ samples) of the vector magnitude V of the QRS-complex. Annotation: Class labels of the majority (left figure); QRS_d (right figure).

8.6   Conclusions

The presented algorithm is an adaptive data analysis procedure for clustering and visualization of large and high dimensional data sets. It can be viewed as a three layered artificial neural network comprising an input layer, a cluster center layer and a representation layer (d input dimensions, a variable number k of cluster center neurons and r 'output neurons' for the r dimensions of the representation centers). There is full connectivity from layer to layer and the input weights of the middle layer 'neurons' represent the corresponding cluster center positions, whereas their output weights represent the corresponding representation center positions. The learning rule (8.1) is basically competitive Hebbian learning, whereas the learning rule (8.6) for the output weights is more complex and non-local.

The adaptivity of this clustering and representation procedure makes it useful for many applications, where the clustering itself is part of a larger program or network operating in a changing enviroment. In such cases, where one wants the system to be able to learn without throwing away the accumulated experience and under human supervision, the procedure described here may turn out to be very useful. Furthermore, it may help in forming hypotheses about the data, which in turn may be substantiated by other statistical methods.

Bibliography on Chapter 8

1. A.J. Bell and T.J. Sejnowski. An information-maximization approach to blind separation and blind deconvolution. *Neural Computation*, 7:1129–1159, 1995.
2. G. Breithardt and M. Borggrefe. Pathophysiological mechanisms and clinical significance of ventricular late potentials. *Eur Heart J*, 7:364–385, 1986.
3. G. Breithardt, M.E. Cain, N. El-Sherif, N. Flowers, V. Hombach, M. Janse, M.B. Simson, and G. Steinbeck. Standards for analysis of ventricular late potentials using high resolution or signal-averaged electrocardiography. *Eur Heart J*, 12:473–80, 1991.
4. C. Darken and J. Moody. Fast adaptive k–means clustering: Some empirical results. In *Proceedings International Joint Conference on Neural Networks*, 1990.
5. J.H. Friedman. Exploratory projection pursuit. *Journal of the American Statistical Association*, 82(397):249–266, 1987.
6. J.H. Friedman and J.W. Tuckey. A projection pursuit algorithm for exploratory data analysis. *IEEE Transactions on Computers*, 9(c-23):881–890, 1974.
7. J.A. Gomes, S.L. Winters, M. Martinson, J. Machac, D. Stewart, and A. Targonski. The prognostic significance of quantitative signal-averaged variables relative to clinical variables, site of myocardial infarction, ejection fraction and ventricular premature beats. *JACC*, 13:377–384, 1989.
8. H.H. Harman. *Modern Factor Analysis*. University of Chicago Press, 1967.
9. M. Höher and V. Hombach. Ventrikuläre Spätpotentiale – Teil I Grundlagen. *Herz & Rhythmus*, 3(3):1–7, 1991.
10. M. Höher and V. Hombach. Ventrikuläre Spätpotentiale – Teil II Klinische Aspekte. *Herz & Rhythmus*, 3(4):8–14, 1991.
11. K. Hornik. Konvergenzanalyse von nn–lernalgorithmen. In G. Dorffner, K. Möller, G. Paaß, and S. Vogel, editors, *Konnektionismus und Neuronale Netze*, volume 272 of *GMD-Studien*, pages 47–62, 1995.
12. P.J. Huber. Projection pursuit. *The Annals of Statistics*, 13(2):249–266, 1985.
13. A. Hyvärinen. Survey on independent component analysis. *Neural Computing Surveys*, 2:94–128, 1999.
14. A.K. Jain and R.C. Dubes. *Algorithms for Clustering Data*. Prentice Hall, Englewood Cliffs, New Jersey, 1988.
15. I.T. Jollife. *Principal Component Analysis*. Springer-Verlag, 1986.
16. M. Kendall. *Multivariate Analysis*. Charles Griffin & Co, 1975.
17. T. Kohonen. *Self-Organizing Maps*. Springer, 1995.
18. U. Kreßel. The Impact of the Learning-Set Size in Handwritten-Digit Recognition. In T. Kohonen, editor, *Artificial Neural Networks*. ICANN-91, North-Holland, 1991.
19. Y. Linde, A. Buzo, and R.M. Gray. An algorithm for vector quantizer design. *IEEE Transactions on Communications*, 28(1):84–95, 1980.
20. S.P. Lloyd. Least squares quantization in PCM. *IEEE Transactions on Information Theory*, 28(2):129–137, 1982.
21. J. MacQueen. Some methods for classification and analysis of multivariate observations. In L.M.LeCam and J.Neyman, editors, *Proceedings of the Fifth Berkeley Symposium on Mathematical Statistics and Probability*, volume I, pages 281–297. Berkeley University of California Press, 1967.
22. D. Michie, D.J. Spiegelhalter, and C.C. Taylor. *Machine Learning, Neural and Statistical Classification*. Ellis Horwood, 1994.
23. J. Moody and C.J. Darken. Fast learning in networks of locally-tuned processing units. *Neural Computation*, 1(2):281–294, 1989.

24. H. Ritter and K. Schulten. Convergence properties of Kohonen's topology converving maps:fluctuations,stability, and dimension selection. *Biological Cybernetics*, 60:59–71, 1988.
25. J.W. Sammon. A nonlinear mapping for data structure analysis. *IEEE Transactions on Computers*, C-18:401–409, May 1969.
26. D.W. Scott. *Multivariate Density Estimation*. John Wiley & Sons, New York, 1992.
27. M.B. Simson. Use of Signals in the Terminal QRS Complex to Identify Patients with Ventricular Tachycardia after Myocardial Infarction. *Circulation*, 64(2):235–242, 1981.

9 Self-Organizing Maps and Financial Forecasting: an Application

Marina Resta

Abstract. Financial markets provide a singular field of analysis and exciting challenges for researchers. During the past decades, strongest assumptions on financial time-series (namely the Random Walk Hypothesis) have been partially discharged, and useful new paradigms (such as chaos, self similarity, self-organized criticality) have been discovered to this field. One of the arguments with respect to random walk is that chaotic time series can be long term random, and still have short term patterns. If, in fact, the marketplace has short term patterns, and some technicians think it does, then there are windows of opportunity that the random walker either denies exist or are unable to be exploited. Self Organizing Maps (SOMs), therefore, offer a powerful tool, suitable to explore financial markets, in order to find significant patterns, and use them as a forecasting tool. Additionally, the possibility to detect features of the market can be employed to find out the statistics of the process under examination, hence to decide whether or not the Random Walk Hypothesis is acceptable. This chapter will explore the use of Self Organizing Maps for such purpose. Both the conditions needed for data to be treated, and the evaluation of results obtained are considered.

9.1 Introduction

The primary aim of this chapter is to survey the recent literature with respect to financial markets behaviour, focusing both on the common and uncommon pitfalls, that should be avoided when it is used to deal with price changes forecasting. It is also used as an alternative approach to this task using neural networks by means of Self Organizing Maps.

It is necessary at this stage to examine the Random Walk Hypothesis [2] due to Louis Bachelier.

The basic idea underlying Random Walk Hypothesis (RWH since now on), is that:

- price changes are independent, and stationary random variables, distributed in a gaussian fashion:

$$Z(t + \tau) - Z(t) \sim \aleph(0, \sigma^2(\tau)) \tag{9.1}$$

where $Z(t + \tau), Z(t)$ are random variables modeling the logarithm of price changes at times $t + \tau$ and t respectively; $\aleph(0, \sigma^2(\tau))$ is a Gaussian process with zero mean, and variance $\sigma^2(\tau)$ is generally proportional to time lag τ.

- price changes reflect all possible information presently available to the markets;
- it is not possible for investors to use information available to gain more than others.

Briefly stating, RWH implies the absence of exploitable excess profit opportunities, both by running any policy of arbitrage and fundamental analysis, as well as by means of any tools of technical analysis.

Over the years the original vision of Bachelier has been integrated into a weaker theoretical framework, more in keeping with the evidence of real markets. The Efficient Market Hypothesis (EMH), developed in earlier sixties, relies on this, and is stated as follows:

"a competitive market may be considered efficient if price changes follow a martingale random process".

The EMH implies the martingale model. A process $Z(t)_{t \in N}$, adapted to a filtration F_t, is said to be a martingale if it has finite first moment and:

$$E[Z(t + \tau)|F_t] = Z(t), \forall \tau > 0 \qquad (9.2)$$

That is the conditional expected value of the random variable $Z(t + \tau)$, with the filtration F_t, i.e. given the values for $Z(t)$ and for all the $Z(i)$, $i = 0, ..., t - 1$, equals the value of random variable at time t. Equation (2) says that if $Z(t)$ follows a martingale, then the best forecast of $Z(t + \tau)$ that could be constructed, based on the current level of information F_t, would just equal to $Z(t)$.

As originally stated, this assumption is weaker than the RWH, but consistent with the random nature of price changes. The basic idea is that asset prices can be dependent but if they follow a process *linear in the mean* (i.e. with white spectrum), then any excess of information can be smoothed through efficient arbitrage policies, which tend to eliminate in the operators any profitable occasions for surplus of gain.

From such a statement, the definitions of informational capital market efficiency follow. According to Fama [12], a market is:

- *efficient in weak form*, if the information includes past prices and returns alone;
- *efficient in semi-strong form*, if information incorporates all public information;
- *efficient in strong form*, if information summarizes all information both public and private.

Strong-form efficiency implies semi-strong efficiency, which in turn allows for weak form efficiency, but the reversal is not ever true.

Additionally, from an operative point of view, those definitions imply respectively that margins exist to both fundamental analysis methods and arbitrage opportunities, or just to arbitraging, or to neither of them, according to the different degree of generalization of their statements.

As common point, they all deny any substance to technical analysis, as method to maintain gains over the market mean.

Technical Analysis (TA) deals with the arguments that (i) using proper tools to smooth data (such as: moving averages, Fourier transforms, Kalman filtering), or (ii) finding significant patterns or recurrent features among prices in the market, it should be possible to figure out useful additional information about movements on the run, hence making advantageous corrections to the position assumed.

The field has been developed from miscellaneous contributions, from the extreme positions of pure graphicists, who scan the markets for significant patterns, and the pure fundamentalists, searching for plausible economical foundations for contingent markets behaviour. There has been space for many intermediate roles, which combine, employing on various degrees of compromise (and imagination!), the use of both graphical and technical indicators, all trying to beat the market. Hence, it is not surprising there is a mixture of scepticism and perplexity accompanying TA, despite of its effective potential. However, the idea to extract patterns from prices sequence and to use them to forecast future behaviour of the market, is quite remarkable, since in this way it should be possible to identify their relevant features.

The significance of this approach has been of interest in recent years. Starting from early seventies strong controversies have arisen between supporters of "conventional" assumptions, and those who found both theoretical and empirical evidences to deny those statements. Mandelbrot [21], for instance, claimed on limitations of EMH and martingale models, pointing on the existence for a class of cases, where useful implementation of arbitrage was impossible.

At the same time, many research studies (see among others: [3], [8]) outlined the evidence of fat tails, as well as anomalous skewness and kurtosis values were exploited in many empirical processes, in contrast with strictly gaussianity assumptions. Additionally, empirical prices tend to exhibit non stationarity, as well as they tend to be better fitted by theoretical processes lacking in moments higher than the first.

Table 9.1. Most diffused non linearity and chaos tests.

Test	Author	Meaning of Results
BDS	Brock, Dechert, LeBaron, Scheinkman [4]	Nonlinearity
Correlation Dimension	Grassberger and Procaccia [13]	Chaos
False Nearest Neighbors	Kennel, Brown and Abarbanel [19]	Chaos
Hinich Bispectrum	Hinich [16], [15]	Lack of 3rd order dependence
Lyapunov Exponent	H. Kantz [18] version	Chaos
NEGM	Nychka, Ellner, Gallant, McCarey [24]	Chaos

Starting from this point, new paradigms such as chaos, self similarity,and self-organized criticality have gained the attention of researchers.

In particular, the interest in chaotic processes has induced the development of a battery of tests, to verify whether or not market price changes may be considered due to a probabilistic or a deterministic process: Table 9.1 reports some of the most diffused tests actually in use, and main purposes they serve to. These tests will be considered further later in the chapter.

Strictly linked to this approach is that of self similarity, and self-organized criticality, whose kernel may be roughly explained as follows. Let us imagine that it is possible to state that market prices follow a rather complicated deterministic rule. Hence, the idea is that such an underlying rule regulates the way the market behaves, with similar patterns replicating over the time, but for a matter of scaling parameters. In this way, it has been introduced the Fractal Hypothesis (FH) for financial markets [22] [5], where the assumption is made that global behaviour comes as a multiplicity of micro patterns replicating periodically in a similar fashion. From a more formal point of view, a process $Z(t)_{t \in N}$ is called *multifractal* if it satisfies:

$$E[|Z(t)|^q] = c(q)t^{\tau(q)+1}, \forall t \in \mathcal{T}, q \in \mathcal{Q} \qquad (9.3)$$

where $\mathcal{T}, \mathcal{Q}$ are non-empty intervals on the real line, with $0 \in T$, $[0, 1] \in Q$ and $c(q)$ and $\tau(q)$ (which is called the *scaling function* of the process), are functions with domain $\mathcal{Q}$. Multifractals are called *uniscaling* when $\tau(q)$ is linear, *multiscaling* otherwise.

The current tendency, is to think financial time series as chaotic ones, as they are long term random, and still have short term patterns thus creating profit opportunities for traders in the short period.

Finally, a unifying vision has been suggested during the past decade [6]. The idea is to formulate the martingale model in terms of differences as follows:

$$E[Z(t + \tau) - Z(t)|F_t] = 0 \qquad (9.4)$$

that is $Z(t)$ is martingale if and only if $Z(t + \tau) - Z(t)$ is a fair game.

From Eq. (4) it comes:

$$Z(t + \tau) = Z(t) + \varepsilon_t \qquad (9.5)$$

where ε_t is the martingale difference. Equation (5) appears to be similar to that implied by equation (1), but, as previously stated, for weaker basics assumptions, since it simply requires the first moment of price changes to exist, where on the other hand RWH imposes conditions over variable higher moments.

In this way, three versions of RWH may be distinguished:

- independently and identically distributed returns;
- independent returns,
- uncorrelated returns

versions. The latter is well accomplishing to the formulation given by equation (5), because prices changes, although uncorrelated, may not be independent over time, exhibiting clusters of volatility (that is what generally happens over real markets) .

This long discussion serves to multiple purposes, since it shows a few critical points:

- the importance of patterns in market prices behavior is still controversal, since one gives more credit to RWH or martingale models rather than to FH. Also in this case, it is not clear at all how self-similarity has to be interpreted from an operative point of view;
- all the techniques described tend to give indications about the nature of the process under examination, but at the moment they are not able to add further information about its future behaviour. Also no methods are known which are able to efficiently implement these techniques into a trading system;
- all the methods mentioned must be carefully managed, because such calculations using time series data can often yield erroneous results. The literature contains many examples of poor or erroneous calculations and it has been shown that some popular methods may produce doubtful results. Limited data set size, noise, nonstationarity and complicated dynamics also serve to compound the problem.

The present work takes into account more than 25000 daily fluctuations x_t from January 2nd 1915 until October 22nd 1999, using:

$$\{x_t\}_{t \in N} : x_t = \frac{(p_{t+1} - p_t)}{p_t} \tag{9.6}$$

where p_t, p_{t+1} are prices at time $t, t+1$ from the Dow Jones index.

Such data will be studied, using testing methodologies, from both chaotic and stochastic processes theory.

An alternative method, which attempts both to detect the nature of data, and to implement such information in a forecasting key, will also be extensively discussed.

This technique relies on the power of Self Organizing Maps [20] as pattern recognition tools applied to financial time series, in order to discover any relevant features.

The outline of the chapter is as follows.

- Sec. 2 is devoted to the study of Dow Jones data (DJ) by means of conventional methods: main linear and non linear tests will be briefly described and hence applied to the time series under examination; advantages and drawbacks of such methods for financial (advanced) practitioners will be outlined.
- Sec. 3 will introduce a novel technique based on Self Organizing Maps which appears to be quite promising, since it can be applied both to detect intrinsic nature of data and also in forecasting. Both aspects will be investigated, and some *instructions* for their use will be suggested. Finally,
- Sec. 4 will draw some conclusions and make some observations for future developments.

9.2 Managing Financial Time Series

Managing time series is apparently a simple task, and consists of just following a few number of rules.

Rule number one: **an important requirement is to have a large dataset**.

With a large number of samples, it is possible to perform a great deal of analysis. As a rule of thumb, the set of data should be partitioned in a reasonable number of subsets, in order to test the robustness of desired hypotheses

(i) over the whole of the sample, as well as

(ii) over the aggregation of the single parts of the sample.

In this study, for instance, data has been partitioned into 10 sets, each set is made of about 2500 elements, and results have been generally studied for both all of two sets.

Another rule is **making exploratory data analysis (EDA) as first step**.

EDA is generally of great help, since it makes possible to have a visual impact on dataset main features. Consider Fig. 9.1.

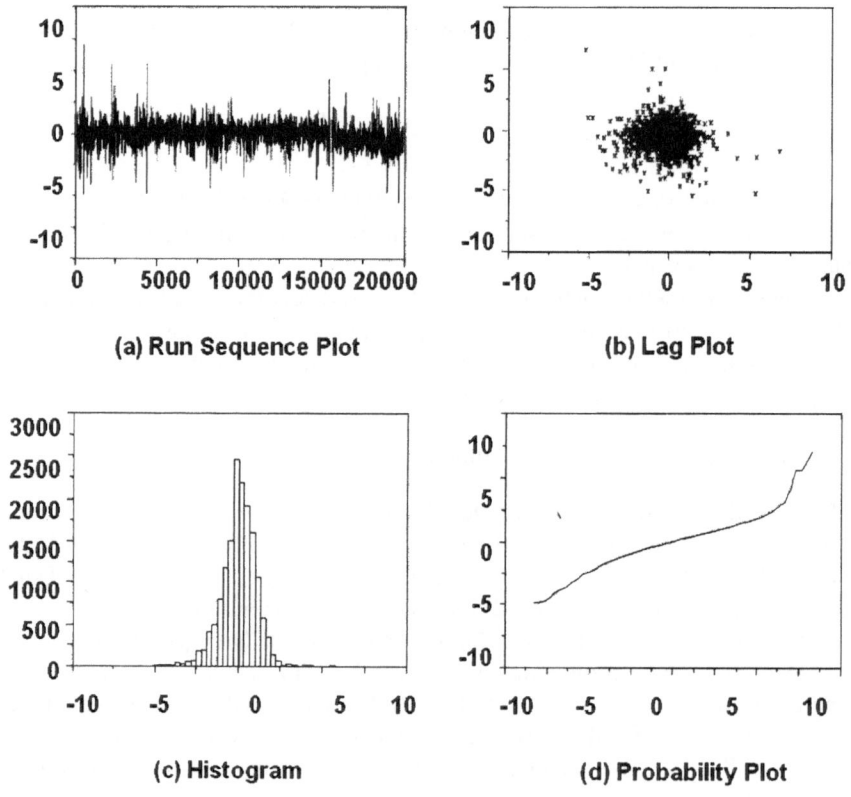

(a) Run Sequence Plot

(b) Lag Plot

(c) Histogram

(d) Probability Plot

Fig. 9.1. EDA on Dow Jones data.

The advantage of such representation stands in the fact that further information is added to general comprehension of the sequence. The lag plot, for instance, is a representation of the relationships between observations $x(t)$ and $x(t + \tau)$, where τ is the lag separating them. If data are normally distributed, then the corresponding lag plot area would be completely covered by points, thus evidencing the absence of any underlying structure. On the other hand, perfect linear relationships produce lag plots with points aligned on a straight line. In this case (where $\tau = 1$ has been used), the basic information which may be obtained from the data is that the generating process is not gaussian, and probably relationships between them are not linear, since points tend to distribute as an irregular shaped cloud around the central section of the plot area. Additionally, the histogram appears to be skewed to the right. A further confirmation supporting the rejection of gaussianity, comes by the probability plot of data. In this plot the order statistics of real sequence is compared to that of normally distributed data, with location and scaling parameters estimated through the empirical ones: in the case of well-fitting data from such theoretical distribution, points in the plot should distribute along the straight line. In this case, departures from the straight line stands for excess of skewness (respect on normal hypothesis) and fat tails.

EDA methods should serve to *first look* purposes, looking deepest inside the data being required, in order to both understand the sample under examination, and use past data to forecast the future. A promising pathway, in such sense, should be to go one step further the evidence coming from the probability plot. Rule number three: **the next rule is that the data distribution should be tested.**

9.2.1   Distribution Tests

Things become more and more complicated! If data are not normally distributed, is it possible to find a well fitting theoretical distribution? A number of tests has been developed for such purposes: in this context Kolmogorov-Smirnov and Anderson-Darling tests have been run on data. An exhaustive description of these procedure is beyond the scope of this work. For more information on this see [7] and [28]. In this case both tests are able to verify whether or not the data comes from specific distributions. Proofs have been made for several different skewed distributions. For the sake of simplicity, only results for the assumption of Lognormal, Exponential, and Gaussian distributions are shown in Table 9.2.

Poor results have been summed, both on the whole sample and in its parts: none of distributions under examination resulted reasonably in fitting data sequences.

Although such tests have advantages, mainly that of being not dependent from sample size, they also have some drawbacks.

Perhaps the most serious limitation is that the distribution must be fully specified. That is, if location, scale, and shape parameters are estimated from the data, the critical region of these tests is no longer valid. Typically it must be determined by simulation.

Hence, although they represent an obliged step to study data, distribution tests are not completely reliable, and further proofs need to be performed.

Rule number four: **monitor the nature of the data.**

Table 9.2. Distribution tests: Normal Kolmogorov-Smirnov goodness of fit.

Test Set	Null Hyp. H0	K-S Value	$\alpha = 10\%$	$\alpha = 5\%$	$\alpha = 1\%$	Conclusions
DJ1	NORMAL	0.690E−01	0.03099	0.03454	0.04140	Reject H0
DJ2	NORMAL	0.829E−01	0.03204	0.03572	0.04281	Reject H0
DJ3	NORMAL	0.962E−01	0.03150	0.03512	0.04209	Reject H0
DJ4	NORMAL	0.751E−01	0.03150	0.03512	0.04209	Reject H0
DJ5	NORMAL	0.1485743	0.03150	0.03512	0.04209	Reject H0
DJ6	NORMAL	0.1689017	0.03150	0.03512	0.04209	Reject H0
DJ7	NORMAL	0.1433285	0.03150	0.03512	0.04209	Reject H0
DJ8	NORMAL	0.939E−01	0.03150	0.03512	0.04209	Reject H0
DJ9	NORMAL	0.414E−01	0.03150	0.03512	0.04209	Reject H0
DJ10	NORMAL	0.840E−01	0.03150	0.03512	0.04209	Reject H0

Table 9.3. Distribution tests: LogNormal Kolmogorov-Smirnov goodness of fit.

Test Set	Null Hyp. H0	K-S Value	$\alpha = 10\%$	$\alpha = 5\%$	$\alpha = 1\%$	Conclusions
DJ1	LOGNORMAL	0.2292848	0.03099	0.03454	0.04140	Reject H0
DJ2	LOGNORMAL	0.968E−01	0.03204	0.03572	0.04281	Reject H0
DJ3	LOGNORMAL	0.897E−01	0.03150	0.03512	0.04209	Reject H0
DJ4	LOGNORMAL	0.2610565	0.03150	0.03512	0.04209	Reject H0
DJ5	LOGNORMAL	0.3889095	0.03150	0.03512	0.04209	Reject H0
DJ6	LOGNORMAL	0.4231583	0.03150	0.03512	0.04209	Reject H0
DJ7	LOGNORMAL	0.3784130	0.03150	0.03512	0.04209	Reject H0
DJ8	LOGNORMAL	0.3414157	0.03150	0.03512	0.04209	Reject H0
DJ9	LOGNORMAL	0.2514822	0.03150	0.03512	0.04209	Reject H0
DJ10	LOGNORMAL	0.2483514	0.03150	0.03512	0.04209	Reject H0

9.2.2 Linearity and Non-Linearity Tests

Since the evidence of non-gaussianity of real data first appeared, a great deal of effort has been spent to build up tools to give a strong theoretical basis to alternative developments:

Following this development, Dow Jones data has been extensively examined by various techniques, which come from Chaos and Non-Linear Processes Theory.

The kernel of this approach may be summarized as follows. Firstly imagine that the data is governed by an unknown process. Then essential features of such generating process can be captured if and only if it is possible to move into its phase space, and to estimate its *correlation dimension*.

From a formal point of view, the Correlation Dimension (CD) is one of the fractal dimensions associated with an inhomogeneous attractor [14], and it estimates the

Table 9.4. Distribution tests: Exponential Kolmogorov-Smirnov goodness of fit.

Test Set	Null Hyp. H0	K-S Value	$\alpha = 10\%$	$\alpha = 5\%$	$\alpha = 1\%$	Conclusion
DJ1	EXP	0.1043695	0.03099	0.03454	0.04140	Reject H0
DJ2	EXP	$0.725E-01$	0.03204	0.03572	0.04281	Reject H0
DJ3	EXP	0.4920000	0.03150	0.03512	0.04209	Reject H0
DJ4	EXP	0.1202800	0.03150	0.03512	0.04209	Reject H0
DJ5	EXP	0.2459789	0.03150	0.03512	0.04209	Reject H0
DJ6	EXP	0.2467793	0.03150	0.03512	0.04209	Reject H0
DJ7	EXP	0.2971990	0.03150	0.03512	0.04209	Reject H0
DJ8	EXP	0.2041992	0.03150	0.03512	0.04209	Reject H0
DJ9	EXP	0.1223868	0.03150	0.03512	0.04209	Reject H0
DJ10	EXP	0.107033	0.03150	0.03512	0.04209	Reject H0

Table 9.5. Distribution tests: Anderson-Darling Normal Test.

Test Set	Null Hyp.H0	Location	Scaling	A-D value	$\theta = 95\%$	Conclusion
DJ1	NORMAL	$0.2653E-01$	1.55690	10.03559	1.321	Reject H0
DJ2	NORMAL	$-0.296E-01$	31.93687	10.03559	1.321	Reject H0
DJ3	NORMAL	$0.55813E-01$	2.063903	10.03559	1.321	Reject H0
DJ4	NORMAL	$0.11002E-01$	1.144464	26.77245	1.321	Reject H0
DJ5	NORMAL	$0.28437E-01$	0.797861	23.94515	1.321	Reject H0
DJ6	NORMAL	$0.61256E-01$	0.669755	14.57517	1.321	Reject H0
DJ7	NORMAL	$0.30032E-01$	0.665021	1.963789	1.321	Reject H0
DJ8	NORMAL	$0.12569E-01$	0.653438	9.711954	1.321	Reject H0
DJ9	NORMAL	$-0.2343E-03$	0.781681	4.658472	1.321	Reject H0
DJ10	NORMAL	$0.5364E-02$	0.973309	3.096840	1.321	Reject H0

average number of data points within a radius r of the data point under examination:

$$CD(r) = \frac{1}{MN} \sum_{k=1}^{M} \sum_{n=1}^{N} \theta(r - ||z(k) - z(n)||) \qquad (9.7)$$

where $||.||$ is the distance between points $z(k)$ and $z(n)$. The Euclidian distance is used, but in principle any distance measure will do. Finally, $\theta(.)$ is the Heaviside function, and it is one when its argument is positive, zero when its argument is negative. This expression counts the number of points of the data set, which are closer than radius r or within a hypersphere of radius r, and then divides by the total number of points NM one looks at. Typically N and M are of order the total number of samples.

Generally, the CD is interpreted as follows. If a signal is stochastic, the Correlation Dimension tends to increase as embedding dimension becomes larger and larger. Otherwise, if the signal is chaotic the CD becomes small.

From the financial point of view, CD is relevant if it is stable at reasonable low levels, of say up to 10 for daily sequences. If CD is small enough, it means that over the short term there should be profitable margin to operate over the market average. On the other hand, if the CD is stable but too high, say in the order of $20 - 30$ for daily data, it means the information is not very appealing for financial studies, which develop appealing trading opportunities in the market.

Figures 9.2 and 9.3 show the behaviour of correlation dimensions over Dow Jones data. The x axes show embedding dimension (ED), and the y axes the corresponding values for the estimation of correlation dimension (ECD). Since ED is available only at discrete values ($ED \in N$) the estimate has been obtained by interpolating through a pure power law.

It can be observed that CD tends to remain at low levels, both in smaller samples of data and in the sequence as a whole. This confirm that the data has a chaotic rather than gaussian nature.

A rule of thumb for the estimation of correlation dimension should be to accompany it with the estimation of false nearest neighbors and Lyapunov exponents, in order to get further confirmation to results. These are explained below.

False Nearest Neighbors is a method of choosing the minimum embedding dimension of a one-dimensional time series. This method finds the nearest neighbor of every point in a given dimension, and then checks to see if these are still close neighbors in one higher dimension. An n-dimensional neighbor is considered false if the distance between the point and its neighbor in the $n + 1$st dimension is greater than the following criterion: "the average distance between the point and its neighbor in each of the previous n dimensions". The percentage of False Near Neighbors should drop to zero when the appropriate embedding dimension has been reached. The appropriate embedding dimension is a bit of a matter of judgement.

The *Lyapunov Exponents* of a system describe the exponential rates of convergence or divergence of nearby trajectories [27], [10]. Each exponent is the average rate of growth of infinitesimal perturbations along any direction. Chaos arises from the exponential growth of infinitesimal perturbations. Perhaps, the most widely accepted definition of chaos is that a chaotic system is one which has *at least* one positive Lyapunov exponent. If the data is embedded in too high of an embedding space, then extra, spurious exponents will be produced. These spurious exponents will be highly unstable with respect to parameter changes but Lyapunov exponents are independent of the embedding procedure. Thus any suitable embedding should, in theory, produce the same Lyapunov spectrum.

When applied to this data, such tests have produced some contradictory results.

As it can be seen from Table 9.6 the minimum percentage of false nearest neighbors tends to have extremely high values at the highest embedding dimension level of 100. Also in these cases it is close to values of stochastic processes (50%). This

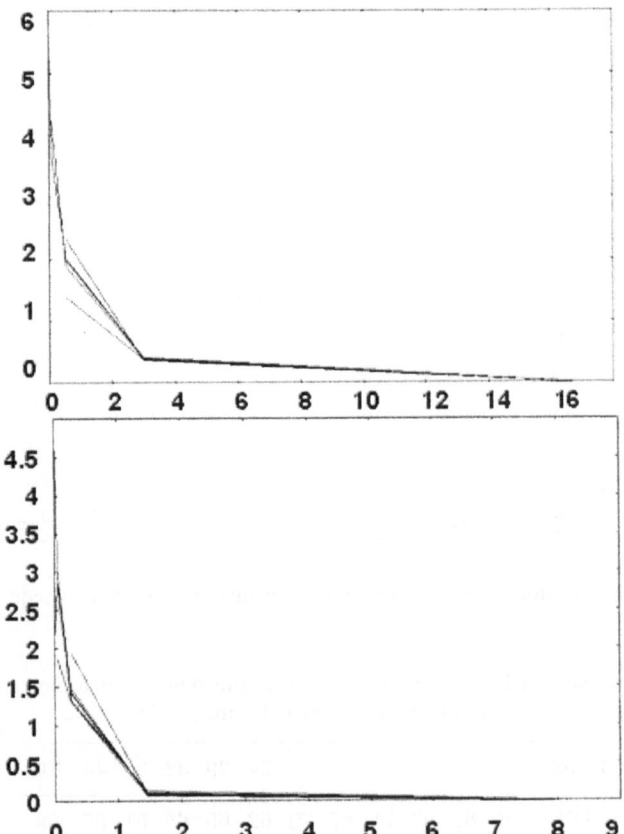

Fig. 9.2. Correlation dimensions on Dow Jones sets at various threshold levels. Results for two sets are shown. Five different values of the threshold in the range among minimum and maximum of the data have been used. Different lines correspond to different threshold settings.

is strongly in contrast to previous tests on Dow Jones data, which accounts for correlation dimension stable, since at relatively low embedding dimension levels.

The behaviour of the Lyapunov Exponents (see Figures 9.4, 9.5), which remains at negative values for each data test set is not consistent with previous observations of Correlation Dimensions.

Since the use of traditional tools for exploring chaos in time series seemed of little help for Dow Jones data, other test methods of general non-linearity, such as Hinich bispectral test [15], and BDS [4] statistics have been examined.

To such purpose, as it can be read from Table 9.7, the sequence has been whitened from any possible linear dependence through a proper AR(p) process acting as a filter, according to the Schwartz (SC) [17] criterion. As further remark, it is notable to observe that although pre-whitening is not necessary for Hinich bispectral test, since it directly verifies for a nonlinear generating mechanism, it has been

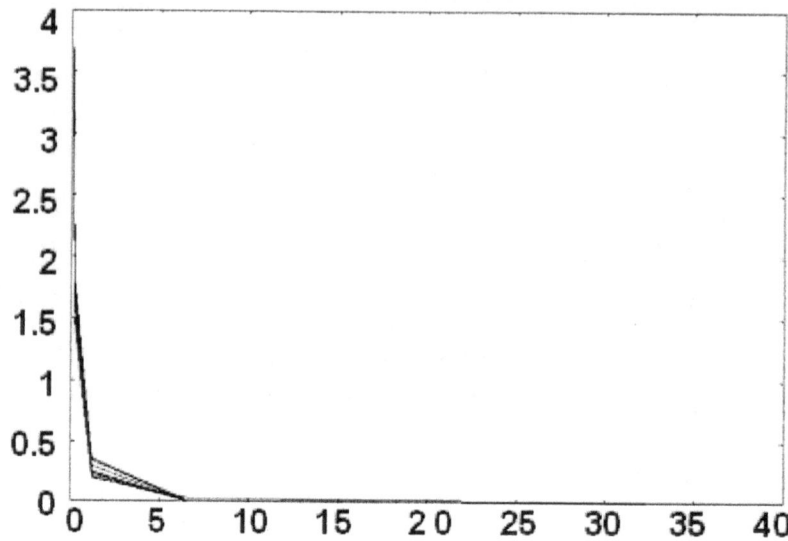

Fig. 9.3. Correlation dimensions of Dow Jones. In this case the whole sequence has been considered.

Table 9.6. False Nearest Neighbors method when embedding dimensions is varied in the range from 1 up to 100. The last line shows results for the whole sample.

Test Set	1	2	5	7	17	22	24	29	32	37	44	100
DJ1	89	80	66	71	67	70	62	65	64	51	57	59
DJ2	85	81	82	76	72	71	70	68	62	61	59	58
DJ3	90	83	80	77	77	74	72	72	72	68	68	69
DJ4	89	84	84	74	72	70	67	66	65	63	63	61
DJ5	90	83	78	68	65	63	61	62	62	62	62	60
DJ6	89	80	80	73	72	68	66	63	59	58	56	55
DJ7	87	82	80	74	66	64	69	68	67	64	60	53
DJ8	88	88	80	74	67	67	63	62	57	54	50	48
DJ9	93	82	77	73	66	64	64	62	60	57	50	47
DJ10	88	79	77	68	70	70	68	65	63	64	61	54
DJ	80	78	71	69	69	64	59	52	49	44	35	30

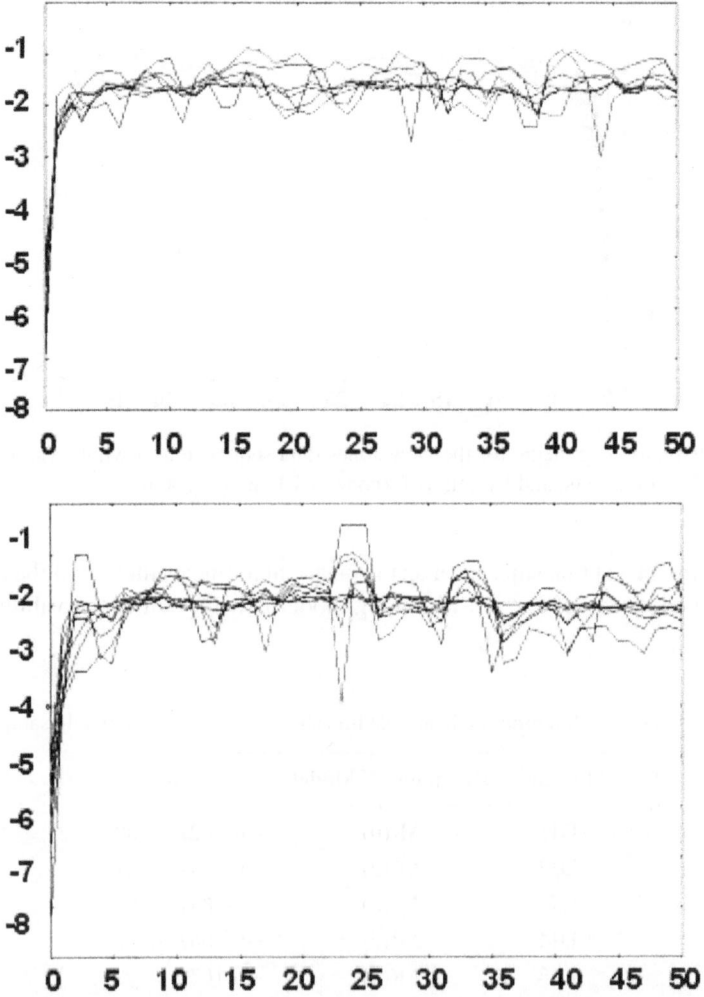

Fig. 9.4. Lyapunov exponents for sample data sets. On x-axis Embedding Dimensions ED are shown, while y-axis shows values obtained for Lyapunov Exponents, LE. Results are reported in two cases, but these statements held for all samples. Various parameter levels are considered, moving among minimum and maximum values five different thresholds have been taken into account. Different lines correspond to different threshold assumptions.

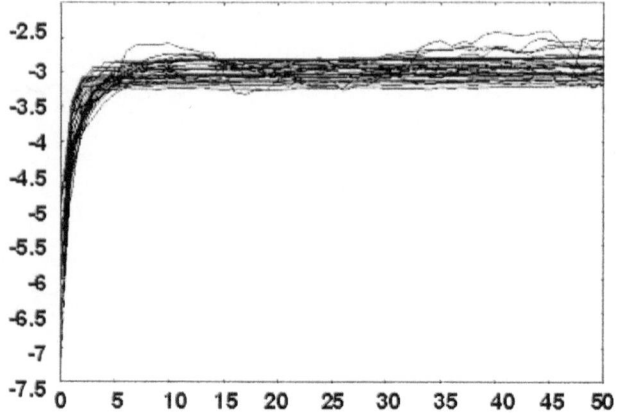

Fig. 9.5. Lyapunov exponents for the Dow Jones data sequence as a whole. Embedding Dimension ED is on x-axis, and Lyapunov Exponents LE are on y-axis.

recently proved [1] that since such test is invariant to linear filtering of the data, and hence, the adequacy of the pre-whitening model is irrelevant to the validity of the test.

Table 9.7. Prewhitening on data. Last line shows results for the whole sample.

Test Set	Minimum SC Model	SC
DJ1	AR(0)	$-0.162E + 02$
DJ2	AR(2)	$-0.714E + 01$
DJ3	AR(0)	$-0.164E + 02$
DJ4	AR(2)	$-0.165E + 02$
DJ5	AR(1)	$-0.167E + 02$
DJ6	AR(1)	$-0.168E + 02$
DJ7	AR(1)	$-0.166E + 02$
DJ8	AR(1)	$-0.167E + 02$
DJ9	AR(1)	$-0.165E + 02$
DJ10	AR(1)	$-0.163E + 02$
DJ	AR(0)	$-0.334E + 01$

At this time, 5000 N-samples were drawn at random from the empirical distribution of the observed N-sample of data. The size of each test has been therefore analyzed using both asymptotic theory and the bootstrap for samples of $N = 2500$ and $N = 25000$. The use of both asymptotic theory and bootstrap may be justified since problems with asymptotically validity of nonlinear tests are known [1].

Results for Hinich bi-spectral test are reported in Tables 9.8, 9.9.

The kernel idea of this test is to interpret the skewness function of data by means of the bispectrum of the series in the frequency domain Ω:

$$B_{xxx}(\omega_1,\omega_2) = \sum_{r=-\infty}^{+\infty} \sum_{s=-\infty}^{+\infty} C_{xxx}(r,s)e^{i2\pi(\omega_1 r+\omega_2 s)} \quad (9.8)$$

where $B_{xxx}(\omega_1,\omega_2)$ is the double Fourier transformation of the third order moments function $C_{xxx}(.)$. By defining the skewness function in terms of the bispectrum, we get:

$$\Gamma(\omega_1,\omega_2) = \frac{|B_{xxx}(\omega_1,\omega_2)|^2}{S_{xx}(\omega_1)S_{xx}(\omega_2)S_{xx}(\omega_1+\omega_2)} \quad (9.9)$$

where $S_{xx}(\omega)$ is the power spectrum of $x(t)$ at frequency ω.

Generally $\Gamma(\omega_1,\omega_2)$ is constant over all frequencies $(\omega_1,\omega_2) \in \Omega$, if the sequence is linear, while $\Gamma(\omega_1,\omega_2)$ is flat at zero over all frequencies is the sequence is Gaussian.

The results have to be interpreted as follows. The significant level is given for each sequence at which the gaussian null hypothesis and the linear hypothesis may be rejected.

Table 9.8. Hinich Bi-Spectral test on data: bootstrap. A number $m = 122$ of sub-samples for each original sets were generated and used to run the test. Resuls are on average. Last line shows results for the whole sample.

Test Set	Ho=Gaussianity	H1=Linearity	Comments
DJ1	0.962	0.939	Strongly reject H0 and H1
DJ2	0.297	0.260	Strongly accept H0 and H1
DJ3	0.352	0.721	Strongly accept H0 and H1
DJ4	0.033	0.030	Strongly accept H0 and H1
DJ5	0.412	0.150	Strongly accept H0 and H1
DJ6	0.374	0.195	Strongly accept H0 and H1
DJ7	0.503	0.691	Strongly accept H0 and H1
DJ8	0.043	0.615	Strongly accept H0 and H1
DJ9	0.213	0.547	Strongly accept H0 and H1
DJ10	0.879	0.994	Strongly reject H0 and H1
DJ	0.828	0.710	Strongly reject H0 and H1

BDS statistics results given in Tables 9.10, 9.11.

The BDS test is based on the Grassberger and Procaccia (1983) [13] correlation integral as the test statistic. In particular, under the null hypothesis of whiteness, the BDS statistic is

Table 9.9. Hinich Bi-Spectral test on data: asymptotic theory. Last line shows results for the whole sample.

Test Set	H0=Gaussianity	H1=Linearity	Comments
DJ1	0.884	0.599	Strongly reject H0 and weakly H1
DJ2	0.000	0.000	Strongly accept H0 and H1
DJ3	0.184	0.254	Strongly accept H0 and H1
DJ4	0.005	0.000	Strongly accept H0 and H1
DJ5	0.222	0.006	Strongly accept H0 and H1
DJ6	0.190	0.013	Strongly accept H0 and H1
DJ7	0.310	0.275	Strongly accept H0 and H1
DJ8	0.029	0.181	Strongly accept H0 and H1
DJ9	0.091	0.123	Strongly accept H0 and H1
DJ10	0.506	0.807	Weakly reject H0 and strongly accept H1
DJ	0.000	0.000	Strongly accept H0 and H1

Table 9.10. BDS test on data: bootstrap. Results for $\epsilon = 2.00$. Last line shows results for the whole sample.

Test Set	Dim. 2	Dim. 3	Dim. 4
DJ1	0.789	0.551	0.596
DJ2	0.000	0.000	0.000
DJ3	0.645	0.650	0.495
DJ4	0.920	0.930	0.810
DJ5	0.002	0.014	0.200
DJ6	0.005	0.020	0.400
DJ7	0.360	0.632	0.900
DJ8	0.550	0.710	0.840
DJ9	0.390	0.000	0.100
DJ10	0.487	0.391	0.424
DJ	0.160	0.410	0.950

$$W(N, m, \epsilon) = \sqrt{N} \frac{C(N, m, \epsilon) - C(N, 1, \epsilon)^m}{\hat{\sigma}(N, m, \epsilon)} \tag{9.10}$$

where $\hat{\sigma}(N, m, \epsilon)$ is an estimate of the asymptotic standard deviation of $C(N, m, \epsilon) - C(N, 1, \epsilon)^m$.

Tables 9.8- 9.11 show significance levels at which the null hypothesis can be rejected.

Table 9.11. BDS test on data: asymptotic theory. Results for $\epsilon = 2.00$. Last line shows results for the whole sample.

Test Set	Dim. 2	Dim. 3	Dim. 4
DJ1	0.797	0.582	0.638
DJ2	0.000	0.000	0.000
DJ3	0.662	0.674	0.536
DJ4	0.880	0.720	1.000
DJ5	0.001	0.013	0.017
DJ6	0.005	0.026	0.047
DJ7	0.370	0.310	0.600
DJ8	0.550	0.710	0.840
DJ9	0.027	0.000	0.000
DJ10	0.501	0.391	0.438
DJ	0.009	0.340	0.840

It can be easily see that results from both tests cannot be read in a unique sense, since there are parts of the sequence where the gaussianity of the data can be rejected, as well as its linearity. On the other hand, there are cases where results are not so clear, typically when results are closer to 0.5). In this case much is left to personal intepretation.

It can be seen that managing a time series is not as easy as may be originally imagined. This is especially true in the case of a financial sequence. It is often not enough to study data using sophisticated methods, such as chaos and non-linear dynamics theory, since they are of little help in typical financial analysis. That is studying data is an effort to integrate new knowledge into trading and hence operative systems.

To this aim perhaps, the most important rule which has to be taken in mind is rule number five: **let data speak for itself**.

9.3 Let Data Speak for Itself

As widely known, Self Organizing Maps [20], are neural networks, which rely on the key concept of *competitive learning*.

Competitive learning is an adaptive process in which neurons become gradually sensitive to different input categories in a specific domain of the input space. Such training is done by means of competition between neurons: when a data input is presented to the trained net, the best matching unit wins, and it is able to provide further information about the sample under examination.

The Self-Organizing algorithm goes one step further, generalizing previous idea of *winner taking all* to that of *winner taking the most*. According to this principle, when a pattern extracted from input space is presented to the net, information about it is not only retrieved by the best neuron which is able to represent it, but also by its closest neighbors, according to a proper similarity criterium. As main result, this procedure allows neurons in a net to organize themselves, so that connectivity structures are formed, which are *topology preserving* respect on input data, and corresponds to a particular Delaunay Triangulation (or its dual Voronoi Tesselation) of the subspace where they are contained [23].

In this way it is possible to extract relevant features of data by means of a relatively low number of neurons, reducing high dimensional relationships to more simplified ones into a (generally) bi-dimensional grid of neurons.

Although extensively employed in technical fields, Self Organizing Maps have not been used as widely in financial forecasting [9], [25], [26]: in this section a procedure will be described which makes use of variants of Self Organizing Maps to build up a system that appears to (i) detect intrinsic nature of (financial) data, (ii) translating information into really simple rules to assume buy/sell position into the market.

9.3.1   A Brief Description of the Algorithm

Since an impressive number of papers has already been devoted to describe Self Organizing Maps and their variants, we limit to enumerate basic steps on which the procedure in use relies.

Let M a map consisting of K neurons arranged in an array form. Assume that each neuron is associated to a vector $\underline{w}$, , initially set at random, defined into a proper space R^d. The algorithm consists of four instructions which are iteratively repeated.

- **STEP 1.** A sample $\underline{x}$ is chosen from input space and presented to M.
- **STEP 2.** The net is ordered according to similarity criteria. If i, j are two neurons of the net, hence:

$$p(i, t + 1) > p(j, t + 1) \tag{9.11}$$

$$\Longleftrightarrow$$

$$d(\underline{x}, \underline{w}_i) > d(\underline{x}, \underline{w}_j) \vee d(\underline{x}, \underline{w}_i) = d(\underline{x}, \underline{w}_j) \wedge (i > j)$$

where, $p(i, t + 1)$ is the position in M of neuron i at time $t + 1$.
- **STEP 3.** Weights adaptation: reference vectors are modified according to their proximity to input sample:

$$\underline{w}_i(t + 1) = \{ \begin{smallmatrix} \underline{w}_i(t) + h(p(i,t),\alpha)(\underline{x}(t) - \underline{w}_i(t)), p(i,t) \leq \theta; \\ \underline{w}_i(t), otherwise \end{smallmatrix} \tag{9.12}$$

where $h(nr, \alpha) = e^{-(nr-1)\alpha}$, and $\theta \in N$, with $0 < \theta < K$.

Weights may be modified into two different ways: they are maintained unvarying, whether or not corresponding neuron position p exceeds a fixed threshold θ, which controls neighbor amplitude. Each time an input is presented to the net, the whole neighborhood of leader neuron, instead of this one by itself, will be modified.

- **STEP 4.** If the whole dataset has been presented to the net, then the procedure stops, otherwise it comes back to stage one, and all steps are repeated.

Figure 9.6 shows the way neural space is modified as effect of input patterns presentation. It is easy to observe that the procedure doesn't converge to a steady state, being dynamically modified as newest information from input space is added. This aspect does not appear as a bottleneck, as it might figure out at first view: on the contrary, dealing with financial data requires a procedure continuously in evolution.

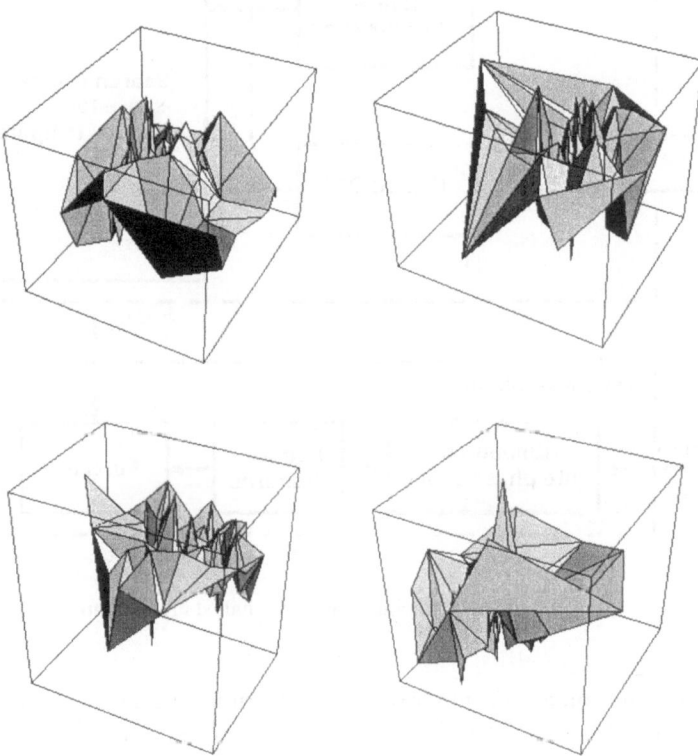

Fig. 9.6. Evolution of neural space during the learning phase. From top to bottom in clockwise sense: neurons at initial step; after 100 runs; after 1000 runs, and after 2000 runs.

9.3.2   Case Study

Neural networks become very valuable for managing financial data. The leading idea is to verify if (i) structural patterns intrinsic to data exist, and if (ii) they can be used through Self Organizing Maps, as patterns classifiers.

A dedicated-financial oriented system has been created, called the *Artificial Technical Analyst* (ATA), whose architecture is shown in Figure 9.7.

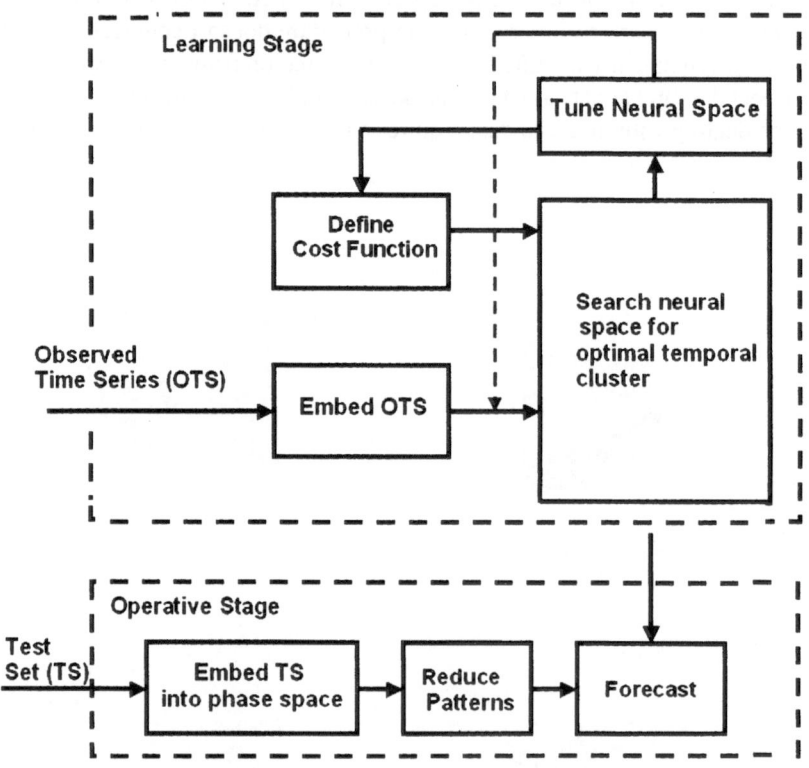

Fig. 9.7. ATA, Artificial Technical Analyst architecture.

The system is organized into two stages: a learning and an operative phase.

During the first stage, data is presented to the system, and properly embedded, so that sequential patterns are created. Definition 1 makes clearer the concept of pattern, used along the procedure.

Definition 1. *Given a list l with length dim(l), a pattern $\underline{u}_i$ of length n is the vector whose n sequential components have been extracted from l, starting from the ith.*

By Definition 1, Dow Jones data has been organized into sequential patterns of proper length, and then analyzed by means of a neural-network based system, using the algorithm briefly summarized in previous section.

Once all patterns have been presented to the system, the second stage begins. A group of new unknown data is considered, and embedded according to the dimension n estimated in previous stage. Hence a "reduction phase" starts, where a number $k, k < n$ of elements is dropped from each pattern, so that such a incomplete shape is presented to the system.

The trained net is now used to forecast, by reconstructing the most probable complete shape for the sample, by taking into account information, gathered during the training phase.

The final result is an operative decision, i.e a buy/sell or standby signal, according to the sign of forecasting, generated by the procedure.

This is in principle how the system works. The need for robust responses, in order to move in an operative way into the markets, led us to introduce some improvements to the original procedure, by considering systems made up by more than one net.

The main problem encountered dealing with financial data through Self Organizing Maps, comes from their properties of classifiers. It is known that such neural nets tend to fit data according to the criterium of more promising patterns: a small number of cells is able to represent the whole universe given by input space, and some neurons are recalled more than once.

This means that a net could have a small number of neurons with high power of representation for the whole sample, increasing the probability that single extremely performing cells could monopolize input patterns representation, leading to overall poor results.

This trouble has been eliminated by the introduction of twin principles of *k-neighborhood* and *g-k neighborhood responses* into a finite ensemble of nets.

Definition 2. *A signal from neuron r may be considered a k-neighborhood response if and only if all k-nearest nodes to r confirm it.*

Definition 3. *Let r_i be the winning neuron for the ith net, for $i = 1, ..., G$. A signal may be considered a g-k-neighborhood response if and only if all k-nearest nodes to r_i confirm it in almost earlier $g \leq G$ closest to input nets.*

From the operative point of view, the two principles act on different levels, since k-neighborhood operates on single nets, and g-k neighborhood runs on sets of nets.

However, they share the common aim which is to filter the responses of the nets, by taking into account the plausibility of their feedback to the input, with different degrees of truth, whether or not the answer of leader neuron is confirmed by a qualified majority of k closest nodes to it (in case of k-neighborhood), and extending such concept over a set of nets (in the case of g-k neighborhood). A response will be accepted if and only if there will exist an acceptable number g of nets closest to input pattern, confirming, with almost k nearest nodes, the result of leaders.

In this way, when the second stage of the procedure starts, it is possible to monitor the forecasting activity of the system of nets, with various degrees of robustness, as the level of confirmations increases over single nets, as well as over the whole ensemble. More properly, for each forecasting task a response matrix is build up, where each cell contains a response, with different degrees of sustainability, according to the qualified majority requested (both over single nets, and over the whole set of nets).

Figures 9.8, 9.9 add further explanations to such concept.

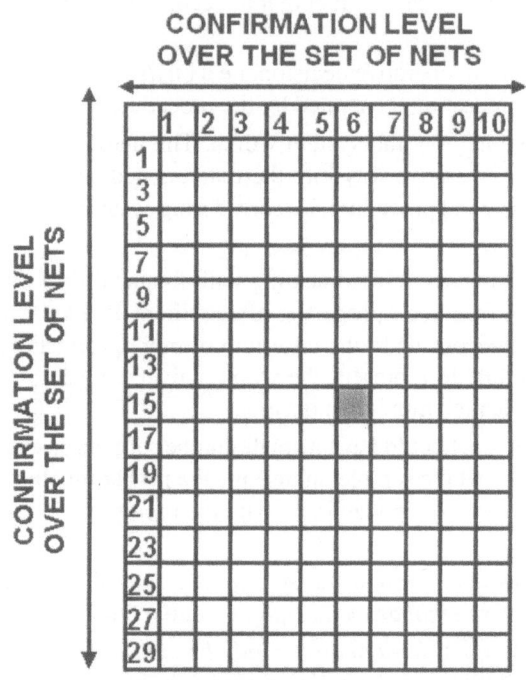

Fig. 9.8. $g - k$ neighborhood at work. In the example a $6 - 15$ response is highlighted.

When applied to real data, this technique may be driven by different requirements. We limit these to three different tasks, linked to practical aspects.

Since Self Organizing Maps are mainly good patterns recognizers, one should to control *what kind of patterns* the procedure knows best. To this purpose, this system has been tested under different operative hypotheses:

- Temporal dependence capabilities (group TS1): as in previous section, the Dow Jones data has been divided into 10 blocks of 2500 elements each. After training, the system has been asked to forecast over additional 500 unknown sequential patterns, extracted from different blocks than the one serving as learning set.

CONFIRMATION LEVEL
OVER THE SET OF NETS

	1	2	3	4	5	6	7	8	9	10
1	1	1	1	1	1	0	0	0	0	0
3	1	-1	1	1	1	0	0	0	0	0
5	-1	-1	1	-1	-1	0	0	0	0	0
7	1	1	-1	0	0	0	0	0	0	0
9	-1	1	-1	0	0	0	0	0	0	0
11	-1	1	-1	0	0	0	0	0	0	0
13	-1	1	1	0	0	0	0	0	0	0
15	-1	-1	0	0	0	0	0	0	0	0
17	1	1	0	0	0	0	0	0	0	0
19	-1	0	0	0	0	0	0	0	0	0
21	0	0	0	0	0	0	0	0	0	0
23	0	0	0	0	0	0	0	0	0	0
25	0	0	0	0	0	0	0	0	0	0
27	0	0	0	0	0	0	0	0	0	0
29	0	0	0	0	0	0	0	0	0	0

(Left axis label: CONFIRMATION LEVEL OVER THE SET OF NETS)

Fig. 9.9. Response Matrix. The x axis shows the confirmation level over the system of nets. The y axis shows the confirmation level into each net.

- Generalization capabilities (group TS2): experiments have been made by extracting at random 3500 patterns from the complete set. After training on 2500 samples, the system has been required to forecast on remaining 1000.
- Control for possible faulty specification (group TS3 and group TS4): previous experiments have been repeated, by forming patterns after data has been mixed at random. Since these patterns are random by construction, in absence of specification errors, the system should be able to perform as well as by tossing a coin. The difference between this experiment and previous ones stands in the fact than in this case, patterns have been formed in a second stage, once the order of single observations in the sequence have been randomized, i.e. both temporal dependence and patterns have been completely altered in a random fashion.

In each of previous cases the task the system was required to forecast, for each unknown pattern, the sign of the last fluctuation composing it, and to assume a position in the market according to it: that is mark long for sign equal to +1, mark short for sign equal to −1, and flat for 0 responses.

In the first simulation, as well as in its random counterpart, a system made on confirmation levels up to 12 neighbors over single nets, and 10 over the set of networks have been used. On the other side, in TS2 and TS4, the confirmation level among nets has been increased up to 10, and that among neighbors over single nets have been reduced to 10.

On following, results are presented and discussed: please note that performances are reported on average (weighted through the number of moves) for simulations TS1, and TS3, being the number of tables to be reported too high (100).

Table 9.12. Results for simulation 1. Average performances of systems: each cell represent the average performance (in all 100 simulations) of the couple (NAL(i), SAL(j)), where NAL is the single Net Acceptance Level, and SAL the System Acceptance Level. For instance, (NAL(6),SAL(7))= 52 means that, moving over markets through operative suggestions (kept at a majority level of sixth nearest neighbors on single nets, and confirmed at least by seven nets), we should get proficiently *on average* 52% of total runs.

System	SAL1	SAL2	SAL3	SAL4	SAL5	SAL6	SAL7	SAL8	SAL9	SAL10
NAL1	52	52	52	52	52	52	52	52	52	52
NAL2	52	52	52	52	52	52	52	52	52	52
NAL3	52	52	52	52	52	52	52	52	52	52
NAL4	52	52	52	52	52	52	52	52	52	52
NAL5	52	52	52	52	52	52	52	52	52	52
NAL6	52	52	52	52	52	52	52	52	52	52
NAL7	53	52	52	52	52	52	52	52	52	52
NAL8	53	52	52	52	52	52	52	52	52	52
NAL9	53	52	52	53	53	53	53	53	53	53
NAL10	53	53	53	53	53	53	53	53	53	53
NAL11	53	53	53	53	53	53	53	53	53	53
NAL12	53	53	53	53	53	53	53	53	53	53

Tables 9.12, 9.13 show average returns held by the system, when tested over different control sets. The results must be interpreted in the following way: Table 9.12 shows average performances over different control blocks; on the other side, Table 9.13 shows the average number of times the system has been in action, i.e. assuming a long - short, or flat position over the market, according to a particular $g − k$ neighborhood response.

At the first view, these performances are not significantly over "standard" 50%, but it s quite interesting to compare them to the results obtained in the "random" counterpart of such simulation (Tables 9.14, and 9.15).

Table 9.13. Results for simulation 1. Average number of moves.

System	SAL1	SAL2	SAL3	SAL4	SAL5	SAL6	SAL7	SAL8	SAL9	SAL10
NAL1	464	457	454	452	450	449	448	447	447	446
NAL2	432	424	422	421	421	420	420	420	419	419
NAL3	407	402	401	400	400	400	399	399	399	399
NAL4	389	385	385	384	384	384	384	384	384	384
NAL5	376	373	372	372	372	372	372	372	372	372
NAL6	364	362	362	361	361	361	361	361	361	361
NAL7	354	352	352	351	351	351	351	351	351	351
NAL8	343	342	342	342	342	342	342	342	342	342
NAL9	334	333	332	332	332	332	332	332	332	332
NAL10	325	324	324	324	324	324	324	324	324	324
NAL11	317	316	316	316	316	316	316	316	316	316
NAL12	310	309	309	309	309	309	309	309	309	309

Table 9.14. Results for simulation 3, as random counterpart of simulation 1. Average performances: values have been rounded at the lower closest integer.

System	SAL1	SAL2	SAL3	SAL4	SAL5	SAL6	SAL7	SAL8	SAL9	SAL10
NAL1	49	49	49	49	49	49	49	49	49	49
NAL2	49	49	49	49	49	49	49	49	49	49
NAL3	49	49	49	49	49	49	49	49	49	49
NAL4	49	49	49	49	49	49	49	49	49	49
NAL5	49	49	49	49	49	49	49	49	49	49
NAL6	49	49	49	49	49	49	49	49	49	49
NAL7	49	49	49	49	49	49	49	49	49	49
NAL8	49	49	49	49	49	49	49	49	49	49
NAL9	49	49	49	49	49	49	49	49	49	49
NAL10	49	49	49	49	49	49	49	49	49	49
NAL11	49	49	49	49	49	49	49	49	49	49
NAL12	49	49	49	49	49	49	49	49	49	49

Table 9.15. Results for simulation TS3, as random counterpart of simulation TS1. Average number of moves.

System	SAL1	SAL2	SAL3	SAL4	SAL5	SAL6	SAL7	SAL8	SAL9	SAL10
NAL1	455	449	445	443	442	441	440	439	439	439
NAL2	424	417	415	414	413	413	412	412	412	412
NAL3	401	396	394	394	394	393	393	393	393	393
NAL4	383	380	379	379	378	378	378	378	378	378
NAL5	370	368	367	367	367	367	367	367	367	367
NAL6	360	358	357	357	357	357	357	357	357	357
NAL7	350	348	348	348	348	348	348	348	348	348
NAL8	340	339	339	339	339	339	338	338	338	338
NAL9	332	331	331	331	331	331	331	331	331	331
NAL10	324	323	323	323	323	323	323	323	323	323
NAL11	317	316	316	316	316	316	316	316	316	316
NAL12	310	310	310	309	309	309	309	309	309	309

While the average number of moves remains similar in both TS1, and TS3, it may be noticed that average results in the latter case is below 50%.

A question arises as which measure are the results different to those obtained moving at random in the market. Some statistical results of these tests are presented below.

It is possible to make two different tests:

- a test for independence, where the results for both TS1, and TS3 are compared over different blocks where they were controlled, and the differences between them are evaluated, to verify whether they come from the same distribution, hence accepting the null hypothesis (independence). A table is obtained, where for each cell (corresponding to a particular $g - k$ neighborhood response), a sentence is reported: when H0 appears,it means the null hypothesis has been accepted. RH0, on the other hand, indicates the rejection of independence.
- a test for non linearity: starting from results for TS1,proper confidence intervals are estimated, and the number of positions is assumed. Hence, the null hypothesis of randomness which means it is impossible to obtain from the patterns any additional information about the process, will be accepted, when results fall into the estimated confidence interval. If it does not, it will be discarded. Table 9.17, shows percentage values associated for each systems, to credibility in rejecting the null hypothesis, at the 95% confidence level.

It may seem remarkable to observe that, results are "not so far" from standard 50%, average performances of the system make possible to reject the null hypothesis in the test for independence in almost all cases.

Table 9.16. Test for independence. S(i) refers on ith system of nets, $i = 1, ..., 10$ while CB(j) indicates the jth block ($j = 1, ..., 10$) where systems were controlled.

System	CB1	CB2	CB3	CB4	CB5	CB6	CB7	CB8	CB9	CB10
S1	RH0	RH0	RH0	RH0	RH0	RH0	RH0	RH0	RH0	RH0
S2	RH0	RH0	RH0	RH0	RH0	RH0	RH0	RH0	RH0	RH0
S3	H0	RH0	RH0	RH0	RH0	RH0	RH0	RH0	RH0	RH0
S4	RH0	RH0	RH0	H0	RH0	RH0	RH0	RH0	RH0	RH0
S5	RH0	RH0	RH0	RH0	RH0	RH0	RH0	RH0	RH0	RH0
S6	RH0	RH0	RH0	RH0	RH0	RH0	RH0	RH0	RH0	RH0
S7	RH0	RH0	RH0	RH0	RH0	RH0	RH0	H0	RH0	RH0
S8	RH0	RH0	RH0	RH0	RH0	RH0	RH0	RH0	RH0	RH0
S9	RH0	RH0	RH0	RH0	RH0	RH0	RH0	H0	RH0	RH0
S10	RH0	RH0	RH0	RH0	RH0	H0	RH0	RH0	RH0	RH0

Table 9.17. Results for simulation 1: non-linearity test.

System	CB1	CB2	CB3	CB4	CB5	CB6	CB7	CB8	CB9	CB10
S1	0	19	100	0	0	98	74	0	61	10
S2	0	9	0	9	90	0	0	0	59	10
S3	0	0	0	0	0	0	100	7	95	0
S4	100	0	29	0	0	0	0	10	0	0
S5	100	100	15	19	0	0	100	0	100	100
S6	0	0	0	7	0	0	8	60	0	100
S7	0	100	42	33	90	100	51	0	0	100
S8	83	28	4	0	40	15	85	0	90	0
S9	0	0	2	25	0	8	100	0	0	100
S10	16	0	0	0	0	22	0	0	0	0

In the second test (see Table 9.17), the systems are revealed able to scan the data, and discover some intrinsic features differing from pure random walks. The number of times the null hypothesis has to be accepted maintains very low.

However, such results simply state that once a confidence interval has been set, neural system results may be at lower or higher values.

As additional step, the same statistics have been employed to rule out, under the profile of returns, whether or not the system may be profitable, when employed to discover temporal patterns and trade (Table 9.17).

The results shown in Table 9.18 show that, although our neural systems are able to detect on data whatever different from pure randomness, their overall results in

Table 9.18. Results for simulation 1: independence test and profile of returns.

System	CB1	CB2	CB3	CB4	CB5	CB6	CB7	CB8	CB9	CB10
S1	0	0	100	0	0	0	74	0	61	10
S2	0	0	0	0	90	0	0	0	59	10
S3	0	0	0	0	0	0	100	0	95	0
S4	100	0	29	0	0	0	0	10	0	0
S5	100	100	0	8	0	0	100	0	100	100
S6	0	0	0	0	0	0	8	0	0	100
S7	0	100	42	0	90	100	51	0	0	100
S8	0	0	0	0	0	15	85	0	90	0
S9	0	0	2	0	0	0	100	0	0	100
S10	16	0	0	0	0	22	0	0	0	0

terms of gain are not completely satisfying. As they allow for higher returns than random walkers only in a reduced number of test blocks.

Such drawbacks could be due to a wrong specification of learning coefficients, as well as embedding dimension or both. Additionally, it may be argued that generalization should be improved by letting the systems to choose among a larger set of data, which has been done in the second experiment. This may be seen in the following tables, which give results for TS2 and its random countepart TS4.

Table 9.19. Results for simulation TS2.

System	SAL1	SAL2	SAL3	SAL4	SAL5	SAL6	SAL7	SAL8	SAL9	SAL10
NAL1	57	57	58	58	58	58	58	58	58	58
NAL2	58	59	58	58	58	58	58	58	58	58
NAL3	59	59	59	59	59	59	59	59	60	60
NAL4	60	60	60	60	60	60	60	60	60	60
NAL5	60	60	60	60	60	60	60	60	60	60
NAL6	60	60	60	60	60	60	60	60	60	60
NAL7	59	59	59	59	59	59	59	59	59	59
NAL8	59	59	59	59	59	59	59	59	59	59
NAL9	59	59	59	59	59	59	59	59	59	59
NAL10	60	59	59	59	59	59	59	59	59	59

At a first examination, results from generalization capabilities appear better than previous ones, and the null hypothesis in this case is rejected at a confidence level of 95%.

Table 9.20. Results for simulation TS4, as random counterpart of simulation TS2.

System	SAL1	SAL2	SAL3	SAL4	SAL5	SAL6	SAL7	SAL8	SAL9	SAL10
NAL1	49	51	49	49	51	51	48	49	49	48
NAL2	48	50	48	49	49	50	50	49	49	50
NAL3	49	48	48	51	49	51	51	50	50	50
NAL4	49	50	48	50	48	49	49	50	51	48
NAL5	48	51	50	50	49	51	51	48	50	48
NAL6	49	50	50	48	51	49	50	50	49	50
NAL7	48	48	48	51	51	48	51	50	50	50
NAL8	51	49	49	51	48	48	51	48	51	49
NAL9	51	49	48	48	49	51	51	50	48	48
NAL10	49	51	49	51	50	51	50	48	50	49

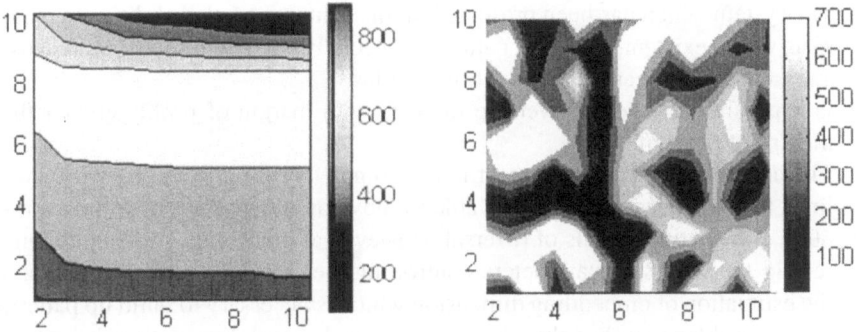

Fig. 9.10. Number of positions assumed in TS2, and TS4.

It appears reasonable from these results to think that systems of neural networks functions well as generalization capabilities, given a large enough training set. If this holds, then they are able to detect, where existent, patterns characterizing a sequence of data (as in case of Dow Jones).

An additional confirmation has come from misspecification experiment (TS4), where, the nets have been driven over patterns created *at random* from the original series.

As it can be seen in Table 9.20, results maintain close to 50%. At the same time, the statistics accepted the null hypothesis at a confidence level of 95%.

Differences appear, also, in the number of moves the system has run in TS2 and TS4, as it can be seen in Figure 9.10, where contour plots are used to represent the distribution of number of positions assumed in the two different runs.

9.4 Conclusions

In this work more recent techniques of time series analysis have been reviewed.

Although very promising approaches have emerged over the past 20 years, a number of problems may be highlighted, especially when they have been employed with financial data. Problems might arise when such techniques are used to provide operative signals to take positions over the market.

Following this evidence, an alternative approach has been suggested, which relies on the idea of letting data to speak for itself.

The kernel idea of this approach is to take advantage, if possible, of emergent patterns in financial time series, employing them to develop trading strategies.

An architecture based on Self Organizing Maps has been implemented, employing the power of such neural nets as classifiers, in order to try to provide answers to a number of questions.

- Is it possible to forecast? More recent developments, indicate that short term patterns should be exploitable.
  The system which has been presented in this context revealed abilities to recognize, where existent, temporal patterns, especially when it has been trained over a reasonably large and diverse set of samples.
- Is it possible to operate over the markets with margin of profits greater than moving at random?
  The answer given in this context is not completely exhaustive. It may be outlined that patterns classifiers are able to move on the markets over the average 50%, anticipating signals of reversal. However, a limit for this procedure might rely in the values of parameters controlling the learning phase, as well as in the estimation of embedding dimension which is necessary to build up patterns. Such variables may hugely affect overall results, and, hence, they have to be managed very carefully, verifying the stability of results over a range of parameters values.

The results rule out ready to be used, statistics which allow operators to evaluate immediately the proficiency of the neural strategy they are running.

In this context the opportunity has been taken to build up tests for both non linearity and independence. Small steps have been made in these directions, although there is a need for more accurate theoretical framework, the early results appear promising.

As final remark, it must be outlined that the procedure suggested may be able to develop operative signals without any human action, except for that of setting the dimension and the number of nets to use.

Bibliography on Chapter 9

1. Ashley, R., D. M. Patterson (2000) A Nonlinear Time Series Workshop: A Toolkit For Detecting And Identifying Nonlinear Serial Dependence, Kluwer Academeic Publishers
2. Bachelier, L. (1900) Thorie de la spculation reprinted in Cootner, P.H. ed., The Random Character of Stock Market Prices, 17-78, 1964 Cambridge MA, MIT Press
3. Bollerslev, T. (1992) Generalized autoregressive conditional heteroskedasticity, Journal of Econometrics, 31, 307-327
4. Brock, W. A., Dechert, W. D., LeBaron, B., Scheinkman, J. A. (1996) A Test for Independence Based on the Correlation Dimension, Econometric Reviews 15, 197-235.
5. Calvet, L., Fisher, A., Mandelbrot, B.(1997) Large Deviations and the Distribution of Price Changes, Cowles Foundation Discussion Paper n. 1165
6. Campbell, J. Y., Lo, A. W. , MacKinlay, A. C. (1997) The Econometrics of Financial Markets, Princeton University Press, Princeton, NJ.
7. Chakravart, Laha, Roy (1967) Handbook of Methods of Applied Statistics, Vol. I, John Wiley 392-394
8. Cochrane, J.(1997) Time Series for Macroeconomic and Finance, Working Paper, University of Chicago
9. Cottrell, M., De Bodt, E., Gregoire, M. (1997) A powerful Tool for Fitting and Forecasting Deterministic and Stochastic Processes: the Kohonen classification, Proceedings of ICANN'97
10. Eckmann, J.-P., Ruelle, D. (1985) Ergodic theory of chaos and strange attractors, Rev. Mod. Phys. 57, 617.
11. Edwards R. D., Magee, J. (1957) Technical Analysis of Stock Trends, Magee
12. Fama, E. F., (1970) Efficient Capital Markets: A Review of Theory and Empirical Work, Journal of Finance 25, 383-417
13. Grassberger, P., Procaccia, I., (1983) Characterization of Strange Attrac- tors, Physical Review Letters 50, 346-349
14. Grassberger, P. (1988) Finite sample corrections to entropy and dimension estimates, Phys. Lett. A 128, 369.
15. Hinich, M. J., (1982) Testing for Gaussianity and Linearity of a Stationary Time Series, Journal of Time Series Analysis 3, 169-176
16. Hinich, M. J., Patterson, D. (1989) Evidence of Nonlinearity in the Trade by-Trade Stock Market Return Generating Process, in: W. A. Barnett, J. Geweke, and K. Shell, eds., Economic Complexity: Chaos, Bubbles, and Nonlinearity, Cambridge University Press, Cambridge, UK
17. Judge, G., Griffiths, W., Hill, C., Ltkepohl, H. L, Lee, T. C. (1985) The Theory and Practice of Econometrics John Wiley and Sons: New York.
18. Kantz, H., (1994) A robust method to estimate the maximal Lyapunov exponent of a time series, Phys. Lett. A 185, 77
19. Kennel,M. B., Brown, R., Abarbanel, H. D. I. (1992) Determining embedding dimension for phase-space reconstruction using a geometrical construction, Phys. Rev. A 45, 3403
20. Kohonen ,T. (1997) Self-Organizing Maps, Springer series in information science
21. Mandelbrot, B. (1967) The variation of certain speculative prices. Journal of Business, 36, 394-419
22. Mandelbrot, B., Fisher, A., Calvet, L. (1997) A Multifractal Model of Asset Returns, Cowles Foundation Discussion Paper n. 1164
23. Martinetz, T., Schulten, K. (1994) Topology Representing Networks, Neural Networks, Vol. 7, No. 3 1994.

24. Nychka, D., Ellner, S., Gallant, R., McCarey, D. (1992) Finding Chaos in Noisy Systems, Journal of the Royal Statistical Society B 54, 399-426.
25. Resta, M. (2000) TRN: picking up the challenge of non linearity testing by means of Topology Representing Networks, Neural Network World
26. Resta, M. (1998) An Hybrid Neural Network System for Market Trading Strategies, in Kohonen, T., Deboeck, G.J. Edrs, Visual Data Explorations in Finance with Self Organizing Maps, Finance Series, Springer Verlag London, 106-116
27. Sano, M., Sawada, Y. (1985) Measurement of the Lyapunov spectrum from a chaotic time series, Phys. Rev. Lett. 55, 1082.
28. Stephens, M. A. (1974) EDF Statistics for Goodness of Fit and Some Comparisons, Journal of the American Statistical Association, vpl. 69, 730-737

10 Unsupervised and Supervised Learning in Radial-Basis-Function Networks

Friedhelm Schwenker, Hans A. Kestler and Günther Palm

Abstract. Learning in radial basis function (RBF) networks is the topic of this chapter. Whereas multilayer perceptrons (MLP) are typically trained with backpropagation algorithms, starting the training procedure with a random initalization of the MLP's parameters, an RBF network may be trained in different ways. We distinguish one-, two-, and three phase learning. A very common learning scheme for RBF networks is two phase learning. Here, the two layers of an RBF network are trained separately. First the RBF layer is calculated, including the RBF centers and scaling parameters, and then the weights of the output layer are adapted. The RBF centers may be trained through unsupervised or supervised learning procedures utilizing clustering, vector quantization or classification tree algorithms. The output layer of the network is adapted by supervised learning. Numerical experiments of RBF classifiers trained by two phase learning are presented for the classification of 3D visual objects and the recognition of hand-written digits. It can be observed that the performance of RBF classifiers trained with two phase learning can be improved through a third backpropagation-like learning phase of the RBF network, adapting the whole set of parameters (RBF centers, scaling parameters, and output layer weights) simultaneously. This, we call three phase learning in RBF networks. A practical advantage of two and three phase learning in RBF networks is the possibility to use unlabeled training data for the first training phase.

Support vector (SV) learning in RBF networks is a special type of one phase learning, where only the output layer weights of the RBF network are calculated, and the RBF centers are restricted to be a subset of the training data.

Numerical experiments with several classifier schemes including nearest neighbor classifiers, learning vector quantization networks and RBF classifiers trained through two phase, three phase and support vector learning are given. The performance of the RBF classifiers trained through SV learning and three phase learning are superior to the results of two phase learning.

10.1 Introduction

The radial basis function (RBF) network model is motivated by the locally tuned response observed in biologic neurons. Neurons with a locally tuned response characteristic can be found in several parts of the nervous system, for example cells in the auditory system selective to small bands of frequencies [10,34] or cells in the visual cortex sensitive to bars oriented in a certain direction or other visual features within a small region of the visual field (see, for instance [32]). These locally tuned neurons show response characteristics bounded to a small range of the input space.

The theoretical basis of the RBF approach lies in the field of interpolation of multivariate functions. Here, multivariate functions $f : \mathbb{R}^d \to \mathbb{R}^m$ are considered. We assume that m is equal to 1 without any loss of generality. The goal of interpolating a set of tupels $(\mathbf{x}^\mu, y^\mu)_{\mu=1}^M$ with $\mathbf{x}^\mu \in \mathbb{R}^d$ and $y^\mu \in \mathbb{R}$ is to find a function

$F : \mathbb{R}^d \to \mathbb{R}$ with $F(\mathbf{x}^\mu) = y^\mu$ for all $\mu = 1, \ldots, M$, where F is an element of a predefined set of functions $\mathcal{F}$, typically $\mathcal{F}$ is a linear space. In the RBF approach the interpolating function F is a linear combination of basis functions:

$$F(\mathbf{x}) = \sum_{\mu=1}^{M} w_\mu h(\|\mathbf{x} - \mathbf{x}^\mu\|) + p(\mathbf{x}) \tag{10.1}$$

where $\| \cdot \|$ denotes the Euclidean norm, $w_1, \ldots, w_M$ are real numbers, h a real valued function, and p a polynomial $p \in \Pi_n^d$ (polynomials of degree at most n in d variables). The degree of the polynomial term has to be fixed in advance. The interpolation problem is to determine the real coefficients $w_1, \ldots, w_M$ and the polynomial term $p := \sum_{l=1}^{D} a_l p_j$ where $p_1, \ldots, p_D$ is the standard basis of Π_n^d and $a_1, \ldots, a_D$ are real coefficients. The function F has to satisfy the conditions:

$$F(\mathbf{x}^\mu) = y^\mu, \quad \mu = 1, \ldots, M \tag{10.2}$$

and

$$\sum_{\mu=1}^{M} w_\mu p_j(\mathbf{x}^\mu) = 0, \quad j = 1, \ldots, D. \tag{10.3}$$

Sufficient conditions for the unique solvability of the interpolation problem were given by several authors [24,16,33]. The function h is called a *radial basis function* if the interpolation problem has a unique solution for any choice of data points. In some cases the polynomial term in formula (10.1) can be omitted, and then the interpolation problem is equivalent to the matrix equation

$$\mathbf{Hw} = \mathbf{y} \tag{10.4}$$

where $\mathbf{w} = (w_1, \ldots, w_M)$, $\mathbf{y} = (y^1, \ldots, y^M)$, and $\mathbf{H}$ is a $M \times M$ matrix defined by

$$\mathbf{H} = (h(\|\mathbf{x}^\nu - \mathbf{x}^\mu\|))_{\mu,\nu=1,\ldots,M}. \tag{10.5}$$

Provided the inverse of $\mathbf{H}$ exists, the solution $\mathbf{w}$ of the interpolation problem can explicitly calculated and has the form:

$$\mathbf{w} = \mathbf{H}^{-1}\mathbf{y}. \tag{10.6}$$

Examples of radial basis functions h often used in applications are:

$$h(r) = e^{-r^2/2\sigma^2} \tag{10.7}$$
$$h(r) = (r^2 + \sigma^2)^{1/2} \tag{10.8}$$
$$h(r) = (r^2 + \sigma^2)^{-1/2} \tag{10.9}$$

Here, σ is a positive real number which we call the *scaling parameter* or the *width* of the radial basis functions. The most popular and widely used radial basis function is the *Gaussian basis function*

$$h(\|\mathbf{x} - \mathbf{c}\|) = exp(-\|\mathbf{x} - \mathbf{c}\|^2/2\sigma^2) \qquad (10.10)$$

with peak at center $\mathbf{c} \in \mathbb{R}^d$ and decreasing as the distance from the center increases. Throughout this chapter we restrict ourselves to this type of radial basis function.

The solution of the exact interpolating RBF mapping passes through every data point $(\mathbf{x}^\mu, y^\mu)$. In the presence of noise the exact solution of the interpolation problem is typically a function oscillating between the given data points. An additional problem with the exact interpolation procedure is that the number of basis functions is equal to the number of data points and so calculating the inverse of the $M \times M$ matrix $\mathbf{H}$ becomes intractable in practice.

In applications where one has to deal with many thousands of noisy data points, an approximative solution to the data is more desirable than an interpolative one. Broomhead and Lowe [4] proposed to reduce the number of basis functions in order to reduce the computational complexity. This technique produces a solution by approximating instead of interpolating the data points. Furthermore, in [4] an interpretation of the RBF method as an artificial neural network model is given. It consists of three neural layers: a layer of input neurons feeding the feature vectors into the network, a hidden layer of RBF neurons, calculating the outcome of the basis functions, and a layer of output neurons, calculating a linear combination of the basis functions. Under some additional conditions imposed on the basis function h the set of RBF networks with free adjustable prototype vectors are shown to be universal approximators, so that any continous function can be approximated with arbitrary precision [29]. This implies that RBF networks with adjustable prototypes can also be used for classification tasks [31].

In the classification scenario the RBF network has to perform a mapping from a continuous input space $\mathbb{R}^d$ into a finite set of classes $Y = \{1, \dots, L\}$, where L is the number of classes. In the training phase the parameters of the network are determined from a finite training set

$$S = \{(\mathbf{x}^\mu, y^\mu) \mid \mathbf{x}^\mu \in \mathbb{R}^d, y^\mu \in Y, \mu = 1, \dots, M\}, \qquad (10.11)$$

here each feature vector $\mathbf{x}^\mu$ is labeled with its class membership y^μ. In the recall phase further unlabeled observations $\mathbf{x} \in \mathbb{R}^d$ are presented to the network which estimates their class memberships $y \in Y$. In our classification scenario utilizing RBF networks the number of output units corresponds to the number of classes, and the class memberships $y \in Y$ are encoded through a 1-of-L coding into a binary vector $\mathbf{z} \in \{0, 1\}^L$ through the relation $\mathbf{z}_i^\mu = 1$ iff $y^\mu = i$. To simplify the notation we do not distinguish between these two representations of the classmembership. In this context it should be mentioned that other coding schemes can be used but are not very common in pattern recognition applications. Using the 1-of-L encoding scheme a RBF network with K basis functions is performing a mapping $F : \mathbb{R}^d \rightarrow$

Input   **RBF-layer, prototypes**   **Output**

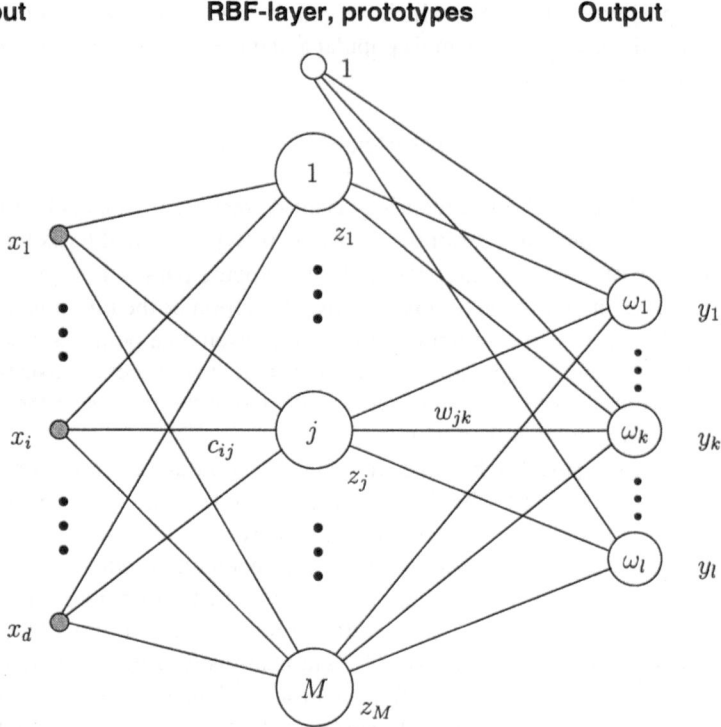

Fig. 10.1. A radial basis function network with M RBF units in the hidden layer and l output units. The input x is a d-dimensional feature vectors. Each unit of the output layer gets an additional bias input.

$\mathbb{R}^L$,

$$F_i(\mathbf{x}) = \sum_{j=1}^{K} w_{ji} h(\|\mathbf{x} - \mathbf{c}_j\|) + w_{0i}, \quad i = 1, \dots, L, \tag{10.12}$$

where the w_{0i} denote the biases, which may be absorbed into the summation by including an extra basis function $h_0 = 1$ whose activation is set equal to 1 on the whole input space $\mathbb{R}^d$. Categorization is performed by assigning the input vector $\mathbf{x}$ the class of the output unit with maximum activation:

$$\text{class}(\mathbf{x}) = \underset{i \in \{1, \dots, L\}}{\text{argmax}} \ F_i(\mathbf{x}). \tag{10.13}$$

Typically, an RBF as a neural network model differs from the RBF as an interpolation method in some ways:

1. The number of basis functions is typically much less than the number of data points, and the basis functions are located in representative prototypes $\mathbf{c}_j \in \mathbb{R}^d$ which are not restricted to be data points.

2. Instead of a global scaling parameter $\sigma \in \mathbb{R}$ for all basis function, each basis function has its own scaling parameter $\sigma_j \in \mathbb{R}$.
3. In some RBF network models the so-called Mahalanobis distance is used instead of the Euclidean distance. In general a Mahalanobis distance between two points $\mathbf{x}, \mathbf{y} \in \mathbb{R}^d$ is defined by a positive definite matrix R and is given through

$$\|\mathbf{x} - \mathbf{y}\|_{\mathbf{R}} = \sqrt{(\mathbf{x} - \mathbf{y})^T \mathbf{R} (\mathbf{x} - \mathbf{y})} \tag{10.14}$$

here T denotes the transpose of a matrix. Typically $\mathbf{R}$ is the inverse of the covariance matrix of the input data points $\mathbf{x}^\mu, \mu = 1, \ldots, M$. The Mahalanobis distance becomes the Euclidean distance if $\mathbf{R}$ is equal to the identity matrix $\mathbf{I}$. In this type of RBF network every basis function has its own matrix $\mathbf{R}_j$, usually defined as the inverse of the covariance matrix of the data points with respect to the center $\mathbf{c}_j$. Such an architecture contains d parameters for each center $\mathbf{c}_j$ plus $d(d+1)/2$ parameters for each matrix $\mathbf{R}_j$. In some approaches the matrix $\mathbf{R}_j$ is simplified to be a diagonal matrix for every center.

To simplify the notation we set $h_j(\mathbf{x}) = h(\|\mathbf{x} - \mathbf{c}_j\|_{\mathbf{R}_j}^2), j = 1, \ldots, K$ and the RBF network (10.12) becomes

$$F_i(\mathbf{x}) = \sum_{j=0}^{K} w_{ji} h_j(\mathbf{x}), \quad i = 1, \ldots, L \tag{10.15}$$

With these modifications the process of adjusting the parameters is usually treated as a typical neural network training process. In many applications the first RBF layer and the output layer are trained separately. This has led to a bad reputation of RBF networks in some application areas, which is due to the impression that the performance of RBF networks after these two training phases is worse than, for example, that of multilayer perceptrons (MLP) networks [25]. However, also a combined training of the whole network in the style of backpropagation has been proposed [31] which leads to a better performance comparable to MLP networks. We distinguish three learning or training schemes for RBF networks.

One Phase Learning. With this learning procedure only the output layer weights $\mathbf{w}$ are adjusted through some kind of supervised optimization, e.g. minimizing the squared difference between the network's output and the desired output value. Here, the centers $\mathbf{c}_j$ are sub-sampled form the set of input vectors $\mathbf{x}^\mu$ (or all data points are used as centers) and typically all scaling parameters are set equal to a predefined real number σ.

Support vector learning is a special example of one phase learning, here only the output layer weights are adjusted, the location of the kernel centers are restricted to the data points $\{\mathbf{x}^\mu \in \mathbb{R}^d : \mu = 1, \ldots, M\}$ and the scaling parameter is fixed in advance (see Sec. 10.5).

Two Phase Learning. Here the two layers of the RBF network are trained separately, first the RBF centers $\mathbf{c}_j$ and the scaling parameters are determinted, and subsequently the output layer is adjusted (see Sec. 10.2).

Three Phase Learning. After the initalization of the RBF networks utilizing two phase learning the whole architecture is adjusted through a further optimization procedure (see Sec. 10.3).

The chapter is organized as follows: In Sec. 10.2 we introduce the classical two stage training of the two layers. Backpropagation learning for RBF networks is reviewed in Sec. 10.3. In Sec. 10.4 a comparision of the different supervised and unsupervised learning rules is given. Support vector learning as a special type of supervised learning in RBF networks is reviewed in Sec. 10.5. In Sec. 10.6 a brief description of the different classifiers used in the evaluation is given. Then we show results for these classifier schemes for the classification of 3D visual objects and for the recognition of hand-written digits. Finally we conclude in Sec. 10.7.

10.2   Two Phase Learning for RBF Networks

In a multilayer perceptron (MLP) network all parameters are usually adapted simultaneously by an optimization procedure. This training procedure is supervised, since it minimizes an error function measuring the difference between the network output and the teacher signal that provides the correct output. In contrast to training a MLP network, learning in an RBF network can be done in two stages:

1. Adjusting the parameters of the RBF layer, including the RBF centers $c_j \in \mathbb{R}^d$, $j = 1, \dots, K$, and the scaling parameters given through scalars, vectors, or matrices $\mathbf{R}_j \in \mathbb{R}, \in \mathbb{R}^d$, or $\in \mathbb{R}^{d^2}$ respectively. Here d is the dimension of the input space.
2. Calculating of the output weights $\mathbf{w}_j \in \mathbb{R}^L$ for $j = 1, \dots, L$ of the network, L is the number of classes.

To determine the centers for the RBF networks, typically unsupervised training procedures from clustering are used [26], whereas in the original use for interpolation the centers are simply the data points. If the RBF network has to perform a classification task, supervised training procedures to determine the RBF centers are also applicable, because the target values of the input vectors are given. In this section we present unsupervised and supervised algorithms to initialize the RBF centers. We then describe heuristics to calculate the scaling parameters of the basis functions and discuss supervised training of the output layer weights.

10.2.1   Unsupervised Vector Quantization to Calculate the RBF Centers

Clustering techniques are typically used when the data points have to be divided into natural groups and no teacher signal is available. Here, the aim is to determine a small but representative set of centers or prototypes from a larger data set in order to minimize some quantization error.

Unsupervised Competitive Learning. A competitive neural network consists of a single layer of K neurons. Their synaptic weight vectors $c_1, \ldots, c_K \in \mathbb{R}^d$ divide the input space into K disjoint regions $\mathcal{R}_1, \ldots, \mathcal{R}_K \subset \mathbb{R}^n$, where each set $\mathcal{R}_j$ is defined by

$$\mathcal{R}_j = \{ x \in \mathbb{R}^d \mid \|x - c_j\| = \min_{i=1,\ldots,K} \|x - c_i\| \}. \tag{10.16}$$

Such a partition of the input space is called a Voronoi tesselation where each weight vector c_j is a representative prototype vector for region $\mathcal{R}_j$.

When an input vector $x \in \mathbb{R}^n$ is presented to the network, all units $j = 1, \ldots, k$ determine their Euclidean distance to x: $d_j = \|x - c_j\|$.

Competition between the units is realized by searching for the minimum distance: $d_{j^*} = \min_{j=1,\ldots,K} d_j$. The corresponding unit with index j^* is called the winner of the competition and this winning unit is trained through the *unsupervised competitive learning* rule

$$\Delta c_{j^*} = \eta_t (x^\mu - c_{j^*}) \tag{10.17}$$

where c_{j^*} is the closest prototype to the input x^μ. For convergence the learning rate η_t has to be a sequence of positive real numbers such that $\eta_t \to \infty$ as the number of data points presentations t grows up to ∞, $\sum_{t=1}^{\infty} \eta_t = \infty$ and $\sum_{t=1}^{\infty} \eta_t < \infty$.

k-means Clustering. One of the most popular methods in cluster analysis is the k-means clustering algorithm. The empirical quantization error defined by

$$E(c_1, \ldots, c_K) = \sum_{j=1}^{K} \sum_{x^\mu \in \mathcal{C}_j} \|x^\mu - c_j\|^2, \tag{10.18}$$

is minimal, if each prototype c_j is equal to the corresponding center of gravity of data points $\mathcal{C}_j := \mathcal{R}_j \cap \{x^1, \ldots, x^M\}$. Starting from a set of initial seed prototypes, these are adapted through the learning rule

$$c_j = \frac{1}{|\mathcal{C}_j|} \sum_{x^\mu \in \mathcal{C}_j} x^\mu, \tag{10.19}$$

which is called *batch mode k-means clustering*. The iteration process can be stopped if the sets of data points within each cluster $\mathcal{C}_j$ in two consecutive learning epochs are equal. Incremental optimization of E can also be realized utilizing learning rule (10.17) or

$$\Delta c_{j^*} = \frac{1}{N_{j^*} + 1} (x^\mu - c_{j^*}) \tag{10.20}$$

N_{j^*} counts how often unit j^* was the winning unit of the competition. The topic of incremental clustering algorithm has been discussed in [7].

Self-Organizing Feature Map. The self-organizing feature map (SOM) is an alternative unsupervised learning structure for data clustering and vector quantization [14]. In addition, the SOM training algorithm implements a type of dimensionality reduction of the input feature space. Here each unit of the network, represented through the synaptic weight vectors c_j, is mapped to a fixed location g_j of a predefined discrete topological space $\mathcal{G}$. Typically, $\mathcal{G}$ is a finite $1D$, $2D$ or $3D$ grid.

Based on a *neighborhood function* $N : \mathcal{G} \times \mathcal{G} \rightarrow \mathbb{R}_+$ the units are adapted through an unsupervised competitive learning rule:

$$\Delta c_j = \eta_t N(g_j, g_{j^*})(x^\mu - c_j) \qquad (10.21)$$

with learning rate η_t and j^* the index of the winning unit. A typical example for the neighborhood function is the Gaussian kernel function:

$$N(g_j, g_{j^*}) = \exp(-\|g_j - g_{j^*}\|^2/2\sigma^2).$$

The prototypes $c_1, \dots, c_K$ trained through batch mode k-means, incremental k-means, SOM-learning or the general unsupervised competitive learning rule can serve as inital locations of centers of the basis functions in RBF networks.

10.2.2 Supervised Learning to Calculate the RBF Centers

LVQ Learning. It is assumed that a classification or pattern recognition task has to be performed by the RBF network. A training set of feature vectors x^μ is given each labeled with a target classification y^μ. In this case supervised learning may be used to determine the set of prototype vectors $c_1, \dots, c_K$.

The LVQ learning algorithm has been suggested by Kohonen [13] for vector quantization and classification tasks. From the basic LVQ 1 version, LVQ2, LVQ3 and OLVQ1 training procedures have been derived. OLVQ1 denotes the optimized LVQ algorithm. Presenting a vector $x^\mu \in \mathbb{R}^d$ together with its classmembership the winning prototype j^* is adapted according to the LVQ1-learning rule:

$$\Delta c_{j^*} = \eta_t(y_{j^*}^\mu - \frac{1}{2}z_{j^*}^\mu)(x^\mu - c_{j^*}). \qquad (10.22)$$

here z^μ is the binary output vector of the network and y^μ is a binary target vector coding the classmembership for feature input vector x^μ. In both vectors z^μ and y^μ exactly one component is equal to 1, all others are 0. The difference $s_{j^*}^\mu = 2(y_{j^*}^\mu - \frac{1}{2}z_{j^*}^\mu)$ is equal to 1 if the classification of the input vector is correct and -1 if it is a false classification by the class label of the nearest prototype. In the LVQ1, LVQ2 and LVQ3 algorithms, η_t is a positive decreasing learning rate. For the OLVQ1 algorithm the learning rate depends on the actual classification by the winning prototype, and is not decreasing in general. It is defined by

$$\eta_{t+1} = \frac{\eta_t}{1 + s_{j^*}^\mu \eta_t}. \qquad (10.23)$$

For a detailed treatment on LVQ learning algorithms see [14]. After LVQ training the prototypes $c_1, \dots, c_K$ can be used as the initial RBF centers [40].

10.2.3   Calculating the Kernel Widths

The setting of the kernel widths is a critical issue in the transition to the RBF network [2]. When the kernel width $\sigma \in \mathbb{R}$ is too large the estimated probability density is over-smoothed and the nature of the underlying true density may be lost. Conversely, when σ is too small there may be an over-adaptation to the particular data set. In addition very small or large σ tend to cause numerical problems with gradient descent methods as their gradients vanish.

In general the Gaussian basis functions $h_1, \dots, h_K$ have the form

$$h_j(\mathbf{x}) = exp(-(\mathbf{x} - \mathbf{c}_j)^T \mathbf{R}_j (\mathbf{x} - \mathbf{c}_j)) \tag{10.24}$$

where each $\mathbf{R}_j$, $j = 1, \dots, K$, is a positive definite $d \times d$ matrix. Girosi[11] called this type of basis function a hyper-basis-function. The contour of a basis function, more formally the set $H_j^\alpha = \{\mathbf{x} \in \mathbb{R}^d \mid h_j(\mathbf{x}) = \alpha\}$, $\alpha > 0$, is a hyperellipsoid in $\mathbb{R}^d$. Depending on the structure of the matrices $\mathbf{R}_j$ four types of hyperellipsoids appear:

A. $\mathbf{R}_j = \frac{1}{2\sigma^2}\mathbf{I}$ with $\sigma^2 > 0$. In this case all basis functions h_j have a radial symmetric contour all with same constant width, and the Mahalanobis distances reduces to the Euclidean distance multipled by a fixed constant scaling parameter. This is the original setting of RBF in the context of interpolation and support vector machines.

B. $\mathbf{R}_j = \frac{1}{2\sigma_j^2}\mathbf{I}$ with $\sigma_j^2 > 0$. Here the basis functions are radially symmetric, but are scaled with different widths.

C. $\mathbf{R}_j$ are diagonal matrices, but the elements of the diagonal are not constant. Here the contour of a basis function h_j is not radially symmetric – in other words the axes of the hyperellipsoids are parallel to the axes of the feature space, but of different length.
   In this case $\mathbf{R}_j$ is completely defined by a d-dimensional vector $\sigma_j \in \mathbb{R}^d$:

$$\mathbf{R}_j = \mathbf{I}(\frac{1}{2\sigma_{1j}^2}, \dots, \frac{1}{2\sigma_{dj}^2}) = \text{diag}\,(\frac{1}{2\sigma_{1j}^2}, \dots, \frac{1}{2\sigma_{1j}^2}). \tag{10.25}$$

D. $\mathbf{R}_j$ is positive definite, but not a diagonal matrix. This implies that shape and orientation of the axes of the hyperellipsoids are arbitary in the feature space.

Different schemes for the initial setting of the kernel widths may be used. Real- and vector-valued kernel widths are considered. In all cases a parameter $\alpha > 0$ has to be set heuristically.

1. All σ_j are set to the same value σ, which is proportional to the average of the p minimal distances between all pairs of prototypes. First all distances $d_{lk} = \|\mathbf{c}_l - \mathbf{c}_k\|$ with $l = 1, \dots, K$ and $k = l + 1, \dots, K$ are calculated and then re-numbered through an index mapping $(l, k) \rightarrow (l - 1)K + (k - 1)$. Thus, there is a permutation τ such that the distances are arranged as an increasing sequence with $d_{\tau(1)} \leq d_{\tau(2)} \leq \cdots \leq d_{\tau(K(K-1)/2)}$ and $\sigma_j = \sigma$ is set to:

$$\sigma_j = \sigma = \alpha \frac{1}{p} \sum_{i=1}^{p} d_{\tau(i)}. \tag{10.26}$$

2. The kernel width σ_j is set to the mean of the distance to the p nearest prototypes of c_j. All distances $d_{lj} = ||c_l - c_j||$ with $l = 1, \ldots, K$ and $l \neq j$ are calculated and re-numbered through a mapping $(l, j) \rightarrow l$ for $l < j$ and $(l, j) \rightarrow l - 1$ for $l > j$, then there is a permutation τ such that $d_{\tau(1)} \leq d_{\tau(2)} \leq \cdots \leq d_{\tau(K-1)}$ and σ_j is set to:

$$\sigma_j = \alpha \frac{1}{p} \sum_{i=1}^{p} d_{\tau(i)} \tag{10.27}$$

3. The distance to the nearest prototype with a different class label is used for the initialization of σ_j:

$$\sigma_j = \alpha \min\{||c_i - c_j|| : class(c_i) \neq class(c_j), i = 1, \ldots, K\} \tag{10.28}$$

4. The kernel width σ_j is set to the mean of distances between the data points and the corresponding cluster $\mathcal{C}_j$:

$$\sigma_j = \alpha \frac{1}{|\mathcal{C}_j|} \sum_{x^\mu \in \mathcal{C}_j} ||x^\mu - c_j|| \tag{10.29}$$

In the case of vector-valued kernel parameters, the widths $\sigma_j \in \mathbb{R}^d$ may be initially set to the variance of each input feature based on all data points in the corresponding cluster $\mathcal{C}_j$:

$$\sigma_{ij}^2 = \alpha \frac{1}{|\mathcal{C}_j|} \sum_{x^\mu \in \mathcal{C}_j} (x_i^\mu - c_{ij})^2 \tag{10.30}$$

Typically, the calculation of the kernel widths is unsupervised without using the target information (cases 1,2, and 4), in the classification scenario the class labels of the prototypes may be used (case 4). In general, the location and the shape of the kernels represented by the centers c_j and the scaling matrices R_j can be calculated using re-estimation techniques known as the expectation-maximization (EM) algorithm [35].

10.2.4 Training the Output Weights of the RBF Network

Provided that the centers c_j and the scaling parameters, given by the matrices R_j, of the basis functions have been determined, the weights of the output layer can be calculated [12,2]. We assume K basis functions in the hidden layer of the RBF network. Let (x^μ, y^μ), $\mu = 1, \ldots, M$ be the set of training examples with feature vector $\mathbf{x}^{\mu} \in \mathbb{R}^d$ and target $\mathbf{y}^{\mu} \in \mathbb{R}^m$, $H_{\mu j} = h_j(\mathbf{x}^{\mu})$ the outcome of the j-th basis function with the μ-th feature vector $\mathbf{x}^{\mu}$ as input, and $Y_{\mu j}$ the j-th component of the μ-th target vector $\mathbf{y}^{\mu}$. Given these two matrices $\mathbf{H} = (H_{\mu j})$ and $\mathbf{Y} = (Y_{\mu j})$ the matrix of the output layer weights $\mathbf{W}$ is the result of a minimization of the error function:

$$E(\mathbf{W}) = \|\mathbf{HW} - \mathbf{Y}\|^2. \tag{10.31}$$

The solution is given explicitly in the form $\mathbf{W} = \mathbf{H}^+\mathbf{Y}$ where $\mathbf{H}^+$ denotes the pseudo inverse matrix of $\mathbf{H}$ which can be defined as

$$\mathbf{H}^+ = \lim_{\alpha \to 0+} (\mathbf{H}^T\mathbf{H} + \alpha\mathbf{Id})^{-1}\mathbf{H}^T. \tag{10.32}$$

Provided that the inverse matrix of $(\mathbf{H}^T\mathbf{H})$ is defined, the pseudo inverse matrix becomes simply $\mathbf{H}^+ = (\mathbf{H}^T\mathbf{H})^{-1}\mathbf{H}^T$. The solution $\mathbf{W}$ is unique and can also be found by gradient descent optimization of error function defined in (10.31). This leads to the delta learning rule for the output weights

$$\Delta w_{jp} = \eta \sum_{\mu=1}^{M} h_j(\mathbf{x}^{\mu})(y_p^{\mu} - F_p(\mathbf{x}^{\mu})), \tag{10.33}$$

or its incremental version

$$\Delta w_{jp} = \eta_t h_j(\mathbf{x}^{\mu})(y_p^{\mu} - F_p(\mathbf{x}^{\mu})). \tag{10.34}$$

After this final step of calculating the output layer weights, all parameters of the RBF network have been determined.

10.3   Backpropagation Learning in RBF Networks

As described in Sec. 10.2 learning in an RBF network can simply be done in two separate learning phases: calculating the RBF layer and then the output layer. This is a very fast training procedure but often leads to RBF classifiers with bad classification performance [25]. We propose a third training phase of RBF networks in the style of backpropagation learning in MLPs, performing an adaptation of all types of parameters simultaneously. We give a brief summary of the use of error-back-propagation in the context of radial basis function network training, for a more detailed treatment see [12,2,43].

If we define as the error function of the network a differentiable function like the sum-of-squares error E,

$$E = \frac{1}{2} \sum_{\mu=1}^{M} \sum_{p=1}^{L} \left(y_p^{\mu} - F_p^{\mu}\right)^2, \tag{10.35}$$

with F_p^μ and y_p^μ as the actual and target output values respectively, and consider a network with differentiable activation functions then a necessary condition for a minimal error is that its derivatives with respect to the parameters center location $\mathbf{c}_j$, kernel width $\mathbf{R}_j$, and output weights $\mathbf{w}_j$ vanish. In the following we consider the case that $\mathbf{R}_j$ is a diagonal matrix defined by a vector $\sigma_j \in \mathbb{R}^d$.

An iterative procedure for finding a solution to this problem is gradient descent. Here, the full parameter set $\mathbf{U} = (\mathbf{c}_j, \sigma_j, \mathbf{w}_j)$ is moved by a small distance η in the direction in which E decreases most rapidly, i.e. in the direction of the negative gradient $-\nabla E$:

$$\mathbf{U}^{(\tau+1)} = \mathbf{U}^{(\tau)} - \eta \nabla E(\mathbf{U}^{(\tau)}). \tag{10.36}$$

For the RBF network (10.15) for the Gaussian basis function we obtain the following expressions for the adaptation rules or the network parameters:

$$\Delta w_{jp} = \eta \sum_{\mu=1}^{M} h_j(\mathbf{x}^\mu)(y_k^\mu - F_k^\mu) \tag{10.37}$$

$$\Delta c_{ij} = \eta \sum_{\mu=1}^{M} h_j(\mathbf{x}^\mu) \frac{x_i^\mu - c_{ij}}{\sigma_{ij}^2} \sum_{p=1}^{L} w_{jp}(y_p^\mu - F_p^\mu) \tag{10.38}$$

$$\Delta \sigma_{ij} = \eta \sum_{\mu=1}^{M} h_j(\mathbf{x}^\mu) \frac{(x_i^\mu - c_{ij})^2}{\sigma_{ij}^3} \sum_{p=1}^{L} w_{jp}(y_p^\mu - F_p^\mu) \tag{10.39}$$

$$\tag{10.40}$$

10.4   Competitive and Gradient Descent Learning

In this section we briefly compare the competitive learning rules (unsupervised: k-means- and SOM-learning, and supervised learning: LVQ learning) with the learning rules for RBF networks derived from gradient descent optimization. In particular the prototype adaptation is considered. The weight updates in all learning rules consisted of two parts—a *presynaptic* and a *postsynaptic* signal.

Presenting an input vector $\mathbf{x}^\mu$ together with its class label $\mathbf{y}^\mu$, the synaptic weight c_{ij} is changed through a *Hebbian learning rule* composed of a presynaptic and a postsynaptic factor.

The different learning rules yield:

	unsupervised	supervised (LVQ)	RBF
presynaptic	$x_i^\mu - c_{ij}$	$x_i^\mu - c_{ij}$	$x_i^\mu - c_{ij}$
postsynaptic	1	$\delta_{jj^*}(y_j^\mu - z_j^\mu/2)$	$h(\mathbf{x}_j^\mu)\sum_p (y_p^\mu - F_p^\mu)w_{jp}$

In all three adaptation rules the presynaptic part is given by the difference vector $(\mathbf{x}^\mu - \mathbf{c}_j)$. This vector is weighted by a postsynaptic signal which is determined by the actual output of network and the desired output $\mathbf{y}^\mu$. In the unsupervised learning algorithms like k-means or SOM there is no target, and so the postsynaptic term is a constant during the whole learning procedure. In LVQ learning the postsynaptic term depends on the difference of two binary variables as for the wellknown perceptron learning rule, therefore it is an element of the set $\{-1, 0, 1\}$. The postsynaptic signal in the RBF-learning rule is real number determined through a weighted sum of the difference vector $(\mathbf{y}^\mu - \mathbf{F}^\mu)$ and the synaptic weight connections form unit j to the output layer. The distance between the prototype vector $\mathbf{c}_j$ and $\mathbf{x}^\mu$ is taken into account in both supervised learning rules. In LVQ-learning only the winning neuron j^* is adapted. In RBF-learning the distance is fed into the decreasing radial basis function h, so the adaptation of the RBF units close to the input is larger than the adaptation rates of the distant RBF units.

10.5 Support Vector Learning in RBF Networks

Here we give a short review on support vector (SV) learning in RBF networks, [6,37,42] for details. The support vector machine (SVM) was initially developed to classify data points of a linear separable data set. In this case a training set consisting of M examples $(\mathbf{x}^\mu, y^\mu)$, $\mathbf{x}^\mu \in \mathbb{R}^d$, and $y^\mu \in \{-1, 1\}$ can be divided up into two sets by a separating hyperplane. Such a hyperplane is determined by a weight vector $\mathbf{w} \in \mathbb{R}^d$ and a bias or threshold $\theta \in \mathbb{R}$ satisfying the separating contraints

$$y^\mu \left[\langle \mathbf{x}^\mu, \mathbf{w} \rangle + \theta \right] \geq 1 \quad \mu = 1, \ldots, M. \tag{10.41}$$

The distance between the separating hyperplane and the closest data points of the training set is called the *margin*. Intuitively, the larger the margin, the higher the generalization ability of the separating hyperplane. The optimal separating hyperplane with maximal margin is unique and can be expressed by a linear combination of those training examples lying exactly at the margin. These data points are called the *support vectors*. This separating hyperplane with maximal margin has the form

$$H(\mathbf{x}) = \sum_{\mu=1}^{M} \alpha_\mu^* y^\mu \langle \mathbf{x}, \mathbf{x}^\mu \rangle + \alpha_0^* \tag{10.42}$$

where $\alpha_1^*, \ldots, \alpha_M^*$ is the solution optimizing the functional

$$Q(\alpha) = \sum_{\mu=1}^{M} \alpha_\mu - \frac{1}{2} \sum_{\mu,\nu=1}^{M} \alpha_\mu \alpha_\nu y^\mu y^\nu \langle \mathbf{x}^\mu, \mathbf{x}^\nu \rangle \tag{10.43}$$

subject to the constraints $\alpha_\mu \geq 0$ for all $\mu = 1, \ldots, M$ and

$$\sum_{\mu=1}^{M} \alpha_\mu y^\mu = 0. \tag{10.44}$$

A vector of the training set $\mathbf{x}^\mu$ is a support vector if the corresponding coefficient $\alpha_\mu^* > 0$. Then the weight vector $\mathbf{w}$ has the form

$$\mathbf{w} = \sum_{\mu=1}^{M} \alpha_\mu y^\mu \mathbf{x}^\mu = \sum_{\mu,\alpha_\mu>0} \alpha_\mu y^\mu \mathbf{x}^\mu$$

and the bias α_0^* is determined by a single support vector $(\mathbf{x}^s, y^s)$:

$$\alpha_0^* = y^s - \langle \mathbf{w}, \mathbf{x}^s \rangle.$$

The SVM approach has been extended to the non-separable situation and to the regression problem. In most applications (regression or pattern recognition problems) linear hyperplanes as solutions are insufficient. For example, in real world pattern recognition problems it is common to define an appropriate set of nonlinear mappings $\mathbf{g} = (g_1, g_2, \dots)$, where the g_j are defined as real valued functions, transforming an input vector $\mathbf{x}^\mu$ to a vector $g(x^\mu)$ which is element of a new feature space $\mathcal{H}$. Then the separating hyperplane can be constructed in this new feature space $\mathcal{H}$ and can be expressed by

$$H(\mathbf{x}) = \sum_{\mu=1}^{M} \alpha_\mu y^\mu \langle g(\mathbf{x}), g(\mathbf{x}^\mu) \rangle + \alpha_0. \tag{10.45}$$

Provided $\mathcal{H}$ is a Hilbert space and K a kernel $K : \mathcal{H} \times \mathcal{H} \to \mathbb{R}$ satisfying the condition of Mercer's theorem an explicit mapping $g : \mathbb{R}^d \to \mathcal{H}$ does not need to be known, because it is implicitly given through

$$K(\mathbf{x}, \mathbf{x}^\mu) = \langle g(\mathbf{x}), g(\mathbf{x}^\mu) \rangle.$$

Kernel function K is representing the inner product between vectors in the feature space $\mathcal{H}$. With a suitable choice of a kernel function the data can become separable in feature space despite being not separable in the input space. Using such a kernel function K the separating hyperplane is given by

$$H(\mathbf{x}) = \sum_{\mu=1}^{M} \alpha_\mu y^\mu K(\mathbf{x}, \mathbf{x}^\mu) + \alpha_0.$$

The coefficients α_μ can be found by solving the optimization problem

$$Q(\alpha) = \sum_{\mu=1}^{M} \alpha_\mu - \frac{1}{2} \sum_{\mu,\nu=1}^{M} \alpha_\mu \alpha_\nu y^\mu y^\nu K(\mathbf{x}^\mu, \mathbf{x}^\nu)$$

subject to the contraints $0 \leq \alpha_\mu \leq C$ for all $\mu = 1, \dots, M$ and

$$\sum_{\mu=1}^{M} \alpha_\mu y^\mu = 0$$

where C is a predefined positive number. An important kernel function satisfying Mercer's condition is the Gaussian kernel function

$$K(\mathbf{x}, \mathbf{y}) = e^{-\frac{\|\mathbf{x}-\mathbf{y}\|^2}{2\sigma^2}}.$$

The separating surface obtained by the SVM approach is a linear combination of Gaussian functions located at the support vectors. The SVM reduces to an RBF network. In contrast to RBF networks described in the previously the centers are now located at certain data points of the training set and the number of centers is automatically determined in this approach. Furthermore, all Gaussians are radially symmetric, all with the same kernel width σ^2.

10.6   Applications

In the following sections we will compare different methods of initialization and optimization on the data sets.

Classsifiers.   For numerical evaluation the following classification schemes were applied:

1NN: Feature vectors are classified through the 1-nearest-neighbour (1NN) rule. Here, the 1NN rule is applied to the whole training set.

LVQ: The 1-nearest-neighbour classifier is trained through Kohonen's supervized OLVQ1 followed by LVQ3 training algorithm (each for 50 training epochs). The 1NN rule is applied to a fraction of the training set (see Sec. 10.2.1).

2-Phase-RBF (data points): A set of data points is randomly selected from the training data set. These data points serve as RBF centers. A single scaling parameter per basis function is determined as the mean of the 3 closest prototypes, see Sec. 10.2.3. The weights of the output layer are calculated through the pseudo inverse matrix solution as described in Sec. 10.2.4.

2-Phase-RBF (k-means): A set of data points is randomly selected from the training data set. These data points are the seeds of an incremental k-means clustering procedure and these k-means centers are used as centers in the RBF network. For each basis function a single scaling parameter is set to the mean of the 3 closest prototypes, and the output layer matrix is calculated through the pseudo inverse matrix solution.

2-Phase-RBF (SOM): A set of data points is randomly selected from the training data set. These data points are the seeds for the SOM learning procedure. For each basis function a single scaling parameter is set to the mean of the 3 closest prototypes, and the output layer matrix is calculated through the pseudo inverse matrix solution.

2-Phase-RBF (LVQ): A set of data points is randomly selected from the training set. These data points are the seeds for the OLVQ1 training algorithm (50 training epochs), followed by LVQ3 training with again 50 training epochs. These proto-types then are used as the centers in an RBF network. A single scaling parameter per basis function is set to the mean of the 3 closest prototypes and the output layer is calculated through the pseudo inverse matrix.

3-Phase-RBF (data points): The **2-Phase-RBF (data points)** network is trained through a third error-backpropagation training procedure with 100 training epochs (see Sec. 10.3).

3-Phase-RBF (k-means): The **2-Phase-RBF (k-means)** network is trained through a third error-backpropagation training procedure with 100 training epochs.

3-Phase-RBF (SOM): The **2-Phase-RBF (SOM)** network is trained through a third error-backpropagation training procedure with 100 training epochs.

3-Phase-RBF (LVQ): The **2-Phase-RBF (LVQ)** network is trained through a third error-backpropagation training procedure with 100 training epochs.

SV-RBF: The RBF network with Gaussian kernel function is trained by support vector learning (see Sec. 10.5). For the optimization the NAG library is used. In multi-class applications (number of classes $L > 2$) a RBF network has been trained through SV learning for each class. In the classification phase the estimate for an unseen exemplar is found through maximum detection among the L classifiers. This is called the *one-against-rest* strategy [38].

10.6.1   Application to Hand-Written Digits

The classification of machine-printed or hand-written characters is one of the classical applications in the field of pattern recognition and machine learning. In optical character recognition (OCR) the problem is to classify characters into a set of classes (letters, digits, special characters (e.g. mathematical characters), characters from different fonts, characters in different sizes, etc.). After some preprocessing and segmentation, the characters are sampled with a few hundred pixels and then categorized into a class of the predefined set of character categories. In this chapter we consider the problem of hand-written digit recognition which appears as an important subproblem in the area of automatic reading of postal addresses.

Data. The data set used for the evaluating of the performance of the RBF classifiers consists of 20,000 hand-written digits (2,000 samples of each class). The digits, normalized in height and width, are represented through a 16×16 matrix G where the entries $G_{ij} \in \{0, \dots, 255\}$ are values taken from an 8 bit gray scale, see Fig. 10.2. Previously, this data set has been used for the evaluation of machine learning techniques in the STATLOG project. Details concerning this data set and the STATLOG project can be found in the final report of STATLOG [25].

Results. The whole data set has been divided into a set of 10,000 training samples and a set of 10,000 test samples (1,000 examples of each class in both data set). The

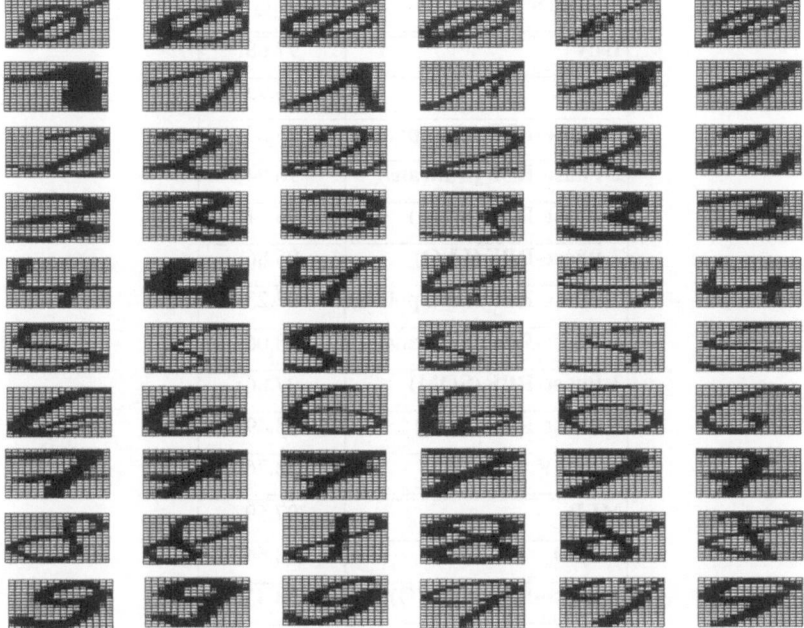

Fig. 10.2. A subset of 60 hand-written digits with 6 exemplars of each class sampled from the training data set is shown.

training set was used to design the classifiers, and the test set was used for performance evaluation. Three different classifiers per architecture were trained, and the classification error was measured on the test set. For this data set we present results for all classifier architectures described before. Furthermore results for multilayer perceptrons **MLP**, and results achieved with the first 40 principal components of this data set for the quadratic polynomial classifier **Poly40**, for the RBF network with SV learning **SV-RBF40**, and for RBF network trained by three phase RBF learning and LVQ prototype initialization **3-Phase-RBF40 (LVQ)** are given.

For the LVQ classifier 200 prototypes (20 per class) are used. The RBF networks initialized through randomly selected data points, through centers calculated utilizing k-means clustering or learning vector quantization also consisted of 200 RBF centers. A 14×14 self organizing map is used with the SOM initalization. The MLP networks consisted of a single hidden layer with 200 sigmoidal units.

Further results for this data set of handwritten digits can be found in the final STATLOG report (see pp. 135-138 in [25]). The error rates for the **1NN, LVQ,** and **MLP** classifiers are similar in both studies. The error rate for the RBF classifier in [25] is close to our results we achieved by **2-Phase RBF** classifiers with an initalization of the RBF centers utilizing **k-means, SOM** and **LVQ**. Indeed, the RBF classifiers considered in [25] were trained in two separate stages. First the RBF centers are calculated through k-means clustering and the pseudo-inverse matrix solution was used to determine the output weight matrix. The performance of the

Classifier	accuracy [%]
1NN	97.68
LVQ	96.99
2-Phase-RBF (data points)	95.24
2-Phase-RBF (k-means)	96.94
2-Phase-RBF (SOM)	96.71
2-Phase-RBF (LVQ)	95.86
3-Phase-RBF (data points)	97.23
3-Phase-RBF (k-means)	98.06
3-Phase-RBF (SOM)	97.86
3-Phase-RBF (LVQ)	98.49
SV-RBF	98.76
MLP	97.59
Poly40	98.64
3-Phase-RBF40 (LVQ)	98.45
SV-RBF40	98.56

Table 10.1. Results for the handwritten digits on the test set of 10,000 examples. The classifiers are trained on a training set (disjoint from the test set) of 10,000 examples. Results are given as the median of three training and test runs.

RBF classifiers can significantly be improved by an additional third optimization procedure in order to fine-tune all network parameters simultaneously. All **3-Phase RBF** classifiers perform better as the corresponding **2-Phase RBF** classifiers. The **3-Phase RBF** classifiers performs as well as other regression based methods like MLPs or polynomials. This is not surprising, as RBFs, MLPs and polynomials are approximation schemes dense in the space of continuous functions.

The **1NN** and **LVQ** classifiers perform surprisingly well, particularly in comparision to RBF classifiers trained only in two phases. The **SV-RBF** and **SV-RBF40** classifiers perform very well in our numerical experiments. We found no significant difference between the classifiers on the 256-dimensional data set and the data set reduced to the 40 principal components.

The error rates for **SV-RBF**, **SV-RBF40**, **Poly40** and for RBF classifiers trained through three phase learning with LVQ prototype initalization **3-Phase-RBF** and **3-Phase-RBF40** are very good. Although the performances of the **SV-RBF** and **3-Phase-RBF** classifiers are approximately identical, the architectures are completely different. The complete **SV-RBF** classifier architecture consisting ten classifiers, where approximately 4,200 support vectors are selected from the training data set.

In contrast the **3-Phase-RBF** classifiers with a single hidden layer containing only 200 representative prototypes distributed over the whole input space.

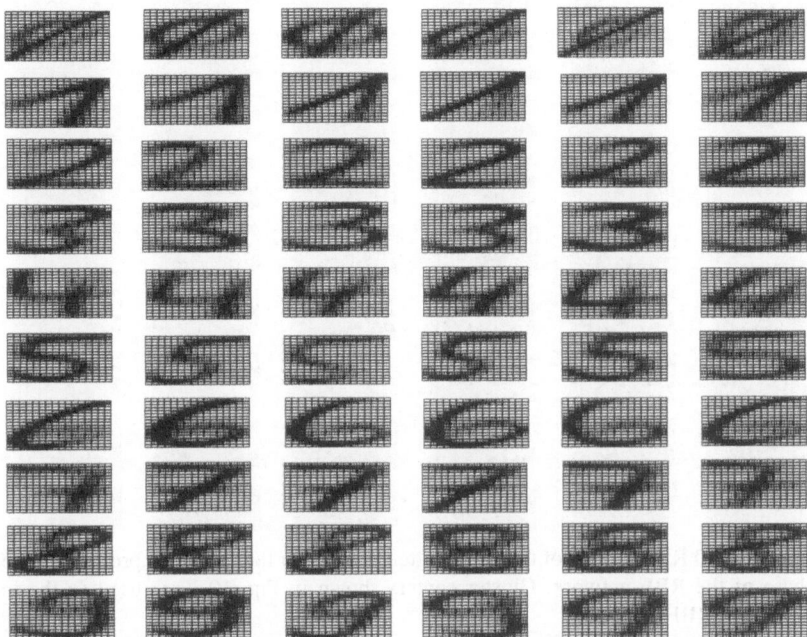

Fig. 10.3. The 60 cluster centers of the handwritten digits after running the incremental k-means clustering algorithm for each of the ten digits separately. For each of the 10 digits $k = 6$ cluster centers are used in this clustering process. The cluster centers are initalized through data points which are randomly selected from the training data set.

An interesting property of RBF networks is that the centers in an RBF network are typical feature vectors and can be considered as representative patterns of the data set, which may be displayed and analyzed in the same way as the data.

In Fig. 10.3 and 10.4, a set of 60 RBF centers is displayed in the same style as the data points shown in Fig. 10.2. Here, for each digit a subset of 6 data points was selected at random from the training set. Each of these 10 subsets serves as seed for the cluster centers of an incremental k-means clustering procedure. After clustering the data of each digit, the union of all 60 cluster centers is used as RBF centers, the scaling parameters are calculated and the output layer weights are adapted in a second training phase as described in Sec. 10.2. These RBF centers are shown in Fig. 10.3.

The whole set of parameters in RBF network is then trained simultaneously by backpropagation for 100 training epochs, see Sec. 10.3. During this third training phase the RBF centers slightly changed their inital locations. These fine-tuned RBF centers are depicted in Fig. 10.4. Pairs of corresponding RBF centers of Fig. 10.3 and 10.4 are very similar. The distance between these pairs of centers before and

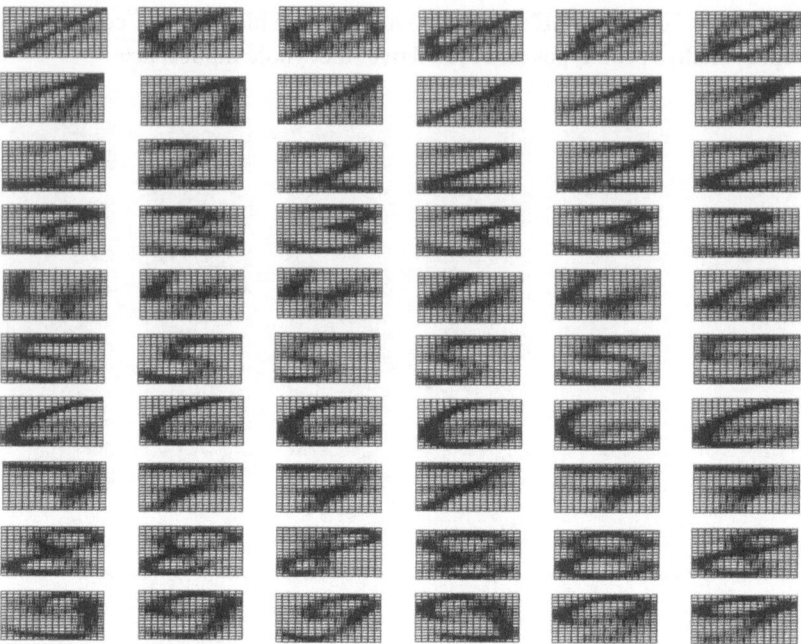

Fig. 10.4. The 60 RBF centers of the handwritten digits after the third backpropagation learning phase of the RBF network. Cluster centers shown in Fig. 10.3 are used as the initial location of the RBF centers.

after the third learning phase was only $\|\Delta \mathbf{c}_j\| = 460$ in the mean, which is significantly smaller than the distances of centers representing the same class (before the third learning phase: 1116 (mean), and after the third learning phase 1153 (mean)) and in particular smaller than the distances of centers representing two different classes (before third learning phase: 1550 (mean), and after the third learning phase: 1695 (mean)). But, calculating the distance matrices of these two sets of centers in order to analyze the distance relations between the RBF centers in more detail, it can be observed that the RBF centers were adapted during this third backpropagation learning phase.

The distance matrices of the centers are visualized as matrices of gray values. In Fig. 10.5 the distance matrices of the RBF centers before (left panel) and after the third learning phase (right panel) are shown. In the left distance matrix many entries with small distances between prototypes of different classes can be observed, particularly between the digits 2, 3, 8, and 9, see Fig. 10.5. These smaller distances between prototypes of different classes typically lead to misclassifications of data points between these classes, therefore such a set of classes is called a *confusion class*. After the third learning phase of this RBF network the centers are adjusted in such a way that these smaller distances between prototypes of different classes disappear, see Fig. 10.5 (right panel).

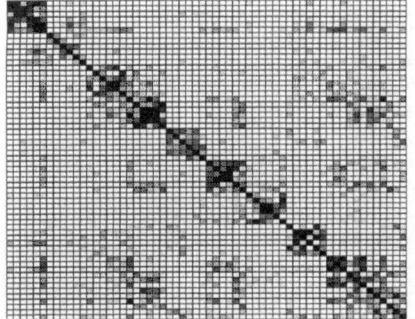 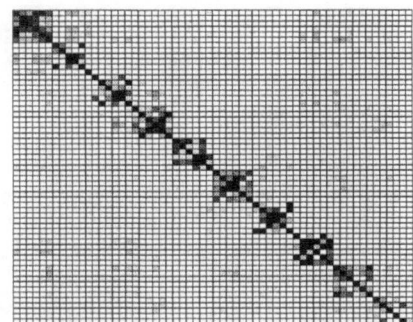

Fig. 10.5. Distance matrices (Euclidean distance) of 60 RBF centers before (left) and after (right) the third learning phase in an RBF network. The centers c_j are sorted by the class memberships in such a way that the centers $c_1, \dots, c_6$ are representing the digit 0, centers $c_7, \dots, c_{12}$ representing the digit 1, etc. Distances $d(c_i, c_j)$ between the centers are encoded through gray values – the smaller the distance $d(c_i, c_j)$ the darker is the corresponding entry in the gray value matrix. In the left distance matrix many small distances from centers of different classes can be observed, in particular the distances between centers of the digits $\{2, 3, 8, 9\}$ are very small. These small distances outside diagonal blocks often lead to mis-classifications. These can not be found in the distance matrix after all three phase learning (right figure).

10.6.2   Application to 3D Visual Object Recognition

The recognition of 3D objects from 2D camera images is one of the most important goals in computer vision. There is a large number of contributions to this field of research from various disciplines, e.g. artificial intelligence and autonomous mobile robots [3,20], artificial neural networks [18,30,36], computer vision and pattern recognition [22,21,41,1,9], psychophysics and brain theory [8,5,19]. Due to the increasing performance of current computer systems and the increasing development of computer vision and pattern recognition techniques several 3D object recognition systems have been developed [28,23,44,27,15]. The recognition of an 3D object consisted the following three subtasks (details on this application may be found in [39]):

1. **Localization of objects in the camera image.**
   In this processing step the entire camera image is segmented into regions, see Fig. 10.6. Each region should contain exactly one single 3D object. Only these marked regions, which we call the regions of interest (ROI), are used for the further image processing steps. Colour-based approaches for the ROI-detection are used.

2. **Extraction of characteristic features.**
   From each ROI within the camera image a set of features is computed. For this, the ROIs are divided into $n \times n$ subimages and for each subimage an orientation histogram with eight orientation bins is calculated from the gray valued image. The orientation histograms of all subimages are concatenated

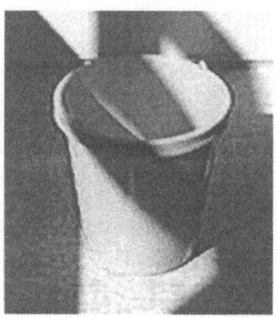

Fig. 10.6. Examples of class bucket of the data set (left) and the calculated region of interest (right).

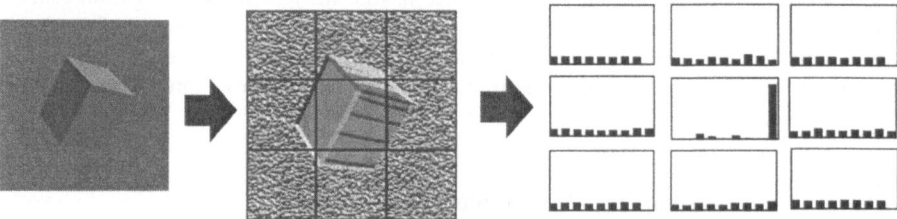

Fig. 10.7. Elements of the feature extraction method. From the grey valued image (left) gradient image (center; absolute value of the gradient) is calculated. Orientation histograms (right) of non–overlapping subimages constitute the feature vector.

into the characterizing feature vector, see Fig. 10.7, here n is set equal to 3. These feature vectors are used for classifier construction in the training phase, and are applied to the trained classifier during the recognition phase.

3. **Classification of the extracted feature vectors.**
   The extracted feature vectors together with the target classification are used in a supervised learning phase to build the neural network classifier. After network training novel feature vectors are presented to the classifier which outputs the estimated class labels.

Data. Camera images were recorded for six different 3D objects (orange juice bottle, small cylinder, large cylinder, cube, ball and bucket) with an initial resolution of 768×576 pixels. To these objects 9 different classes were assigned (bottle lying/upright, cylinders lying/upright). The test scenes were acquired under mixed natural and artificial lighting, see Fig. 10.8. Regions of interest where calculated from 1800 images using the colour blob detection method as described before. These regions where checked and labeled by hand, 1786 images remained for classifier evaluation. Regions of interest are detected using three colour ranges, one for red

Fig. 10.8. Examples of the real–world data set (class 0/1: orange juice bottle upright/lying, class 2/3: large cylinder upright/lying, class 4/5: small cylinder upright/lying, class 6: cube, class 7: ball, class 8: bucket).

(bucket, cylinder, ball), blue (cylinder) and yellow (cylinder, bucket, orange juice). The image in Fig. 10.6 gives an example of the automatically extracted region of interest. Features were calculated from concatenated 5×5 histograms with 3×3 Sobel operator, see Fig. 10.7. These data set of 1786 feature vectors of $\mathbb{R}^{200}$ serve as the evaluation data set.

Results In this application for the LVQ classifiers and the RBF networks initialized through randomly selected data points, through prototypes calculated by clustering or vector quantization 90 centers (10 per class) have been used in the numerical experiments. As in application of the hand-written digits, the **1NN** and particularly the **LVQ** classifiers perform very well. The errror rate of the **LVQ** classifier was lower than all **2-Phase RBF** classifiers, surprisingly also better than the RBF network initalized with the LVQ prototypes and additional output layer training. As already observed in the OCR application the performance of the **2-Phase RBF** classifiers can significantly be improved by an additional third backpropagation-like optimization procedure. All **3-Phase RBF** classifiers perform better as the corresponding **2-Phase RBF** classifiers. In this application the best classification results were achieved with the **SV-RBF** classifier and the **3-Phase-RBF (LVQ)** trained through three phase learning with LVQ prototype initalization.

In Fig. 10.9 the distance matrices of $9 \times 6 = 54$ RBF centers before (left panel) and after (right panel) the third learning phase of the RBF network are shown.

Classifier	accuracy [%]
1NN	90.51 ± 0.17
LVQ	92.70 ± 0.71
2-Phase-RBF (data points)	87.72 ± 0.65
2-Phase-RBF (k-means)	88.16 ± 0.30
2-Phase-RBF (LVQ)	92.10 ± 0.40
3-Phase-RBF (data points)	89.96 ± 0.36
3-Phase-RBF (k-means)	92.94 ± 0.47
3-Phase-RBF (LVQ)	93.92 ± 0.19
SV-RBF	93.81 ± 0.18

Table 10.2. Classification results of the camera images. The mean of five 5-fold cross-validation runs and the standard deviation is given.

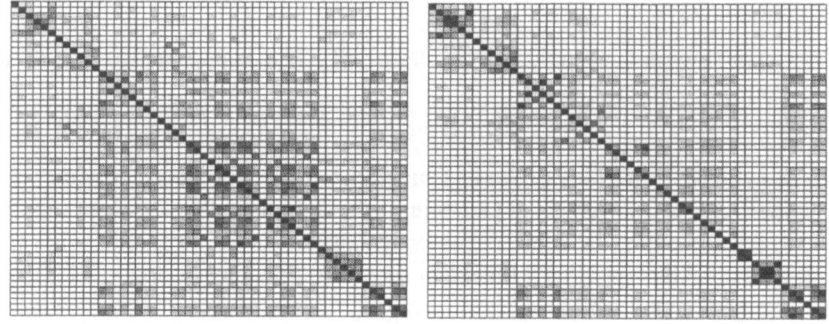

Fig. 10.9. The distance matrices of $9 \times 6 = 54$ RBF centers before (left panel) and after (right panel) the third learning phase in an RBF network are given. Centers c_j are sorted by the class memberships in such a way that the centers $c_1, \ldots, c_6$ are representing class 0, $c_7, \ldots, c_{12}$ representing class 1, etc. The distances $d(\mathbf{c}_i, \mathbf{c}_j)$ are encoded through gray values. In the distance matrix calculated before the third learning phase has been started (left), many small distances for centers of different classes can be observed.

The RBF centers were calculated as for the application to hand-written digits, see Sec. 10.6.1. In both distance matrices a large confusion class can be observed, containing the classes 2, 3, 4, 5, 6, and 8. The centers of class 7 are separated from the centers of the other classes. After the third learning phase of the RBF network these distances between centers of different classes become a little larger. This can be observed in Fig. 10.9 (right panel) where the number of small distances outside the diagonal blocks is reduced.

10.7 Conclusions

In this chapter unsupervised and supervised learning algorithms for RBF networks have been presented and applied to build RBF classifiers for real world applications in pattern recognition. Results for the classification of 3D visual objects and the recognition of hand-written digits are given.

We have discussed three different types of RBF learning schemes: Two phase, three phase and support vector learning. The first step of two phase RBF learning is closely related to density estimation, in particular when unsupervised clustering methods are used. In learning phase two and three an error criterion measuring the difference between the network's output and the target output is minimized. In the context of learning in RBF networks we considered support vector learning as a special type of one phase learning scheme.

Using only the first two phases is very common, see the studies on machine learning algorithms [25,17]. In our applications the performance of classifiers trained by two phase learning was improved through a third learning phase. Therefore, the most economical approach simply uses the first two steps as a fast way of computing a good initialization for the final gradient descent. The performance of RBF networks trained by three phase learning and support vector learning is comparable, but RBF networks trained by support vector learning often lead to a more complex network structure. A practical advantage of two and three phase learning in RBF networks is the possibility to use unlabeled training data and clustering algorithms for the first training phase of the RBF centers. These RBF units can be viewed as smoothed representative data points and actually they can be displayed and interpreted in the same way as data points—an important property of RBF networks in applications where the classifier system has to be built by a non-specialist in the field of classifier design.

Bibliography on Chapter 10

1. R. Basri. Recognition by prototypes. *International Journal of Computer Vision*, 19:147–168, 1996.
2. C. M. Bishop. *Neural Networks for Pattern Recognition*. Clarendon Press, Oxford, 1995.
3. R. Brooks. Model-based three-dimensional interpreations of two-dimensional images. *IEEE Transactions on Pattern Analysis and Machine Intelligence*, 5:140–149, 1983.
4. D.S. Broomhead and D. Lowe. Multivariable functional interpolation and adaptive networks. *Complex Systems*, 2:321–355, 1988.
5. H.H. Bülthoff, S. Edelman, and M.J. Tarr. How are three-dimensional objects represented in the brain? *Cerebral Cortex*, 5:247–260, 1995.
6. N. Cristianini and J. Shawe-Taylor. *An introduction to support vector machines*. Cambridge University Press, 2000.
7. C. Darken and J. Moody. Fast adaptive k–means clustering: Some empirical results. *Proceedings International Joint Conference on Neural Networks*, 1990.
8. S. Edelman and H.H. Bülthoff. Orientation dependence in the recognition of familiar and novel views of three-dimensional objects. *Vision Research*, 32:2385–2400, 1992.
9. S. Edelman and S. Duvdevani-Bar. A model of visual recognition and categorization. *Phil. Trans. R. Soc. London B*, 352:1191–1202, 1997.
10. O. Ghitza. Auditory nerve representation as a basis for speech recognition. In S. Furui and M. Sondhi, editors, *Advances in Speech Signal Processing*, pages 453–485. Marcel Dekker, NY, 1991.
11. F. Girosi, M. Jones, and T. Poggio. Regularization theory and neural network architectures. *Neural Computation*, 7:219–269, 1995.
12. J. Hertz, A. Krogh, and R. G. Palmer. *Introduction to the Theory of Neural Computation*. Addison Wesley, New York, 1991.
13. T. Kohonen. The self–organizing map. *Proc. IEEE*, 78(9):1464–1480, 1990.
14. T. Kohonen. *Self-Organizing Maps*. Springer, 1995.
15. M. Lades, J. Vorbrüggen, J. Buhmann, J. Lange, C. v.d. Malsburg, R. Würtz, and W. Konen. Distortion invariant object recognition in the dynamic link architecture. *IEEE Transactions on Computers*, 42:300–311, 1993.
16. W.A. Light. Some aspects of radial basis function approximation. In S.P. Singh, editor, *Approximation Theory, Spline Functions and Applications*, pages 163–190. Kluwer, 1992.
17. T.-J. Lim, W.-Y. Loh, and Y.-S-. Shih. A comparision of prediction accuracy, complexity, and training time of thirty-tree old and new classification algorithms. *Machine Learning*, pages 1–27, 2000.
18. J.J. Little, T. Poggio, and E.B. Gamble. Seeing in parallel: The vision machine. *International Journal of Supercomputing Applications*, 2:13–28, 1988.
19. N.K. Logothetis and D.L. Scheinberg. Visual object recognition. *Annual Review of Neuroscience*, 19:577–621, 1996.
20. D.G. Lowe. Three-dimensional object recognition from single two-dimensional images. *Artificial Intelligence*, 31:355–395, 1987.
21. D. Marr. *Vision*. Freeman, San Fransisco, 1982.
22. D. Marr and H.K. Nishihara. Representation and recognition of the spatial organization of three dimensional structure. *Proceedings of the Royal Society of London B*, 200:269–294, 1978.
23. B. Mel. Seemore: Combining colour, shape, and texture histogramming in a neurally-inspired approach to visual object recognition. *Neural Computation*, 9:777–804, 1997.

24. C.A. Micchelli. Interpolation of Scattered Data: Distance Matrices and Conditionally Positive Definite Functions. *Constructive Approximation*, 2:11–22, 1986.
25. D. Michie, D.J. Spiegelhalter, and C.C. Taylor. *Machine Learning, Neural and Statistical Classification*. Ellis Horwood, 1994.
26. J. Moody and C. J. Darken. Fast Learning in Networks of locally-tuned Processing Units. *Neural Computation*, 1:284–294, 1989.
27. H. Murase and S. Nayar. Visual learning and recognition of 3d objects from appearance. *International Journal of Computer Vision*, 14:5–24, 1995.
28. C. Papageorgiou and T. Poggio. A trainable system for object detection. *International Journal of Computer Vision*, 38:15–33, 2000.
29. J. Park and I. W. Sandberg. Approximation and Radial Basis Function Networks. *Neural Computation*, 5:305–316, 1993.
30. T. Poggio and S. Edelman. A network that learns to recognize tree-dimensional objects. *Nature*, 343:263–266, 1990.
31. T. Poggio and F. Girosi. Networks for approximation and learning. *Proceedings of the IEEE*, 78:1481–1497, 1990.
32. T. Poggio and F. Girosi. Regularization algorithms for learning that are equivalent to multilayer networks. *Science*, 2247:978–982, 1990.
33. M. J. D. Powell. The Theory of Radial Basis Function Approximation in 1990. In W. Light, editor, *Advances in Numerical Analysis*, volume II, pages 105–210. Oxford Science Publications, 1992.
34. L. Rabiner and B.-H. Juang. *Fundamentals of Speech Recognition*. Prentice Hall, 1993.
35. B.D. Ripley. *Pattern Recognition and Neural Networks*. Cambridge University Press, 1996.
36. B. Schiele and J. Crowley. Probabilistic object recognition using multidimensional receptive field histograms. In *Proc. of the 13th Int. Conf. on Pattern Recognition*, pages 50–54. IEEE Computer Press, 1996.
37. A. Schölkopf, C. Burges, and A. Smola. *Advances in Kernel Methods — Support Vector Learning*. MIT Press, 1998.
38. F. Schwenker. Hierarchical Support Vector Machines for Multi-Class Pattern Recognition. In R.J. Howlett and L.C. Jain, editors, *Knowledge-Based Intelligent Engineering Systems and Aplied Technologies KES 2000*, pages 561–565. 2000.
39. F. Schwenker and H.A. Kestler. 3-D Visual Object Classification with Hierarchical RBF Networks. In R.J. Howlett and L.C. Jain, editors, *Radial Basis Function Neural Networks: Theory and Applications*. Physica-Verlag, 2000 (in press).
40. F. Schwenker, H.A. Kestler, G. Palm, and M. Höher. Similarities of LVQ and RBF learning. In *Proc. IEEE Int. Conf. SMC*, pages 646–651, 1994.
41. S. Ullman. *High-level Vision. Object Recognition and Visual Cognition*. The MIT Press, Cambridge, 1996.
42. V.N. Vapnik. *Statistical Learning Theory*. John Wiley and Sons, 1998.
43. P.D. Wasserman. *Advanced methods in neural computing*. Van Nostrand Reinhold, New York, 1993.
44. S.C. Zhu and A.L. Yuille. Forms: A flexible object recognition and modeling system. *International Journal of Computer Vision*, 20:1–39, 1996.

11 Parallel Implementations of Self-Organizing Maps

Timo D. Hämäläinen

Abstract. This chapter focuses on parallel implementations of the Self-Organizing Map (SOM) featuring different levels of parallelism. The basic arithmetic-logical operations of SOM are first reviewed for a consideration of implementation issues such as number precision, memory consumption and time complexity. Mapping involves *network*, *training set*, *neuron* and *weight* parallelism. Examples of the weight and neuron parallel mappings are given for abstract platforms to conduct general principles. Neuron parallel mapping is considered in great detail as it is the most commonly used approach. A review of implementations is given from supercomputers to VLSI (Very Large Scale Integration) chips with criteria for performance comparison.

11.1 Introduction

Self-Organizing Maps have been studied extensively since the beginning of the 1980's leading to several variations and extensions to the original model [1-3]. First used in academic research, SOMs have grown up and are now widely used in many practical applications in industry and business. Pattern recognition, diagnostics, and classification are examples of the traditional tasks suitable for SOMs. One of the most well-known early applications was the phonetic typewriter, which transcripts spoken phonemes to characters [4]. A new and important application area is visualization and abstraction of large multimedia contents in different text, voice and video databases [5-7]. The capability to map high-dimensional input data to a one- or two-dimensional space makes the SOM a very interesting tool for solving this ever increasing problem.

The original inspiration behind the SOM was a brain-like abstraction and self-organization of data with numerous neurons contributing to the process. Early hardware implementations were fairly modest considering the number of neurons, i.e. the SOM size. Only a maximum of a few hundred neurons was acceptable, as otherwise the computation time would have been unreasonably long. For the same reason, there were limitations to the input data dimensionality and also to the total amount of data. The early implementations made use of the supercomputers of those days. At the beginning of the 1990's dedicated hardware systems began to be used. Almost all special implementations have been parallel due to regular, well-localized arithmetic operations independent of the SOM variant.

Processing power, hard disk capacities and memory sizes have rapidly increased in personal computers since 1995. This has increased the number of customized parallel SOM implementation projects. The problems and applications that were used for dedicated implementations could be then executed on a desktop PC faster and much more cost effectively. For this reason, research work on implementations was directed either to very special, embedded systems or workstation based coarse parallel clustering. The former approach has resulted in several VLSI chips for particular applications, where data is captured and processed on-the-fly without minor local storage of data.

Today most of the research development and applications are based on general purpose computers. New applications, however, require significantly larger maps, the input data is often very high-dimensional and a large data set should be handled in a short time window for a reasonable response time for users. This will, in turn, introduce again parallel implementation needs.

Simultaneously the development of VLSI technology has enabled whole systems to be integrated on a single chip called System-on-Chip (SoC). For SOM implementations this means that the former knowledge of parallel processing could now be used for SoCs. On the other hand, also general-purpose processors include more and more parallel processing features inside for better performance. It can be concluded that the physical platforms are now different, but the implementation principles have mainly been untouched.

This Chapter is partitioned into two logical parts, where the first one concentrates on theoretical issues for parallel mappings and the latter part to practical implementations. In the following, the computation of the basic SOM is first reviewed for understanding the different mapping approaches. The main focus in this Chapter is not on algorithmic speed-up methods that require modifications to the original algorithm. Those are regarded as variants of the original SOM, but the methods for parallel implementations discussed here can well be applied also for them.

After reviewing the SOM flow of computation, three important issues are considered: number precision, memory consumption and time complexity. The latter opens the discussion about the level of parallelism and introduces different kinds of parallel processor topologies. After that, the neuron parallel mapping is considered in more detail, since it has been frequently used and found to be efficient especially for custom parallel computers.

In the implementation part of the Chapter, a classification of the implementations is first given according to key design features. In addition, example implementations are given and evaluated in respect to the theoretical mapping principles. One of basic reasons for parallel implementations is performance, of which measurement and comparison criteria between implementations are given. Conclusions close the Chapter with remarks to the future trends.

11.2 Computation of SOM

The sequence of operations in the basic SOM is fairly straightforward for both training and forward phase. The map is usually initialized with random reference vectors (weights), but in many applications it is useful to preprocess also the weights to improve the map properties and speed up the learning phase [2]. In the following, the map is considered to be two dimensional with X columns and Y rows. Let the input vectors be $\xi = [\xi_1, ..., \xi_m]$, where m refers to the input vector dimension and

$\xi_i \in [0,2^{b-1}]$, where b is the number of bits used for each element of ξ. The weights are referenced to as.

$$\underline{w}_{xy} = [w_1, ..., w_m], 0 \le x \le (X-1), 0 \le y \le (Y-1) \qquad (11.1)$$

The next step is to find the weight vector closest to ξ and mark it as the winner neuron.

$$w_{win} = \min_{xy} \left\| \xi - w_{xy} \right\| \qquad (11.2)$$

The Euclidean distance normally used for the computation is often replaced by the Manhattan distance in implementations for simplicity. Alternatively, the square root operation of the Euclidean norm could be omitted, since only the location of the neuron is significant and can be determined by comparing the squared values. Also other metrics could be used according to implementation and application.

At this point, the next step is to determine the neighbourhood H of the winner. There have been presented several methods and algorithms about the neighbourhood area size, shape and how it is modified in respect of time. Independent of the method, comparison operations are carried out to determine the neighbourhood neurons.

In the basic SOM, the weights of the neurons belonging to the neighbourhood are updated in the next step. This is also called *on-line* weight update method, because the weights are updated at every time $t = 0, 1, ...$ an input vector is given to the network:

$$\begin{cases} w_{xy}(t+1) = w_{xy}(t) + \alpha(\xi(t) - (w_{xy}(t))), & w_{xy} \in H \\ w_{xy}(t+1) = w_{xy}(t), & w_{xy} \notin H \end{cases} \qquad (11.3)$$

where $0 < \alpha(t) < 1$ is the adaptation gain. Very often parallel implementations use the *batch mode* learning, in which a set of T input vectors are presented and weights updated after that. The time can now be expressed by $t = eT + t'$ where $e = 0, 1, \ldots$ is the epoch number, T is the epoch length and $0 \leq t' < T$ the time running inside an epoch. Assuming α is kept constant during the epoch, the weight update for the batch update can be expressed as

$$w_{xy}((e+1)T) = w_{xy}(eT) + \alpha \left(\sum_{t'=0}^{T-1} H(\underline{\xi}(t') - (w_{xy}(eT))) \right) \tag{11.4}$$

where H determines, whether the input vector belongs to the winner neighbourhood or not. In practice, a list of winners and input vectors have to be maintained. Other modifications exist, in which on-line and batch algorithms are combined for the selection of winners or weight updates [8].

The modifications are usually motivated by simpler hardware or much improved raw computing performance, if some computations could be omitted. However, the learning time could still be longer or the convergence worse than using the original algorithm, which must be carefully considered before hardware implementation. It should be noted that another set of modifications rely on optimizing memory or computation without major changes to the basic operations, and have proven not to deteriorate the convergence [7].

As the unsupervised training of SOM proceeds, neurons in the map will be ordered with respect to the topology of input vectors. If the map is to be used for classification, neurons should be *labelled* by broadcasting training vectors with known classification and associating to each winner a corresponding label. Fine-tuning can then be performed e.g. with an LVQ algorithm [2].

11.2.1 Precision

Normal PC or workstation user might not take much care of the number representation e.g. in MATLAB experimentation, because the dynamic range of floating point double precision arithmetic is very large. Even with the batch mode learning, the weights might not abandon the range, but for single precision floating point and fixed point arithmetic a much more careful design is required. Most of the presented digital SOM implementations use fixed point computation due to much faster execution speed over corresponding floating point computation.

Sometimes a question about the need of fixed point numbers is arisen, because the floating point performance is ever increasing. However, new multimedia instructions e.g. Intel MMX (Matrix Math eXtension) [9] and SSE (Streaming SIMD Extensions) [10] make use of fixed point subword parallelism inside a general purpose processor. These are needed to speed up pixel data calculations, but can be used for SOM and other neural networks computations as well. Since the pixel data is usually eight bits wide, the best performance in SOM computation also requires $b = 8$, but this might not be enough for the best convergence of the map.

The minimum number of bits b is restricted by the input vector probability distribution and the number of neurons in the map. The learning rate parameter is found not to be as significant, but caution must be exercised to avoid overflows [11]. If the input vector element satisfies $\Xi_{min} \leq \xi \leq \Xi_{max}$, the resolution or quantization step q between two discrete values of ξ is:

$$q = \frac{\Xi_{max} - \Xi_{min}}{2^b - 1} .$$

(11.5)

For the minimum number of bits required, consider a map of size $X = Y = M$ and assume there is a uniform probability distribution on any square $a \times a$ within the map. For a rectangular neighbourhood with

$$\lim_{t \to \infty} H(t) = 1$$

(11.6)

the asymptotic weight values are almost evenly distributed. It can be deduced that the smallest distance separating any pair of neurons is $a/(2M)$ and, thus, must be

$$q > \frac{a}{2M}$$

(11.7)

according to [11]. This yields to a theoretical minimum.

$$b > \log_2\left(1 + \frac{2M}{a}\right)$$

(11.8)

With $a = 1, M = 20$ six bits would be the absolute minimum. Eight bits is the lowest limit in practise to save implementation costs. A general purpose SOM implementation should also allow longer bit vectors or otherwise the convergence of the map might not satisfy the needs of an arbitrary application.

The number of bits affects also the weight update results. There are two dangers with too few bits that are weights falling outside the range and too coarse resolution for the update. When the learning rate parameter becomes very small the update value might be cut to zero for the winner and its close neighbours. However, the neurons at the edge might be updated by q that in successive iterations deteriorates the mapping. This could be avoided by skipping the update of all neurons when the winner's update value is zero.

Especially batch mode learning might suffer from the overflow of the weight values. One technique is to saturate the weight value near or equal to Ξ_{max} and not allow it to roll over to zero. Another way is to compute the weight iteratively and applying a mean over all modifications inside an epoch. For single precision floating point numbers, block floating point technique could be used to ensure proper range within the epoch.

11.2.2 Memory Consumption

The precision of the weights is also important from the memory consumption point of view. Memory requirement for a map of size M is Mmb in bits. For example, a map of $20x20$ neurons with $m = 4012$ occupies about 12 Mbytes of memory when double-precision floating point numbers are used. With eight bit fixed-point numbers about 1.5 Mbytes is required. For current workstation users this does not seem to be very much, but the repetitive nature of SOM computations makes also memory space an issue.

The new SOM applications may require very large maps and high-dimensional weights. As an example, in [7] the largest map exceeds one million neurons with 500-dimensional input vectors. With eight bits precision this yields about 500 Mbytes of memory space to the weights alone, and to this there must be added temporary parameters and pointer lists for the execution control. It is therefore very important to design the computation steps to minimize data transfers to slower main memory and try to keep all the data for critical kernels in cache [12].

References to the main memory, or even disk swap activity, could be decreased by not storing all intermediate values or packing sparse weights and input vectors like in multimedia computation. This will increase computation, but this is not a significant shortcoming for current computers. The performance bottleneck is now clearly in the communication both between chips and on-chip, which in turn favours computation intensive implementations. An interesting solution for SOM computations would be to use embedded on-chip DRAM technology that is designed for high-end 3D graphics accelerators [13]. This ensures very fast memory accesses to a much larger memory than the current caches.

11.2.3 Time Complexity

Based on the basic SOM operations, the time complexity is next considered for a two dimensional map. The big-O notation is used in the following. The first phase of computing distances requires a copy of the input vector for all neurons. This can be completed in a time,

$$O\left(\left\lceil \frac{b}{B} \right\rceil m\right) = O(m) \tag{11.9}$$

where B is the width of a data path (in bits) for all the units performing computation. In an ideal case all the input vector elements are broadcast simultaneously to all neurons, which can be completed in a unit time $O(1)$.

The computation of distances takes $O(m)$ if carried out in parallel for all neurons. If the total number of neurons is K, the minimum time to find the winner neuron is $O(K)$. The computation of the neighbourhood takes $O(K/P)$ for an arbitrary shaped neighbourhood. Rectangular neighbourhoods, however, can be computed in $O(1)$ using the method given later in this chapter. The updating of neurons takes place in $O(m)$. Thus, the total time for the SOM training is $O(m + K)$ for an ideal case [14].

In practise, the computation and communication limits the execution time more, for which reason basic parallel processing systems are considered next. Fig. 11.1 depicts some examples of the basic topologies and Table 11.1 corresponding complexities for the SOM computation. The presented time complexities best fit to the intuitively natural neuron parallel mapping scheme, but can also be applied to other schemes as well. The step of finding the winner neuron is divided into two parts, since in many implementations a set of neurons are processed locally in some processing units and after that a global winner is voted. Some implementations do not find a global winner at all, but this is more a matter of applied level of parallelism. A step of delivering information about the winner to all other processing units has been added as well.

The steps can also been divided according to the type of operation. Step 1 requires broadcast-type of communication, which is best implemented in bus topology in theory. In practise the bus length and number of connected units is very limited. Tree and hypercube are the next best topologies that do not have a similar limitation [15]. Depending on the mapping method, step 2 involves computation and possibly exchange of data between other processing units. Only the computation is taken into account for Table 11.1. Step 3 is assumed to take place locally, i.e. computation only, and Step 4 requires both communication and computation. The tree topology is best suited for this kind of global reduction operation.

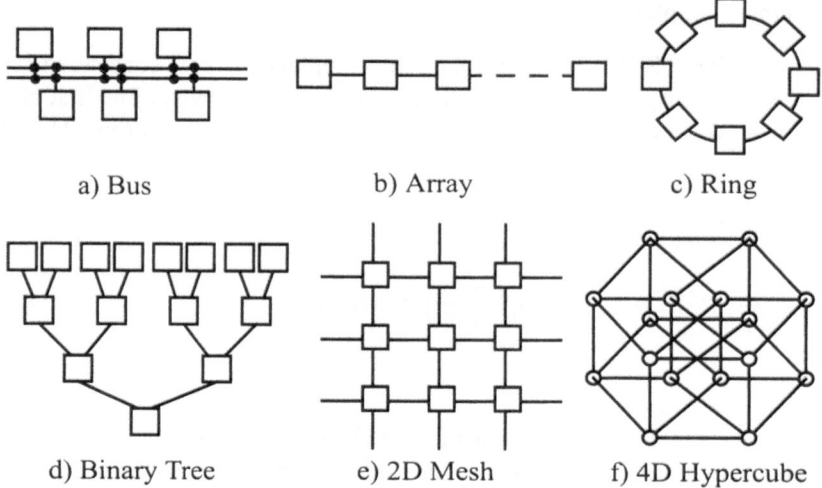

a) Bus   b) Array   c) Ring

d) Binary Tree   e) 2D Mesh   f) 4D Hypercube

Fig. 11.1. Examples of the basic communication topologies for parallel implementations.

Step 5 requires again broadcast type of communication for delivering the winner information. For step 6, computation and possibly communication with some near processing units are required. Step 7 clearly involves computation. A more detailed analysis is required for practical implementations that might mix several topologies and mapping methods. In addition, the figures apply only to ideal systems with no limitations on the number of processing units or communication paths. It should be noted again that several modifications exist that may not contain all of the steps.

The next two sections present the different levels of parallelism and a detailed view of the two most often applied levels: weight and neuron parallelism. In addition, mappings of these levels are given to some example parallel topologies.

Table 11.1. Time complexity for SOM computation on example topologies.

Step	Ideal	Bus	Tree	Ring/Array	d-dim. Mesh	Hypercube
1. Deliver ξ	$O(1)$	$O(m)$	$O(m \log P)$	$O(m P)$	$O(m P^{(1/d)})$	$O(m \log P)$
2. Compute distances	$O(m)$	$O(m (K / P))$	$O(m (K / P))$	$O(m (K / P))$	$O(m (K / P))$	$O(m (K / P))$
3. Find local winner	-	$O(K / P)$	$O(K / P)$	$O(K / P)$	$O(K / P)$	$O(K / P)$
4. Find global winner	$O(K)$	$O(P)$	$O(\log P)$	$O(P)$	$O(P)$	$O(P / (\log P))$
5. Deliver winner	$O(1)$	$O(1)$	$O(\log P)$	$O(P)$	$O(P^{(1/d)})$	$O(\log P)$
6. Compute neighbourhood	$O(1)$	$O(K / P)$	$O(K / P)$	$O(K / P)$	$O(K / P)$	$O(K / P)$
7. Update	$O(m)$	$O(m (K / P))$	$O(m (K / P))$	$O(m (K / P))$	$O(m (K / P))$	$O(m (K / P))$
Total	$O(m + K)$	$O(P + m (K / P))$	$O(m \log P + m (K / P))$	$O(m P + m (K / P))$	$O(P + m (K / P))$	$O(m \log P + m (K / P))$

11.3 Parallel Mapping Approaches

Algorithms can be made parallel in several ways featuring different granularity in execution. For neural networks computation at least five different levels of parallelism are applicable that are network parallelism, training set parallelism, neuron parallelism, weight parallelism and bit parallelism [49]. In the following, the term processing unit refers to any device performing computations, e.g. computer, microprocessor or a VLSI chip.

The network parallelism represents the most coarse granularity, in which one processing unit (typically a computer) takes care of all the computations involved in a whole neural network. Applied to SOMs this level of parallelism means several maps trained simultaneously with the same input data for all the maps, as depicted in Fig. 11.2 (top). This level of parallelism can be used when an optimal map is explored for a problem, because the maps can be of different size and have different parameters. However, there could be used some interaction between the maps like exchange of winner information, but this will not be regarded as pure network parallel implementation.

The training set parallelism differs in a way that identical copies of the SOM algorithm are delivered to a set of computers, but they use different sets from the original input data (Fig. 11.2). This method is particularly useful if the data set is very

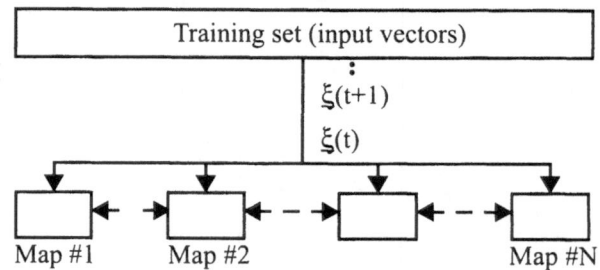

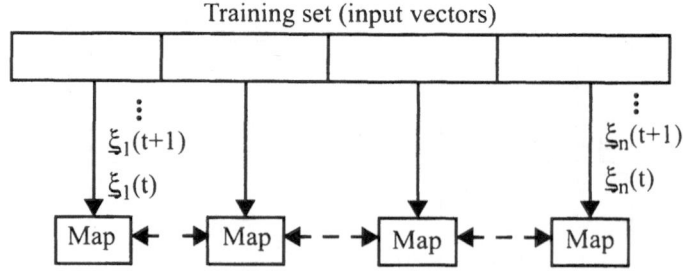

Fig. 11.2. Network parallelism (top) and data set parallelism (bottom).

large and can be divided into smaller parts. There could again be interaction between the maps to globally synchronize the maps periodically. It should be noted that each computer runs a complete map, and if the purpose is only to divide the original map to many processing units the neuron parallel mapping is resulted in.

The layer parallel method is best suited for multi-layer networks, in which layers are pipelined and the outputs of the layers computed simultaneously. For SOM computation, this is useful for pre- and post-processing purposes. The next two levels, namely neuron and weight parallelism, are maybe the most natural levels in the spirit of biological neurons. Both of these levels have been used for dedicated embedded implementations, because they offer fine granularity and can be mapped conveniently to systems with simple processing units. The weight parallel mapping is considered next in more detail, and the neuron parallel mapping is given in Sec. 11.4.

In the weight parallel mapping, one processing unit computes a partial distance in one dimension. This requires to deliver ξ_i to the unit containing the corresponding weight element w_i. After completing calculations, the final distance should be computed by adding partial distances. This procedure is repeated to all weight vectors in the SOM, and for the rest of the SOM computation the system is switched to another mode and processing units given different tasks. It can be summarized that the weights per neuron are processed in parallel, and neurons in sequence one after another.

Pipelining and batch mode learning are very suitable for this level of parallelism. Several input vectors is fed to the system, and many input vectors are undergoing the distance computation at the same time. After the last input vector, the processing units start the winner selection and updating process. For a fully trained map, the weight elements could be permanently stored to each processing unit, which speeds up the computation.

The weight parallel mapping represents very fine granularity in execution, which normally leads to communication intensive operation. In addition, the time to switch the mode from distance computations may cause additional overhead. A benefit in this level of parallelism is that the number of neurons is not limited. The input vector dimension m, however, is typically limited to the number of processing units. Otherwise one processing unit should compute more than one partial distances, which makes the computation less parallel.

In the following, example mappings are given for mesh and tree topologies to give a more detailed view of this level of parallelism. Especially two-dimensional mesh implementations have been very popular, and the tree topology is found to be very effective in global computations. In Fig. 11.3, the mesh of processing units is connected to a controller that initializes the execution and feeds the mesh with input vector elements.

The mesh is used in a systolic manner that means that there are several distance calculations active for different input vectors simultaneously. One row of the mesh forms one neuron, and all the units in a row compute one partial distance d_i for the neuron. In addition, a processing unit receives a partial distance from the left and accumulates it to its own distance. In this way the controller finally receives the total

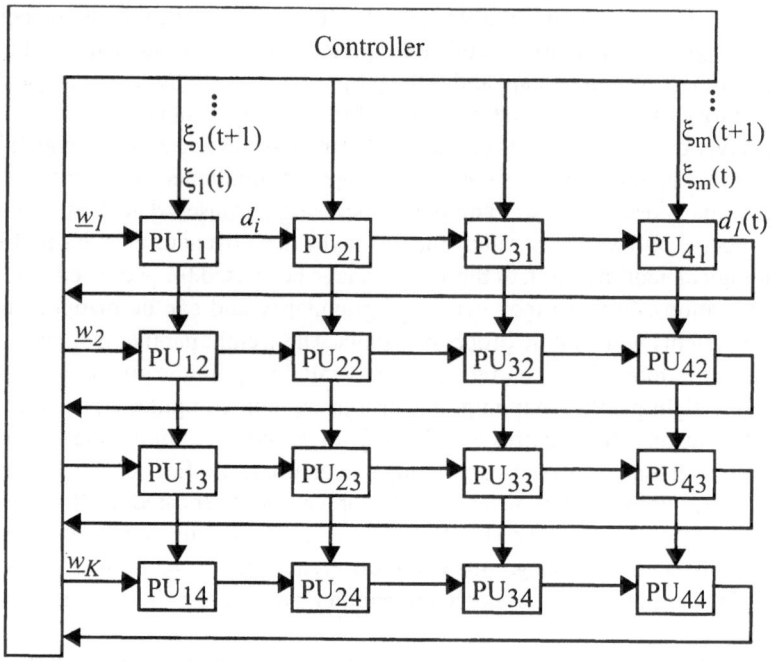

Fig. 11.3. Weight parallel mapping on mesh topology.

distance from the last processing unit. From this point on, the controller could compare the distances to find the winner, or the computation could take place in processing units. Once the winner is found, the weights of the processing units are updated.

An example mapping for the tree topology is depicted in Fig. 11.4. The array of processing units performs the partial distance computation, but now the tree could be used to compute the total distance. The mapping on this topology also prefers batch mode learning and pipelining to reduce overhead. After each epoch, the processing units update the weights according to the list of winners that the controller broadcasts.

Current advancements in implementation technology allow fairly complex and still very fast processing units, which favours computation intensive systems. For this reason, the weight parallel mapping is becoming too fine grained parallel scheme and is therefore losing importance in SOM implementations. The neuron parallel level is thus becoming the lowest reasonable level of parallelism from this point of view. On the other hand, many parallel implementations utilizing several computers or workstations use the neuron parallel mapping, for which reason this mapping scheme is considered in more detail in the following.

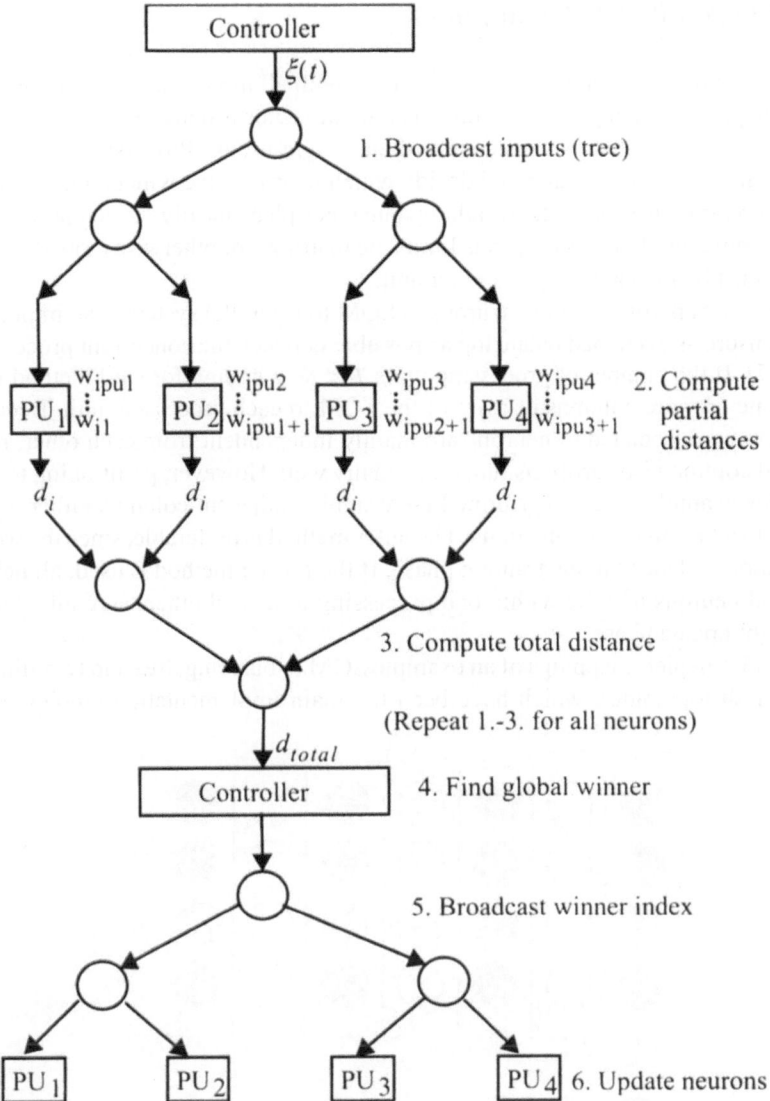

Fig. 11.4. Weight parallel mapping on tree topology.

11.4 Neuron Parallel Mapping

In a pure neuron parallel mapping, one neuron with all the m elements of a weight w_{xy} is mapped to each processing unit. The input vector is delivered to all processing units, which evaluate the distance to their weight vector. Processing units compare distances with each other and decide which neuron is the winner and what are the neighbouring neurons. The weight update takes place locally in each processing unit. The number of processing units limits the map size, or otherwise more than one neuron must be mapped per processing unit.

The way of partitioning the neurons in SOM to a parallel system is an important task to ensure as even load balancing as possible between the concurrent processing units [17]. If the number of processing units $P < K$, a straightforward method is to assign one or more columns or rows of the SOM to each processing unit. Because computations associated to neurons are mainly independent from each other, rowwise and columnwise partitions function equally well. However, partitioning to columns, for example, can be performed by assigning adjacent columns either to the same or to adjacent processing units. The latter method is preferable, since the workload is more balanced in the training phase. If the former method is used, all neighbourhood neurons may lie within one processing units and others stay idle during the weight update phase.

Fig. 11.5 depicts mappings of an example SOM to bus, ring, tree and two-dimensional mesh topologies, which have been the main implementation topologies. It

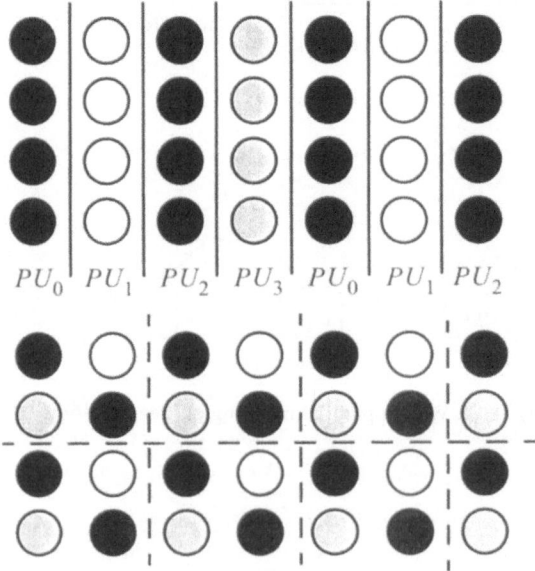

Fig. 11.5. Mapping of a 7x4 SOM to a four-processor (top) bus, ring, array or tree topology (bottom) 2x2 mesh topology.

should be noted that some processing units have to accommodate more neurons than others, if K is not divisible by P. Next the details of the neuron parallel mapping is considered in more detail.

11.4.1 Common Operations

The mapping on bus, ring (array) and tree topologies differ mainly on the way of inter-processor communication. To be strict, the tree topology here is considered to be an array, where the processing units are connected via tree-shape communication network. Common to all these topologies are neuron mapping, indexing, and local computations that are presented in the following. The columnwise mapping is taken as an example case.

Neuron Indexing

Let the processing units be numbered $(0, ..., P\text{-}1)$, such that the first unit is referred to as PU_0 and so on. The original neuron map can be of any size determined by the column number X and the row number Y, and a single neuron is referenced by a coordinate pair (x_{MAP}, y_{MAP}). The origin of the map is on the upper left corner, as shown in Fig. 11.5. Then, the number of columns assigned to each processing unit can be determined as follows [14].

$$\begin{cases} N_{PU_0}...N_{PU_{P-1}} = \dfrac{X}{P}, \text{if } X mod P = 0 \\ N_{PU_0}...N_{PU_{i-1}} = \left\lfloor \dfrac{X}{P} \right\rfloor + 1 \\ \qquad N_{PU_i}...N_{PU_{P-1}} = \left\lfloor \dfrac{X}{P} \right\rfloor \end{cases}, \text{if } X mod P = i \qquad (11.10)$$

where $\lfloor \ \rfloor$ refers to a mathematical floor operation, mod refers to the modulus operation and N_{PU_i} to the number of columns in i:th processing unit. If the number of columns is divisible by the number of processing units ($X mod P = 0$), each

Row index, global and local

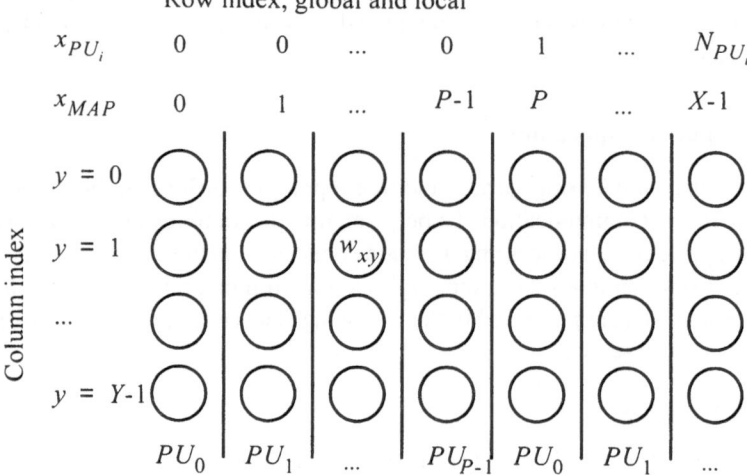

Fig. 11.6. Partition of original neuron map to columns for processing units.

processing unit contains an equal number of columns. Otherwise the first i processing units contain one more column than the rest of processing units.

It should be noted, that the number of neurons in a processing unit is YN_{PU_i}. After partition to columns, global coordinates (x_{MAP}, y_{MAP}) are not very useful, because each processing unit should carry local information. To make processing units independent to each other in this sense, global coordinates are transformed to local coordinates for each processing unit. Due to this, the corresponding program code for processing units could be the same for all of them, and there is no need to initialize them separately. In addition, the only global information needed is the order number of the processing unit. Since the original map is delivered columnwise, the y coordinates are the same in global and local coordinates:

$$y_{PU_i} = y_{MAP}, i \in [0, P - 1], y_{MAP} \in [0, Y - 1].$$ (11.11)

In the following, subscripts are omitted in y coordinates for simplicity. Instead, a local x coordinate for a processing unit number i can be determined as follows.

$$x_{PU_i} = \left\lfloor \frac{x_{MAP} - i}{P} \right\rfloor, i \in [0, P - 1], x_{MAP} \in [0, X - 1] \qquad (11.12)$$

The floor operation is effectively the same as a binary shift operation in the program code. In this case the term $(x_{MAP} - i)$ (fixed point binary number) is shifted P times to the right, and thus the columns are numbered locally $(0 \ldots N_{PU_i})$ in each processing unit, as depicted in Fig. 11.6.

Distance Computation

The distance computation takes place locally on all processing units. To speed up the computation, not all squared terms are needed to compute for all weights in order to find the winner. For the first weight vector w_{00} in the processing unit the corresponding distance d_{00} is first calculated. Since this is the first distance computed so far, this could be marked as the minimum distance in this processing unit: $d_{min} = d_{00}$. After that, a next weight vector is taken, and squared terms are accumulated for it (Euclidean distance). However, if at any time $k < m$,

$$d_{min} < \sum_{j=1}^{k} (\xi_j - w_{xyj})^2, x \in [1, N_{PU_i}], y \in [1, Y], \qquad (11.13)$$

i.e. the cumulated value for squared terms is greater than d_{min} before all terms are computed, this weight may be skipped. This speeds up the computation, but the execution time varies between neurons and processing units. It should be noted that the slowest processing units still dominates the overall execution time.

Neighbourhood Computation

For simplicity, the neighbourhood is assumed to be a rectangle with a radius r, which means that $(2r + 1)^2$ neurons belong to the rectangle. For each neuron in the neighbourhood,

$$\begin{cases} x_{MAP} - r \leq \chi_{MAP} \leq x_{MAP} + r \\ y - r \leq \Upsilon \leq y + r \end{cases} \tag{11.14}$$

using global coordinates of the map. Next an efficient method to compute the neighbourhood in each processing unit is considered. The neurons belonging to the neighbourhood lie within several processing units, so there are sub-rectangles (within assigned columns) of the neighbourhood in one or more processing units, as shown in Fig. 11.7.

One approach to determine whether there are neighbourhood neurons is to compare all local coordinates with the winner's coordinates that are delivered to all

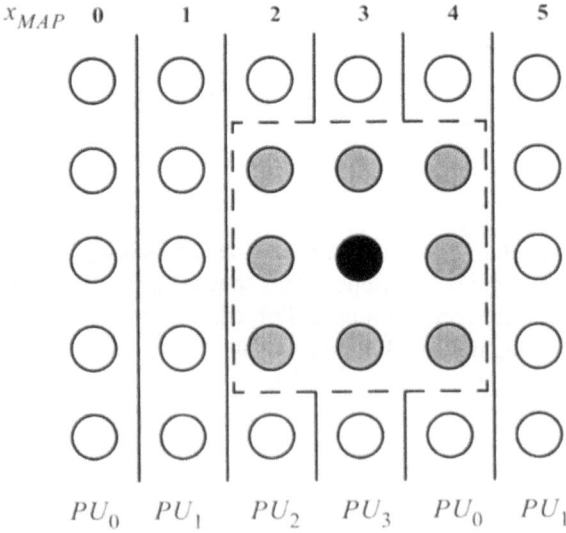

Fig. 11.7. An example of the neighbourhood rectangle falling across several processing units.

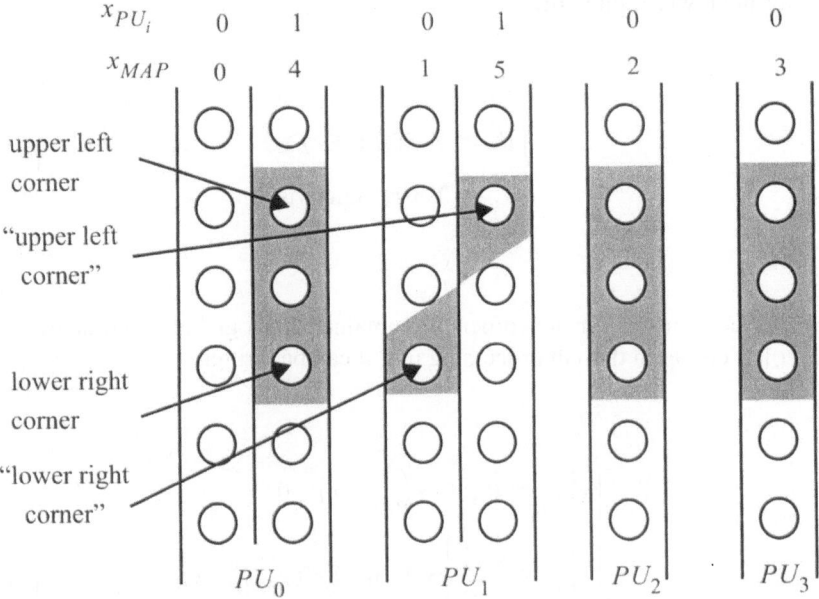

Fig. 11.8. Computation of neighbourhood subrectangles. Note rejected "rectangle" in PU_1.

processing units. However, this is computationally not very efficient, because it requires the more comparison operations the larger the map is.

A more efficient method is to compute only the corners of sub-rectangles in each
processing unit [14]. The benefit of this approach is that all neurons in the neighbourhood can be determined in a constant time span independent of the original map
size or the neighbourhood radius. The procedure is depicted in Fig. 11.8, showing
that each processing unit determines the upper left and lower right corners with respect to the columns assigned to that processing unit.

It should be noted, that a processing unit computes these corners, although none
of its neurons really belong to the neighbourhood of the original map. In this case,
the sub-rectangle will be distorted and not accepted for further operations. The y coordinate for the upper left corner is computed as

$$
\begin{cases}
y^L = 0, & \text{if } (\Upsilon - r) \le 0 \\
y^L = \Upsilon - r, & \text{otherwise,}
\end{cases}
\tag{11.15}
$$

and for the lower right corner

$$
\begin{cases}
y^R = Y, & \text{if } (\Upsilon + r) \geq Y \\
y^R = \Upsilon + r, & \text{otherwise.}
\end{cases}
\tag{11.16}
$$

Local x coordinates for the corners are obtained from global coordinates. For the upper left corner in the i th processing unit it can be written

$$
\begin{cases}
x^L_{PU_i} = 0 \text{ , if } (\chi_{MAP} - r) \leq 0 \\[2mm]
x^L_{PU_i} = \left\lfloor \dfrac{x^*_L}{P} \right\rfloor + 1, \text{ if } x^*_L < (\chi_{MAP} - r) \\[2mm]
x^L_{PU_i} = \left\lfloor \dfrac{x^*_L}{P} \right\rfloor, \text{ otherwise,}
\end{cases}
\tag{11.17}
$$

where

$$
x^*_L = \left\lfloor \frac{\chi_{MAP} - r}{P} \right\rfloor \cdot P + i, \, i \in [0, P-1] \, .
\tag{11.18}
$$

Here x^*_L can be thought as an *aliased image* of the winner neuron coordinate for each local coordinate system in processing units. In Eq. (11.17), a special case is handled in which the winner is at the leftmost edge of the original map, i. e. a part of the neighbourhood is outside the map. The coordinate for the lower right corner is computed similarly:

$$
\begin{cases}
x^R_{PU_i} = (N_{PUi} - 1), \text{ if } (\chi_{MAP} + r) \geq X \\[2mm]
x^R_{PU_i} = \left\lfloor \dfrac{x^*_R}{P} \right\rfloor - 1, \text{ if } x^*_R > (\chi_{MAP} + r) \\[2mm]
x^R_{PU_i} = \left\lfloor \dfrac{x^*_R}{P} \right\rfloor, \text{ otherwise,}
\end{cases}
\tag{11.19}
$$

where

$$
x^*_R = \left\lfloor \frac{\chi_{MAP} + r}{P} \right\rfloor \cdot P + i, \, i \in [0, P - 1] .
\tag{11.20}
$$

It should be noted, that the number of operations in the above procedure is independent of the processing unit count. The next section continues with mapping examples to different communication topologies.

11.4.2 Mapping to Communication Topologies

As shown in Sec. 11.2, the major differences in SOM execution time between parallel implementations are due to the communication topologies. The basic bus topology is only suitable to a few processing units, and an array is often used instead of bus if a large number of processing units is preferred. The ring differs from the array only by the connection of outermost units. To illustrate the use of array and ring topologies, a flow graph of SOM computation on the ring is depicted in Fig. 11.9.

It is assumed that there is a controller that initiates the system by loading weights and feeds input vectors to the processing units during the computation. Each input vector element should be rotated around the ring, and each processing units picks up an element for computing the corresponding distance. If there is not enough memory in processing units also the weights should be rotated, which finally leads to a weight parallel mapping.

Once all partial distances are ready, processing units start to compute distances to another neuron, if there are more than one per processing unit. A local winner is then searched for, and all of those local winners are rotated around the ring to find out the global winner. All processing units keep track of the indices of the candidates for local neighbourhood computation that takes place next. Updating is carried out concurrently in the processing units.

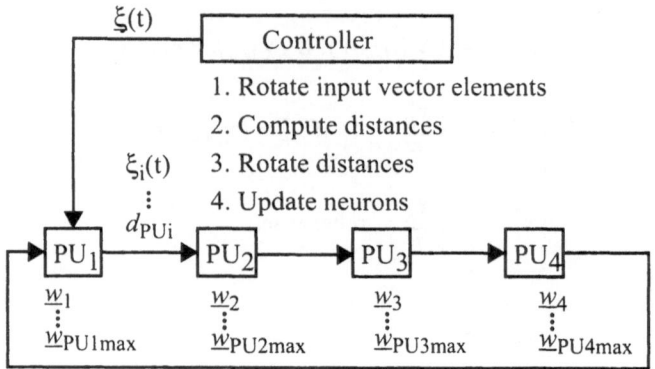

Fig. 11.9. Neuron parallel mapping on ring topology.

Another mapping example is given in Fig. 11.10, where the tree topology is applied. In this example, the nodes forming the trunk of the tree are not used for computing distances, i.e. they do not accommodate any neurons. The nodes are still used to carry out calculations like comparison operations.

The operation starts with a broadcast of input vector, which is very suitable operation to the tree topology. Each processing unit finds a local winner, which are compared with each others in the tree network, which ensures fastest possible search for the global winner. The host receives a global winner and broadcasts the index of the winner back to all processing units, which then check, whether their neuron is the winner and whether they have neurons in the neighbourhood. Selected neurons are updated in each processing unit.

The mesh is intuitively the most natural topology for two-dimensional, neuron parallel mappings of SOM. The execution starts by assigning one neuron per processing unit, and sending input vector elements to the units. There are several possibilities to arrange this, of which one is to receive and forward the elements in each unit. Once the total distance per neuron is computed, the candidate winners could be rotated both vertically and horizontally to find out the global winner. The neighbourhood is straightforward to determine once the winner is found. It should be noted that several modifications exist in implementations to pipeline and speed up the global winner search.

From the topology point of view, an optimal parallel implementation should have efficient data broadcast to all processing units, fast data reduction mechanism for the global winner search and full communication between processing units for fast neighbourhood neuron updates. It could be concluded that a hybrid topology is the best solution for a parallel SOM implementation.

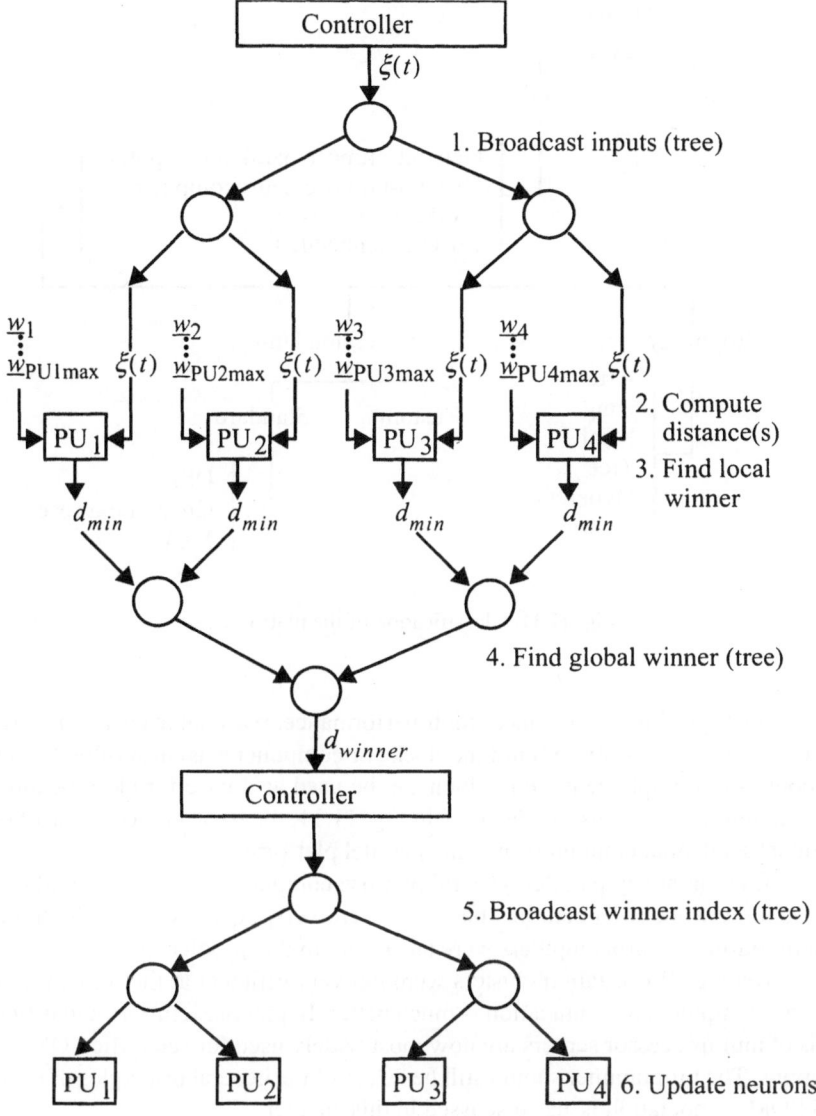

Fig. 11.10. Neuron parallel mapping on tree topology.

11.5 Implementations

Above discussion of SOM computation, basic parallel processing system topologies and levels of parallelism gives a general design space in which real implementations are placed. In addition, there are several practical issues that complicate the design

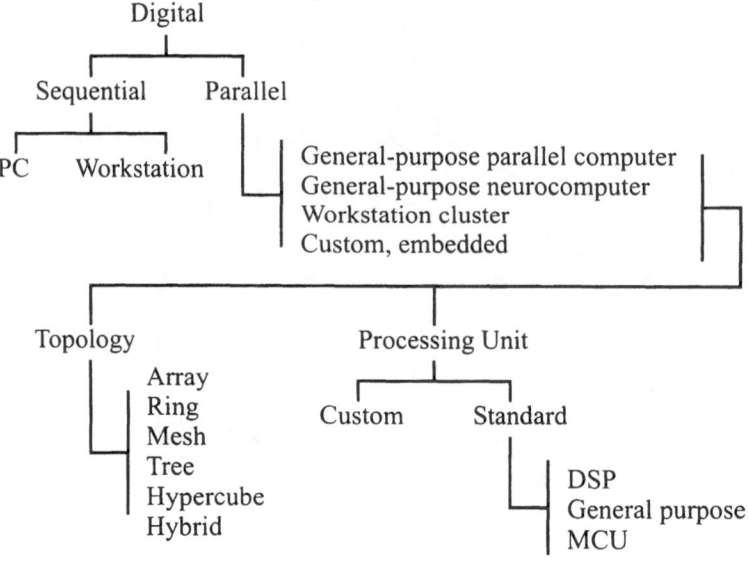

Fig. 11.11. Classification of the platforms.

to fulfil the typical requirements of high-performance, reasonable cost and convenient use [18]. Availability of building blocks or components is most often the starting point for an implementation, which can be used as a base for classification of different implementations. As depicted in Fig. 11.11, the basic choice is use of conventional sequential computers or some parallel platform.

The first massively parallel, general purpose computers were very popular also for neural network computations, but those were very expensive to build and use. General-purpose neurocomputers were an answer to this problem as well as several special systems. Workstation clusters were not very efficient at that time, but now the inter-computer communication is much better. In practise, clusters and different kinds of multiprocessor servers are now most widely used for scientific SOM computations. The programmer should still be aware of the general principles of the parallel SOM computation issues discussed in this chapter.

The type of processing unit is an important choice, and a common solution is to use a high-performance Digital Signal Processor (DSP) or a self-designed VLSI chip for embedded implementations. DSPs are motivated by the efficient multiply-accumulate performance and programmability. For application specific implementations a custom VLSI processing unit is well motivated for optimized performance and cost. Current System-on-Chip implementation could make use of DSP cores or tailored processing units. Independent of the level of system integration, a processing unit should be capable to multi-processor communication.

For a fixed number of processing units, the topology could be chosen quite freely. However, expandability is often required, which is most conveniently implemented in bus, ring and mesh topologies. It should also be noted that the topology itself does not ensure a good implementation, if the speed of communication and computation is not balanced. For this reason, there exists several implementations and proposals that rely on different technologies. Examples of some implementations are given in the following.

11.5.1 General-Purpose Parallel Computers

As stated earlier, general-purpose parallel computers have also been used for SOM computations. The platforms could be divided into massively parallel computers, supercomputers and clusters of workstations.

The first category of parallel computers are designed and used mainly for scientific, engineering and military applications, in which the computation is well localized and thus suitable to parallelization. Example platforms for SOM implementations are Connection Machine [20], MasPar [18], [20], Intel iPSC/860 [18], [19], and WARP [21]. Training set as well as weight or neuron parallel mappings have been applied in these implementations.

Supercomputers are very close to the above group, but the number of processing units is not as high. However, the processing units are more powerful and connected in a high-performance memory architecture. The network or training set parallelism are the most natural levels of parallelism and for this reason often used. SOM implementations have been presented e.g. on 16-processor SGI Power Challenge, as well as 16 and 64-processor SGI Origin2000 [12], [7].

Very popular low-cost parallel SOM implementations have been clusters of workstations or PCs that have normal local area networking and some message passing interface like PVM (Parallel Virtual Machine) [22], [24], [25]. One motivation is to use e.g. classroom PCs for scientific computations when there are no practicals and the computers would stay idle. The computation in this kind of clusters is clearly communication bound and only the most coarse levels of parallelism are reasonable. For example, in [23] a PVM based SOM implementation was experimented on 12 Pentium based PCs with a neuron parallel mapping and fairly small SOM sizes. The speed-up was even worse than using a single PC in some cases.

A more organized way to put PCs carry out parallel computation is Beowulf-clustering [26]. In a typical arrangement, many PCs are mounted in a rack without displays and connected with optical or carefully designed traditional local area network. Some clusters use Myricom switches that allow very fast communication backplane [27].

UNIX workstation manufacturers offer servers that consist of several processors. Also those have been used for neural network computations, but basically the user should carry out parallelization of the algorithm as in other platforms. While the clusters of computers require message passing interface and corresponding design of the programs, multi-processor computers require the program code to be written with threads. The operating system takes care of the execution after that, but very careful design requires understanding of the computer architecture as well.

11.5.2 General-Purpose Neurocomputers

Several general-purpose neurocomputers and accelerator boards to be used with a host computer have been presented. Most of the implementations utilize off-the-shelf building blocks like DSPs, microprocessors, microcontrollers (MCU) as well as programmable logic chips. Transputers have been very popular building blocks for multiprocessor systems, and also several SOM implementations have been presented with different level of parallelism and topology [28], [30-34]. Transputers are practically obsolete what comes to the raw performance, but they are still used for some special cases [29].

SOM implementations have also been presented for general purpose neurocomputers that make use of other kinds of standard processors. Examples of such implementations are Manitoba's system [35], MANTRA [36], MUSIC [37], PARNEU [38], RENNS [39], and TUTNC [14]. These systems are connected to a host computer and most of them could be expanded. All prefer neuron parallel mapping as the native level of parallelism, but typically more than one levels could be implemented. TUTNC and PARNEU are given as examples of these systems in the following.

TUTNC

TUTNC (Tampere University of Neural Computer) [40] is a parallel co-processor system hosted by a personal computer. The host is used to control the co-processor board by sending real time commands and serving data transfers. It also acts as a software development platform and offers windows based, graphical user interface.

The overall architecture of TUTNC is shown in Fig. 11.12 for a prototype system of four processing units. The system consists of a number of identical processing units (PU), a tree shape network of communication units (CU) and a single host interface unit.

The communication network, or tree, is used to transfer data between processing units and the host interface, and it also performs some computational tasks needed in many neural network algorithms. The tree can be configured as an adder tree or it can perform comparison (sorting) operations. An important feature is that each horizontal level of the tree is also a pipeline stage, such that data transfers and computations are pipelined.

The processing units operate in a SPMD (Same Program Multiple Data) mode, which means that the units are synchronized only when a communication operation is required. The units can also execute independently different programs, if required. Each unit contains a very low-cost Texas Instruments' TMS320C25 Digital Signal Processor (DSP), local memory and interface logic in one Xilinx XC4005 FPGA to connect the unit to the tree. Each CU has three bidirectional data ports. The data coming from the bottom port can be either distributed to both branches at the

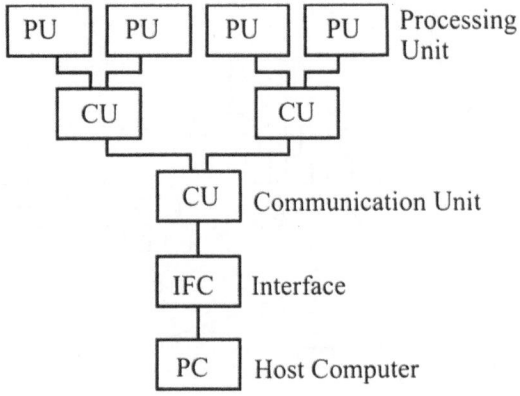

Fig. 11.12. TUTNC architecture.

top of the CU, or only one of them. In this way the data can be either broadcast to all processing units or selectively written to one processing unit.

When data arrives from the top ports, the bottom port can be connected to either of the top ports, the internal adder or comparator. In the first case, the data is transferred from a single processing unit to the host, while in the latter cases the network computes a sum of all numbers sent from the processing units, or finds a minimum or maximum values of them. In the prototype system one CU is implemented in XC4005 FPGA chip and the system consists of three CUs and four processing units.

In SOM computations, the TUTNC can be used either in weight or neuron parallel mode. For the neuron parallel mapping, the performance was found to be computation-bound, i.e. the communication network architecture could support a significantly larger number of processing units. In practise, the number of processing units should never exceed the number of neurons, or otherwise some of the units stay idle. The measured and estimated performance for a larger configuration is given in [14].

PARNEU

The tree topology of TUTNC was found to be very suitable for general neural network computation, but the expandability is quite complex. In practise, a new board should be manufactured for a larger number of processing units or special bridging boards. To overcome this, PARNEU (Partial Tree Shape Neurocomputer) was designed and implemented [41].

The architecture of PARNEU is presented in Fig. 11.13. PARNEU can operate in a stand-alone mode, but for convenient user interface, it is currently connected to a host computer. The master processing unit (MPU) controls the operations inside the PARNEU system.

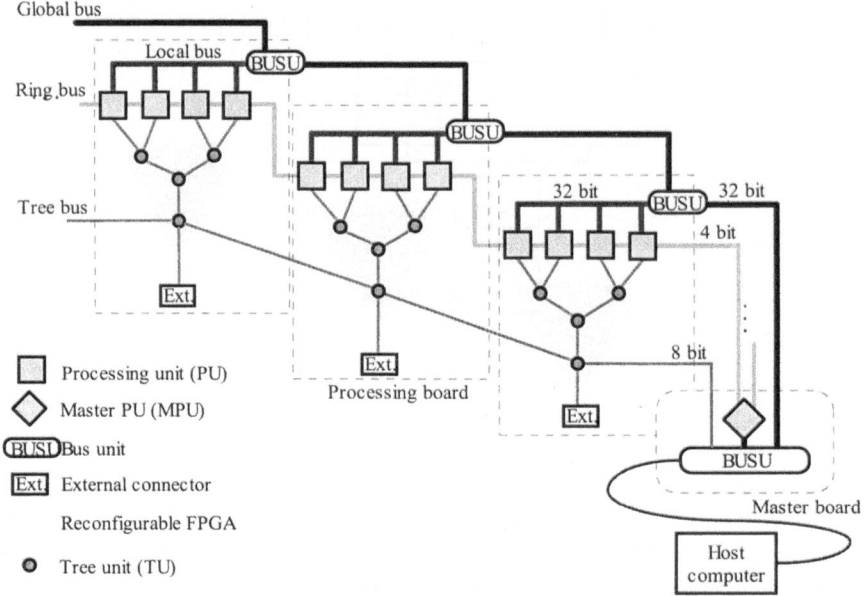

Global bus

Local bus   BUSU

Ring bus

Tree bus

BUSU

32 bit   BUSU   32 bit

4 bit

Ext.

Processing unit (PU)

Master PU (MPU)

BUSU Bus unit

Ext. External connector

Reconfigurable FPGA

Tree unit (TU)

8 bit

Ext.

Processing board

Ext.

BUSU

Master board

Host
computer

Fig. 11.13. PARNEU architecture.

The rack mounted PARNEU system consists of modular processing boards each carrying four processing units (PUs) that are connected to the MPU and further to the host computer. The internal communication structure is very flexible. A global bus (GB) connects all processing boards to the MPU for efficient broadcast operation. A serial ring bus connects each processing unit to adjacent units. Thus, data can be circulated between units is a systolic way. The third method to move data is a reconfigurable partial tree structure that is formed from the FPGAs located in each card.

The active tree network can perform global reduction operations like summation and comparison operations as well as pre- or post-processing operations. In these cases, the data transfer is programmed and operations in the partial tree network are required consequences. Compared to other systems, the combination of active partial tree network and global bus is a unique feature that allows very fast pipelined operations.

Processing units are implemented with Analog Devices ADSP-21062 floating-point digital signal processors. Each unit has internal 256 kilobytes SRAM memory that allows reasonable weight dimensions for SOM computation. The system allows MIMD-type programming, but the master-slave configuration and Single-Program Multiple-Data (SPMD) programming style are mainly used.

The maximum performance of PARNEU scales linearly as new processors are added, and good speed-ups have been obtained in practise for SOM computations

[38]. In addition, the architecture of PARNEU is freely expandible also in practise due to the backplane design [42].

Custom Neurochips and Accelerators

Custom neurochips are well motivated for embedded systems, but many such chips are also used to build up an accelerator board for a computer or even larger scale neurocomputers. While neurocomputers have mainly been designed and implemented by research laboratories, many custom chips are commercially available.

Several custom neural network chip designs have been presented during the last decade, but a smaller number of them are really implemented and used in practise. This is naturally a matter of implementation cost, which is significantly higher for ASIC (Application Specific Integrated Circuit) compared to standard chips. For this reason, programmable logic chips have also been used to custom implementations. Examples of commercial chips used to parallel SOM implementation are CNAPS [43], MA-16 [46] and SAND [44], [45]. All of these are placed in an accelerator board, and utilize weight or neuron parallelism. Examples of university and research laboratory implementations are [47-51]. Most of the chips speed-up either synaptic computations or the winner selection.

11.5.3 Performance

Two widely used performance measures for neural networks are Connections Per Second (CPS) and Connection Updates Per Second (CUPS). While CPS is used to measure forward mode performance, CUPS shows how fast the weights can be updated during the learning phase. Unfortunately, the reported CPS and CUPS values might not be comparable between implementations. Referring to the discussion about neuron parallel mapping, reliable measures could be defined as follows

$$
\begin{cases}
\text{CPS} = \dfrac{XYm}{T_B + T_{LW} + T_{GW}} = \dfrac{\text{total number of weigths}}{\text{total forward time}} \\[2em]
\text{CUPS} = \dfrac{min\{XYm, m(2r+1)^2\}}{T_B + T_{LW} + T_{GW} + T_{CN} + T_U} = \dfrac{\text{really updated weigths}}{\text{total training time}}
\end{cases}
\tag{11.21}
$$

where T_B denotes time for an input vector delivery, T_{LW} is the time for searching a local winner in a processing unit, T_{GW} is the time needed to find the global winner, T_{CN} refers to the time of neighbourhood computation in a processing unit and T_U is the time for the weight update.

Weights to be updated can be determined by multiplying the number of neighbourhood neurons with input vector dimension m. If the neighbourhood rectangular is larger than the map, all the weights are updated. T_{LW}, T_{CN} and T_U should be measured from the slowest processing unit, to which most neurons is assigned.

It should also be noted that the CUPS value depends on the neighbourhood size, which, in turn is dependent on the radius r as well as the location of the winner neuron in the map. As the neighbourhood radius is decreased during execution, T_U decreases. For these reasons, the CUPS value should be reported with all this information for fair comparisons.

A normalized measure of the performance would be reasonable that shows the effect of topology, processing unit and mapping approach comparable to others. The technology evolves all the time and most important is to analyse the factors contributing to the performance instead of reporting peak performance. Unfortunately the implementation details for such an analysis is not available for most of the systems. Another possibility is to develop some benchmarking problems like those for PC processors. The real performance and applicability are finally dependent on whether the requirements have been fulfilled or not for the application. The wall clock time is the most obvious measure and well motivated for the new, large SOM implementations on workstations.

Table 11.2. Performance figures for sample SOM implementations.

System	PUs	Neurons	m	MCUPS
PC (400 MHz, Intel PII) [38]	1	1024	32	47.2
Manitoba's system [35]	1	-	-	2
TUTNC [14]	4	256	40	1
WARP [21]	10	1024	128	12,5
RENNS [39]	15	256	4	8
Transputer [28]	30	14400	100	2.4
PARNEU [38]	32	1024	32	152
REMAP3 [16]	128	2048	128	17.5
CNAPS [43]	256	512	128	183
MANTRA 1 [36]	400	60	76	13.9
MasPar MP-1 [18], [20]	4096	-	-	17.2
CM-2 [20]	16 000	16384	100	48

Table 11.2 summarizes some of the reported performance figures for SOM implementations. For the reasons given above, the Table should be considered as a general view to the order of obtainable speed. Given CUPS values are given with all weights included independent of how many are really updated.

11.6 Conclusions

General principles and examples of past and current implementations for the Self Organizing Map have been considered. The technological development has changed the design space from communication intensive, fine-grain parallel implementations towards a more coarse grained levels of parallelism for the applications requiring large maps and massive input data. Although the platforms are much more powerful today, the problem of partitioning the elementary SOM computations in an optimal way is still present as previously. The difference is that the while the former implementations attempted to speed-up basic arithmetic computations, the current problem is memory organization and inter-processor communication.

In addition to the above high-end line, in which many of the scientific experiments fall, there is the space of embedded implementations like neuromorphic vision chips that have opposite requirements. Low power consumption combined to small physical size favours System-on-Chip approach. Current VLSI technology allows to integrate previous neurocomputers even in one chip, which in turn implies the need of previous knowledge of building parallel SOM implementations. The platforms are different, but the principles the same for current and future designs.

Bibliography on Chapter 11

1. T. Kohonen, *Self-Organization and Associative Memory*, Springer-Verlag, Berlin, 1980.
2. T. Kohonen, *Self-Organizing Maps*, Springer-Verlag, Berlin, 1995.
3. T. Kohonen, S. Kaski (Eds.), *Kohonen Maps*, Elsevier, Amsterdam, 1999.
4. T. Kohonen, "The 'Neural' Phonetic Typewriter, *Computer*", Vol. 21, 1988, 11-22.
5. X. Ling, D. Soergel, G. Marchionini, "A self-organizing semantic map for information retrieval", in: *Proceedings of 14th annual international conference on R&D in information retrieval*, 1991, 262-269.
6. T. Kohonen, S. Kaski, K. Lagus, T. Honkela, "Very large two-level SOM for the browsing of newsgroups", in: *Proceedings of International Conference on Artificial Neural Networks*, 1996, 269-274.
7. T. Kohonen, S. Kaski, K. Lagus, J. Salojärvi, J. Honkela, V. Paatero, and H. Saarela, "Self organization of a massive document collection", *IEEE Transactions on Neural Networks*, Vol. 11, No. 3, 2000, 574 -585.
8. P. Ienne, P. Thiran, T. Vassilas, "Modified self-organizing feature map algorithms for efficient digital hardware implementation", *IEEE Transactions on Neural Networks*, Vol. 8, No. 2, 2000, 315-330.
9. A. Peleg, U. Weiser, "MMX technology extension to the Intel architecture," *IEEE Micro*, Vol. 16, No. 4, 1996, 42-50.
10. T. Thakkar and T. Huff, "The internet streaming SIMD extensions," *IEEE Computer*, Dec. 1999, 26-34.
11. P. Thiran, V. Peiris, P. Heim, B. Hochet, "Quantization effects in digitally behaving circuit implementations of Kohonen networks", *IEEE Transactions on Neural Networks*, Vol. 5, No. 3, 1994, 450–458.
12. A. Rauber, P. Tomisch,D. Merkl, "parSOM: a parallel Implementation of the self organizing map exploiting cache effects - making the SOM fit for interactive high-performance data analysis", In: *Proceedings of the IEEE-INNS-ENNS International Joint Conference on Neural Networks*, Vol. 6, 2000, 177-182.
13. P.W. Diodato, "Embedded DRAM: more than just a memory", *IEEE Communications Magazine*, Vol. 38, No. 7, 2000, 118-126.
14. T. Hämäläinen, H. Klapuri, J. Saarinen and K. Kaski, "Mapping of SOM and LVQ Algorithms on a Tree Shape Parallel Computer System", *Parallel Computing*, Vol. 23, 1997, 271-289.
15. D. Bertsekas, J. Tsitsiklis, *Parallel and Distributed Computation: Numerical Methods*, Prentice-Hall, USA, 1989.
16. T. Nordström and B. Svensson, "Using and designing massively parallel computers for artificial neural networks", *Journal of Parallel and Distributed Computing*, Vol. 14, No. 3, 1992, 260-285.
17. C.H. Wu, R.E. Hodges, "Parallelizing the self-organizing feature map on multiprocessor systems", *Parallel Computing*, No. 17, 1991, 821-832.
18. V. Demian, J-C. Mignot, "Implementation of the self-organizing feature map on parallel computers", *Computers and Artificial Intelligence*, No. 1, 1996, 63-80.
19. E. Schikuta, C. Weidmann, "Data parallel simulation of self-organizing maps on hypercube architectures", in: *Proceedings of Workshop on Self-Organizing Maps, WSOM'97*, Helsinki, Finland, 1997, 142-147.
20. K. Obermayer, H. Ritter and K. Schulten, "Large-scale simulations of self-organizing neural networks on parallel computers: application to biological modelling", *Parallel Computing*, Vol. 13, No. 3, 1990, 381-404.
21. R. Mann and S. Haykin, "A parallel implementation of Kohonen feature maps on the Warp systolic computer", in: *Proceedings of International Joint Conference on Neural Networks*, Vol. 2, 1990, 84-87.

22. M. Yasunaga, K. Tominaga, Jung Hwan Kim, "Parallel self-organization map using multiple stimuli", in: *Proceedings International Joint Conference on Neural Networks*, Vol. 2, 1999, 1127-1130.

23. N. Bandeira, V.J. Lobo, F. Moura-Pires, "Training a self-organizing map distributed on a PVM network", in: *Proceedings of IEEE Joint Conference on Neural Networks*, Vol. 1, 1998, 457-461.

24. J.S. Lange, P. Schonmeier, H. Freiesleben,"Parallelization of analyses using self-organizing maps with PVM", *Nuclear Instruments and Methods in Physics Research A*, No. 389, 1997, 274-76.

25. H. Guan, Chi-kwong Li, To-yat Cheung, Songnian Yu, "Parallel design and implementation of SOM neural computing model in PVM environment of a distributed system", in: *Proceedings of Advances in Parallel and Distributed Computing*,1997, 26-31.

26. T. Bollinger, "Linux in practice: an overview of applications", *IEEE Software*, Vol. 16 No. 1, 1999, 72-79.

27. N. Boden, D. Cohen, R. Felderman, A. Kulawik, C. Sietz, J. Seizovic, W. Su, "Myrinet - A Gigabit-per-Second Local Area Network", IEEE Micro, Feb 1995, 29-36.

28. H. Simeon and A. Ultsch, *"Kohonen Networks on Transputers: Implementation and Animation"*, in: *Proceedings of International Neural Network Conference*, Vol. 2, 1990, 643-646.

29. J.S. Lange, C. Fukunaga, M. Tanaka, A. Bozek, "Transputer Self-Organizing Map Algorithm for Beam Background Rejection at the Belle Silicon Vertex Detector", *Nuclear Instruments & Methods In Physics Research Section A* No. 420, 1999, 288-309.

30. J.M. Auger, "Parallel implementation on transputer of Kohonen's algorithm", in: *Proceedings of Computing with Parallel Architectures: T. Node*,1991, 215-226.

31. M.E. Azema-Barac, "A generic strategy for mapping neural network models on transputer-based machines", in: G. L. Reijns, J. Luo (Eds.), *Transputing in numerical and neural network applications*, IOS Press, 1992, 244-249.

32. H. Kihl, J.P. Urban, J. Gresser, S. Hagmann, "Neural network based hand-eye positioning with a Transputer-based system", in: *Proceedings of High-Performance Computing and Networking - International Conference and Exhibition*, 1995, 281-286.

33. R. Togneri and Y. Attikiouzel, "Parallel Implementation of the Kohonen Algorithm on Transputer", in: *Proceedings of International Joint Conference on Neural Networks*, Vol. II, 1991, 1717--1722.

34. S.A. Wilde, K.M. Curtis, "A transputer based self-organizing neural network for speech synthesis parameter arbitration", in: *Proceedings of Transputer Applications and Systems - 1993 World Transputer Congress*, 1993, 1242-1253.

35. H.C. Card, G.K. Rosendahl, D.K. McNeill, R.D. McLeod, "Competitive Learning Algorithms and Neurocomputer Architecture", *IEEE Transactions on Computers*, Vol. 47, No. 8, 1998, 847-858.

36. T. Cornu, P. Ienne, D. Niebur, P. Thiran and M. Viredaz, "Design, Implementation and Test of a Multi-Model Systolic Neural Network Accelerator", *Scientific Programming*, Vol. 5, No. 1, 1996, 47-61.

37. U. Müller, A. Gunzinger, W. Guggenbühl, "Fast Neural Net Simulation with a DSP Processor Array", *IEEE Transactions on Neural Networks*, Vol. 6, No. 1, 1995, 203-213.

38. P. Kolinummi, P. Pulkkinen, T. Hämäläinen, J. Saarinen, "Parallel implementation of Self-Organizing map on the partial tree shape neurocomputer", *Neural Processing Letters*, Vol. 12, No. 2, 2000, 171-182.

39. G. Myklebust and J.G. Solheim, "Parallel self-organizing Maps for actual applications", in: *Proceedings IEEE International Conference on Neural Networks*, Vol. II, 1995, 1054-1059.

40. T. Hämäläinen, J. Saarinen and K. Kaski, "TUTNC: A general purpose parallel computer for neural network computations", *Microprocessors and Microsystems*, Vol. 9, No. 8, 1995, 447-465.

41. P. Kolinummi, P. Hämäläinen, T. Hämäläinen, J. Saarinen, "PARNEU: General-purpose partial tree computer", *Microprocessors and Microsystems*, Vol. 24, No. 1, 2000, 23-42.

42. P. Kolinummi, T. Hämäläinen, J. Saarinen, "Chained Backplane communication architecture for scalable multiprocessor systems", *Journal of Systems Architecture* Vol. 46, No. 11, 955-972.

43. D. Hammerström and N. Nguyen, "An Implementation of Kohonen's self-organizing map on the Adaptive Solutions neurocomputer", in: T. Kohonen, K. Mäkisara, O. Simula and J. Kangas (Eds.), *Artificial Neural Networks*, North-Holland, Amsterdam, Vol. 1, 1991, 715-720.

44. T. Fischer, W. Eppler, H. Gemmeke, G. Kock, T. Becher, "The SAND neurochip and its embedding in the MiND system", in: *Proceedings of Artificial Neural Networks*, 1997, 1235-1240.

45. W. Eppler, T. Fischer, H. Gemmeke, T. Koder, R. Stotzka, "Neural chip SAND/1 for real time pattern recognition ", *IEEE Transactions on Nuclear Science*, Vol. 45, No. 4, 1819-1823.

46. U. Ramacher, "SYNAPSE - A Neurocomputer That Synthesizes Neural Algorithms on a Parallel Systolic Engine", *Journal of Parallel and Distributed Computing*, Vol. 14, No. 3, 1992, 306-318.

47. B. Hochet, V. Peiris, S. Abdo, M. Declercq, "Implementation of a Learning Kohonen Neuron Based on a New Multilevel Storage Technique", *IEEE Journal of Solid-State Circuits*, Vol. 26, No. 3, 1991, 262-267.

48. J. Choi and B. J. Sheu, "A high precision VLSI winner-take-all circuit for self-organizing neural networks" *IEEE Journal of Solid-State Circuits*, Vol. 28, May 1993, 579-584.

49. S.Rüping, M. Porrmann, U. Rueckert, "SOM Accelerator System", *Neurocomputing*, Vol. 21, No.1-3, 1998, 31-50.

50. X. Fang, P. Thole, J. Göppert and W Rosenstiel, "A Hardware Supported System for a Special Online Application of Self-Organizing Map", In: *Proceedings of the International Conference on Neural Networks*,1996, 956-961.

51. J. Lubkin, G. Cauwenberghs, "VLSI implementation of fuzzy adaptive resonance and learning vector quantization", in: *Proceedings of the Seventh International Conference on Microelectronics for Neural, Fuzzy and Bio-Inspired Systems*, 1999,147-54.